CISM COURSES AND LECTURES

The series presents lecture notes, monographs, edited works and proceedings in the field of Mechanics, Engineering, Computer Science and Applied Mathematics.
Purpose of the series is to make known in the international scientific and technical community results obtained in some of the activities organized by CISM, the International Centre for Mechanical Sciences.

INTERNATIONAL CENTRE FOR MECHANICAL SCIENCES

COURSES AND LECTURES - No. 449

PHASE CHANGE WITH CONVECTION: MODELLING AND VALIDATION

EDITED BY

TOMASZ A. KOWALEWSKI
ACADEMY OF SCIENCES WARSAW

DOMINIQUE GOBIN
CAMPUS UNIVERSITAIRE ORSAY

Springer-Verlag Wien GmbH

The publication of this volume was co-sponsored and co-financed by the UNESCO Venice Office - Regional Bureau for Science in Europe (ROSTE) and its content corresponds to a CISM Advanced Course supported by the same UNESCO Regional Bureau.

This volume contains 135 illustrations

Originally published by Springer-Verlag Wien New York in 2004
SPIN 10984598

In order to make this volume available as economically and as rapidly as possible the authors' typescripts have been reproduced in their original forms. This method unfortunately has its typographical limitations but it is hoped that they in no way distract the reader.

ISBN 978-3-211-20891-5 ISBN 978-3-7091-2764-3 (eBook)
DOI 10.1007/978-3-7091-2764-3

PREFACE

Solid-liquid phase-change phenomena are present in a large number of industrial applications (materials processing, crystal growth, casting of metal – matrix composites, heat storage, food conservation, cryosurgery) and natural processes (iceberg evolution, magma chambers, crust formation). It is generally recognized that the dynamics of such phase change processes are largely influenced by natural convection. Numerical simulation of these non-linear, moving boundary, thermal and fluid flow problems is not a trivial task. The phenomena occurring during solidification of different materials or in different configurations are so diverse that no unique, simple description covering all ranges and scales is possible at the moment. Therefore, the modelling of phase change processes present in practical configurations requires the profound understanding of limitations of the existing physical and numerical models. Validation and verification of the numerical models before their implementation to practical industrial situations becomes an important issue if quantitative modelling of the real physical problems is a target.

The CISM course on Phase Change with Convection (PCC02) given in Udine in September 2002, followed PCC99 Workshop[1]*, which aimed to create a common platform for different groups working on modelling phase change problems. The main outcome of the workshop was to emphasize necessity for deeper understanding both the physical background and mathematical problems concerning modelling phase changes problems. The aim of the CISM course was to present a review of modelling phase change problems and of recent methods of numerical and experimental analysis used, with a particular focus on solidification coupled to convective flow. Special attention was given to the validation and verification of numerical codes and to the applications to practical problems.*

The lecture notes prepared for the course are addressed to advanced students and scientists from engineering and applied sciences, as well as to physicists and mathematicians interested in the fundamentals of the field. There is large number of publications, workshops and reviews concerning modelling of solidification problems and only a minute fraction of these problems could be addressed during our short course. Nevertheless, we hope that theoretical background, discussion of physical phenomena, and practical examples of tailoring numerical codes given in our lectures will help readers to understand limitations of numerical models used for various engineering applications.

Tomasz Kowalewski
Dominique Gobin

---

[1] *ESF Workshop on "Phase Change with Convection" held in Warsaw 1999, web page: http://fluid.ippt.gov.pl/pcc99*

CONTENTS

Solidification microstructure, dendrites and convection

Gustav Amberg[*‡]

[*] Department of Mechanics, KTH, Stockholm, Sweden

1 Introduction

One crucial step in almost all materials processes is solidification in one form or other. The conditions under which the melt resolidifies will be crucial for the final microstructure of the material. The size and morphology of the individual grains that make up a polycrystalline material, the homogeneity of a monocrystal, the actual phase that is formed, as well as its local composition, is determined by the interplay between local heat and mass transfer and the thermodynamics of the phase change. Even though the microstructure of the material may change considerably during subsequent cooling and following process steps, the foundation has been laid at the point of solidification. Since local heat and mass transfer governs the phase change, it is obvious that any melt convection at all will be paramount in determining the structure of the material, thus making this an area of important applications that should interest fluid mechanists.

In this chapter we will describe this following an increasing geometrical complexity, from a planar interface via individual dendrites to a mushy zone, and discuss different effects of convection as we go along. We will start by discussing the instabilities that may appear on a planar solidification interface. These instabilities are the first manifestations of what may grow into for instance a cellular or dendritic microstructure. Even though these instabilities are present also in the absence of convection, density differences etc that are induced by the instability may change its properties. We will then proceed to describe how the growth of individual fully developed dendrites are affected by convection. In order to describe fully developed dendrites we will also have to outline the phase field method, which has become a popular tool recently. Finally we will discuss the phenomenon of chimney formation in directional solidification of a mushy layer. This is a convective instability occurring in the two phase mush that may be taken as a prototype for well known defects in solidified materials.

Throughout this chapter there will be frequent references to (Kurz and Fisher, 1992) , which is the standard textbook on fundamental solidification. The recent book by Davis (2001) also describes solidification from a fundamental point of view, but also includes some recent results on convection effects. A recent survey of current research trends is given in Boettinger et.al. (2000).

2 Stability of a solidification front

A generic example of solidification of a pure liquid would be the unidirectional solidification of an undercooled sample initially at a temperature below the freezing point. The simplest mathematical description of this would assume a planar phase change boundary, with a constant given freezing temperature at the solidification interface, and a constant latent heat release expressed as a discontinuity of the normal temperature gradient at the interface. This evokes a picture of the solidification front as a smooth interface advancing over the domain. In reality solidification starts with nucleation, either homogeneous nucleation in the melt, or heterogeneous nucleation originating typically on solid particles suspended in the melt or on crucible walls. Still, it would be possible in principle that the solid would grow from a crucible wall to form eventually a smooth planar solid liquid interface. Similarly small nuclei in the melt could grow into smooth globular solids.

This however is very much the exception rather than the rule when dealing with metals and crystalline materials. The reason is that a planar solidification interface advancing into an undercooled melt is subject to a fingering instability: if a bump is formed on the solidification front, the local temperature gradient ahead of it will increase, and thus cool the front more efficiently there, causing an amplification of the disturbance. Furthermore, this mathematical problem is in fact ill-posed, since the growth rate of a disturbance of the planar shape of the interface will grow unboundedly with the wavenumber of the disturbance. The assumption responsible for this is that the temperature is assumed constant on the interface. A more realistic description of solidification is obtained by recognizing that the temperature at the interface depends on the local curvature of the material, as well as the speed of the front. Also the interface kinetics are highly anisotropic due to the anisotropic properties of the crystalline solid that is formed.

In a binary mixture, the interface temperature also depends on the local composition and it is possible to make a close analogy between solidification of a pure material and the approximately isothermal solidification of a supersaturated system. The basic instability of a planar or spherical front was first investigated by Mullins and Sekerka (1963), Mullins and Sekerka (1964), and has since been studied extensively in different contexts, for instance effects of natural and forced convection in the melt, Davis (1990). In what follows we will outline some of this work.

2.1 Mathematical model

In this section a mathematical model for a solidifying binary alloy, including curvature and kinetic effects, will be formulated. All equations in this section will be represented in non-dimensional form and relative to a coordinate system fixed in the solid. Length and fluid velocity have been scaled with a reference length l and velocity U. Time has been nondimensionalized with the convective timescale l/U, and, as appropriate for these low Reynolds number flows, the pressure has been scaled with $\rho_0 \nu U/l$, where ρ_0 is a reference value of the density, and ν is the kinematic viscosity of the melt. Temperature T, is scaled so that θ represents the scaled temperature in the relation $T = T_0 + \Delta T\ \theta$, where T_0 is a suitable reference temperature and ΔT is a characteristic temperature difference, for instance the initial undercooling of the melt. Similarly, concentration of solute C is replaced by a scaled concentration c, according to $C = C_0 + \Delta C\ c$.

With this scaling, the conservation of heat takes the following form in the liquid phase:

$$Re Pr\,(\frac{\partial \theta}{\partial t} + \bar{u} \cdot \nabla \theta) = \nabla^2 \theta, \tag{2.1}$$

where t is the non-dimensional time, $\bar{u}$ is the non-dimensional flow velocity, $Re = Ul/\nu$ is the flow Reynolds number, $Pr = \nu/\kappa_l$ is the Prandtl number. ν and κ_l are the kinematic viscosity and the heat diffusivity in the liquid, respectively.

In the solid phase, which is assumed to be stationary, the convective term vanishes:

$$Re Pr\,\frac{\partial \theta}{\partial t} = \kappa'_s \nabla^2 \theta, \tag{2.2}$$

κ'_s denotes the ratio κ_s/κ_l, i.e. the nondimensional thermal diffusivity in the solid.

Similarly, conservation of solute in the liquid yields:

$$Re Sc(\frac{\partial c_l}{\partial t} + \bar{u} \cdot \nabla c_l) = \nabla^2 c_l, \tag{2.3}$$

in the solid:

$$Re Sc \frac{\partial c_s}{\partial t} = D'_s \nabla^2 c_s. \tag{2.4}$$

Here Sc denotes the Schmidt number ν/D_l, where D_l is the mass diffusivity of solute. D'_s is the nondimensional mass diffusivity in the solid, i.e. the ratio D_s/D_l.

The motion in the melt is governed by the Navier-Stokes equations, in the usual way. In the Boussinesq approximation, valid for small relative density variations, this is:

$$Re(\frac{\partial \bar{u}}{\partial t} + \bar{u} \cdot \nabla \bar{u}) = -\nabla p + \nabla^2 \bar{u} - (\frac{Ra}{Re Pr}\,\theta + \frac{Ra_c}{Re Sc}\,c)\,\bar{e}_g. \tag{2.5}$$

Here an expression for the density has been used, $\rho = \rho_0(\Delta T \beta_T \theta + \Delta C \beta_c c)$, where the thermal and solutal expansion coefficients, β_T and β_c, are assumed constant. The two Rayleigh numbers $Ra = \rho_0 g \beta_T \Delta T l^3/(\kappa_l \nu)$ and $Ra_c = \rho_0 g \beta_c \Delta C l^3/(D_l \nu)$ characterize the thermal and solutal buoyancy, respectively.

In accordance with the Boussinesq approximation, the melt flow is assumed to be divergence free:

$$\nabla \cdot \bar{u} = 0, \tag{2.6}$$

For the present purposes we will also disregard solidification shrinkage, i.e. assume that $\rho_s = \rho_l = \rho_0$. While there are certainly many phenomena where the melt motion induced by a change in volume due to a phase change is crucial, the convective motions that are the main focus here are thought of as being caused by external agents, such as gravity induced convection, stirring etc, and are presumed to be considerably larger than that induced by solidification shrinkage.

The details of phase change now enters as boundary conditions between the solid and the liquid regions. Conservation of heat over the solid-liquid interface implies:

$$V_n = \frac{\Delta}{Re Pr}\,(\kappa'_s \nabla \theta_s - \nabla \theta_l) \cdot \bar{n} \tag{2.7}$$

where V_n is the dimensionless normal velocity of the interface into the liquid and $\Delta = c_p\Delta T/L_f$ is the dimensionless undercooling (or inverse Stefan number). Indices l and s are used to denote the temperature on the liquid and solid sides of the interface, respectively. c_p and L_f are the specific heat and the latent heat respectively. $\bar{n}$ is the normal of the interface pointing into the liquid, see figure 1.

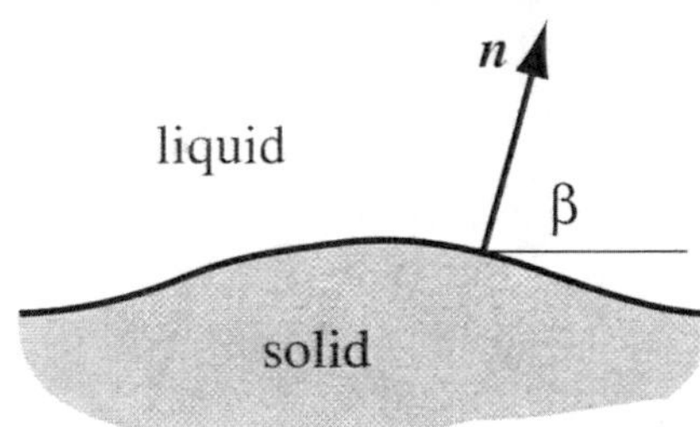

Figure 1. Sketch of a portion of the solidification interface

The corresponding balance of solute over the solidification interface is as follows, where again indices l and s are used to distinguish values on the liquid and solid sides of the interface:

$$V_n(c_l - c_s) = \frac{1}{ReSc}(D'_s\nabla c_s - \nabla c_l) \cdot \bar{n} \tag{2.8}$$

Due to different solubilities in the solid and liquid phases, the concentrations in the liquid and the solid should be different. A common approximation is to assume a constant partition ratio k_p, i.e.:

$$k_p = c_s/c_l \tag{2.9}$$

Using this definition equation (2.8) becomes:

$$V_n(1 - k_p)c_l = \frac{1}{ReSc}(D'_s\nabla c_s - \nabla c_l) \cdot \bar{n} \tag{2.10}$$

What remains in order to have a complete system of equations is information on the phase change kinetics. This takes the form of a dependency of the interface temperature on interface shape, interface velocity and composition. The kinetics is typically anisotropic, which makes the interface temperature depend also on the local angle between a chosen lattice orientation direction and the interface normal, via a dependency of the interface surface energy on this angle:

$$\sigma = \sigma_0\eta(\beta). \tag{2.11}$$

Here σ_0 is the average interfacial energy and the η-function represents the interfacial energy dependency on the angle, β. A typical four-fold symmetry in two dimensions model is obtained with

$$\eta(\beta) = 1 + \epsilon\, cos(4(\beta - \beta_0)), \tag{2.12}$$

where the value of ϵ determines the degree of anisotropy. β_0 denotes the angle where the surface energy is maximal. With this expression and a value of $\beta_0 = 0$, the horizontal and vertical directions in figure 1 will be the preferred directions for growth.

The relation giving the temperature at the interface is thus:

$$\theta = \Gamma[\eta(\beta) + \frac{d^2\eta(\beta)}{d\beta^2}]\chi - \frac{V_n}{\mu'\zeta(\beta)} - c. \tag{2.13}$$

The first term on the right hand side gives the capillary undercooling, i.e. the depression of temperature that should be present at a convex tip. χ is the dimensionless curvature of the interface (negative for a convex solid region). This term is proportional to the surface energy through the nondimensional parameter $\Gamma = T_m\sigma_0/(l\rho_0 L_f \Delta T)$, and includes the anisotropy discussed above. T_m is the melting temperature of the pure material. In what follows we will assume that anisotropy is small enough so that the bracketed term is always positive, i.e. with the anisotropy function in equation (2.12) this would mean that $|\epsilon| \leq 1/15$. If this is not the case the interface will develop sharp corners, as discussed in chapter 5 in Davis (2001). In this case the mathematical model as written here is ill posed, and must be amended to account for sharp corners.

The second term in equation (2.13) is the kinetic undercooling, i.e. the departure from equilibrium that occurs at elevated interface velocities. $\mu' = \mu\Delta T/U$, with μ denoting the dimensional kinetic coefficient, which determines the magnitude of the kinetic undercooling. A function $\zeta(\beta)$ has been included in this term to allow for anisotropic kinetics.

The third term represents the constitutional undercooling, i.e. the dependence of interface temperature on composition of the alloy. Here we have used a common simple approximation of the phase diagram, where the relation between liquid concentration and temperature in equilibrium is assumed to be linear:

$$T = T_m + mC. \tag{2.14}$$

Here the slope of the liquidus, m, is typically negative so that we expect a decreasing interface temperature with increasing solute content. Choosing reference temperature and concentration that satisfy this liquidus relation, i.e. $T_0 = T_m + mC_0$, and $\Delta T = -m\Delta C$, this term gets the particularly simple form given in equation (2.13).

2.2 1D solidification, base state

We will now consider an idealized directional solidification experiment, which in principle is possible to realize by translating a sample in a temperature gradient. If the sample has small transverse dimensions the temperature field may be considered as controlled by the ambient and equal to a prescribed gradient. The translation will then cause the sample to solidify directionally at the prescribed translational speed V. Our main objective now is to study the stability of a planar interface in this configuration.

The temperature field is thus assumed to be prescribed as $T = T_0 + zG_T$, increasing linearly in the direction of growth. This 'frozen temperature approximation' (FTA) implies an assumption of high thermal diffusivity ($\kappa_l/D_l >> 1$) and small latent heat that is quite possible to realize experimentally. The condition on latent heat is obtained from equation (2.7), as $\Delta/(RePr) = \kappa_l c_p G_T/(L_f V) >> 1$. We will also assume that the mass diffusion in the solid is very small, i.e. $D_s' \approx 0$. To begin with we will assume that there is no melt motion, i.e. $\bar{u} = 0$.

In choosing the scales for nondimensionalization, we expect the process to be governed by mass diffusion ahead of the solidification front, and the relevant length scale is thus $l = D_l/V$. The velocity scale is taken as the speed of the front, $U = V$. Furthermore the temperature is nondimensionalized by taking $T_0 = T_m$ and $\Delta T = G_T D_l/V$. To be consistent with this, the scaled concentration is defined so that $C_0 = 0$ and $\Delta C = -\Delta T/m = -G_T D_l/(Vm)$. The coordinate system is changed to one that translates with the solidification front, at speed V in the z-direction.

Before proceeding with the stability, we need to calculate the one-dimensional undisturbed concentration profile. With the assumptions above, together with the assumption that concentration depends only on the z-coordinate in the direction of growth, the equation (2.3) simplifies as follows:

$$-\frac{dc_B}{dz} = \frac{d^2c_B}{dz^2} \tag{2.15}$$

The boundary conditions at the interface are obtained from equations (2.8) and (2.13). Notice that the constant temperature gradient assumed in the FTA, together with the scaling used implies that the nondimensional temperature is simply equal to the nondimensional z-coordinate, $\theta = z$. The as yet undetermined position of the interface is denoted by h_0.

$$(1 - k_p)c_B = -\frac{dc_B}{dz} \text{ at } z = h_0 \tag{2.16}$$

$$z = -c_B \text{ at } z = h_0 \tag{2.17}$$

Far ahead of the front, the melt is assumed to have a uniform concentration C_∞, which corresponds to a rescaled concentration $c_\infty = -mC_\infty V/(G_T D_l)$ (remember that the slope of the liquidus m is typically negative so that c_∞ is indeed positive).

$$c_B = c_\infty \text{ at } z = \infty \tag{2.18}$$

Equation (2.15), together with boundary conditions (2.16), (2.17) and (2.18) now determines the concentration profile, and also the position of the interface h_0. The solution is:

$$c_B = c_\infty((\frac{1}{k_p} - 1)e^{h_0 - z} + 1) \tag{2.19}$$

$$h_0 = -c_B(h_0) = -\frac{c_\infty}{k_p} \tag{2.20}$$

For a value of the partition ratio k_p different from unity there is thus an accumulation of solute ahead of the front, see figure 2. Notice that the concentration in the solid is obtained from the definition of partition ratio, equation (2.9), as $c_s = c_B(h_0)k_p = c_\infty$. This of course follows from conservation of solute and the assumption of a steady state. Also, the interface is displaced to lower values of z, i.e. lower temperature, by the accumulation of solute.

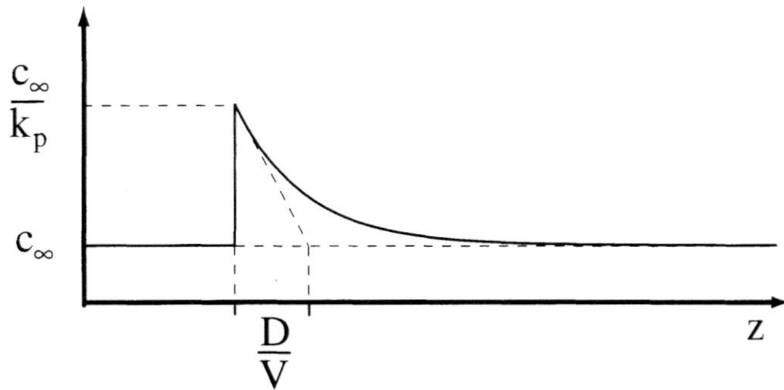

Figure 2. Sketch of the base state concentration profile in directional solidification

The most important feature of this base state profile is however that it determines a base gradient in the concentration at the front. This gradient will be important for the stability properties, and is as follows in nondimensional terms:

$$\frac{dc_B}{dz}(h_0) = c_\infty \frac{1-k_p}{k_p} = \frac{mC_\infty V}{D_l G_T} \frac{1-k_p}{k_p} \tag{2.21}$$

Or, with c_∞ as defined above, the dimensional concentration gradient becomes:

$$G_c = \frac{\Delta C}{l} \frac{dc_B}{dz}(h_0) = -\frac{C_\infty V}{D_l} \frac{1-k_p}{k_p} \tag{2.22}$$

2.3 Morphological instability

We will now study the stability of the one-dimensional solution obtained above. For simplicity we will be satisfied with two-dimensional disturbances, and will assume that all variables are independent of the y-coordinate.

In two dimensions we introduce the interface shape as $z = h(x,t)$. The normal velocity of the interface V_n can be related to $h(x,t)$ by considering the kinematics of the interface, the result being

$$V_n = \frac{1 + \frac{\partial h}{\partial t}}{\sqrt{1 + (\frac{\partial h}{\partial x})^2}} \tag{2.23}$$

The constant unit term in the nominator accounts for the translation of the coordinate system with the constant speed V, which, with the present choice of scaling, becomes unity in this nondimensional equation.

The curvature χ that appears in equation (2.13) is expressed in terms of the interface shape as

$$\chi = \frac{\frac{\partial^2 h}{\partial x^2}}{(1 + (\frac{\partial h}{\partial x})^2)^{3/2}} \tag{2.24}$$

Proceeding with the same choice of nondimensional scales as above, we now split the concentration into a base profile and a disturbance as follows:

$$c_l = c_B(h_0 + \zeta) + \tilde{c}(x, \zeta, t) \tag{2.25}$$

Here we also introduced the translated coordinate $\zeta = z - h_0$, which is zero at the interface.

A similar split is introduced for the interface shape $h(x,t)$:

$$h(x,t) = h_0 + \tilde{h}(x,t) \tag{2.26}$$

Here c_B and h_0 is the solution to the one dimensional problem in the previous section given by equations (2.19, 2.20).

We will now consider small concentration disturbances, $\tilde{c} << c_B$, and an almost flat interface $\partial\tilde{h}/\partial x << 1$, and linearize the problem. In the absence of melt flow, the concentration equation (2.3) is linear in itself, giving an equation for the disturbance as

$$\frac{\partial \tilde{c}}{\partial t} - \frac{\partial \tilde{c}}{\partial z} = \nabla^2 \tilde{c} \tag{2.27}$$

We will require that disturbances decay far ahead of the front, so that

$$\tilde{c} \to 0 \text{ as } \zeta \to \infty \tag{2.28}$$

The boundary conditions at the solidification interface are given in equations (2.13,2.10). In (2.13) we will consider an isotropic material, so that $\eta(\beta) \equiv 1$. We will also disregard the kinetic undercooling, i.e. $\mu' \to \infty$. In (2.10) we are again assuming that the mass diffusivity in the solid is much less than that in the liquid, i.e. $D'_s \approx 0$.

The interfacial conditions (2.13,2.10) are to be applied at the actual interface position $h_0 + \tilde{h}$. This additional nonlinearity is linearized by expanding the conditions in Taylor series around $z = h_0$, or $\zeta = 0$. The resulting conditions are

$$\tilde{c} = -\tilde{h} + \frac{1-k_p}{k_p} c_\infty \tilde{h} + \Gamma \frac{\partial^2 \tilde{h}}{\partial x^2} \qquad \text{at } \zeta = 0 \tag{2.29}$$

$$\frac{\partial \tilde{h}}{\partial t} + k_p \tilde{h} + \frac{k_p}{c_\infty} \tilde{c} = -\frac{k_p}{c_\infty(1-k_p)} \frac{\partial \tilde{c}}{\partial \zeta} \qquad \text{at } \zeta = 0 \tag{2.30}$$

Equations (2.27,2.30) constitute a linear homogeneous problem with constant coefficients. A fundamental solution must thus have the form

$$\tilde{c}(x, \zeta, t) = c_1 \, e^{\lambda \zeta} e^{\sigma t + iax} \tag{2.31}$$

$$\tilde{h}(x,t) = h_1 \, e^{\sigma t + iax} \tag{2.32}$$

By inserting $\tilde{c}$ according to eq (2.31) in equation (2.27) a quadratic equation for the characteristic exponent λ is obtained. The solution is $\lambda = -1/2 - 1/2\sqrt{1 + 4(\sigma + a^2)}$. The sign in front of the radical is chosen so that the condition far ahead of the front, eq (2.28), is satisfied.

When this solution is inserted into the interfacial conditions (2.29,2.30), two linear homogeneous equations are obtained for (c_1, h_1). In order for a nonzero equation to exist the determinant of this 2x2 system must be zero, which yields an equation for the growth rate σ. The solution to this is:

$$\sigma = -\frac{1}{2M^2}(\Gamma a^2(M(2k_p+1)+R)-(1+\Gamma a^2)^2+M(2k_p+1)+R(1-M))) \quad (2.33)$$

where

$$R = \sqrt{(1+\Gamma a^2)^2+4M(Ma^2-k_p(1+\Gamma a^2))}$$

Here the nondimensional undercooling c_∞ has been replaced by the morphological number $M = c_\infty(1-k_p)/k_p = -(1-k_p)mC_\infty V/(G_T D_l k_p)$.

We will first investigate this expression in the special case of vanishing surface energy, i.e. $\Gamma = 0$. The result is

$$\sigma = -\frac{k_p}{M} - \frac{1}{2M^2}(1-M)(R-1) \quad (2.34)$$

where R is now

$$R = \sqrt{(1+4M(Ma^2-k_p)}$$

From this we see that when $M=1$, $\sigma = -k_p$, independent of a. Further investigation of the expression in equation (2.34) shows that $\sigma < 0$ whenever $M < 1$, i.e. that the planar front is stable and all shape perturbations will decay. In order to interpret this condition in physical terms, we rewrite it in the following way:

$$M = c_\infty(1-k_p)/k_p = -(1-k_p)mC_\infty V/(G_T D_l k_p) = \frac{mG_c}{G_T} < 1$$

Here the base profile concentration gradient at the front defined in equation (2.22), G_c, has been used. The condition for stability $M < 1$ is thus equivalent to

$$mG_c < G_T. \quad (2.35)$$

Consider a perturbation dh of the interface position h. In keeping with the frozen temperature approximation that has already been invoked, i.e. heat diffusivities much larger than mass diffusivities, the temperature of the interface must be equal to the background temperature also at the perturbed position, i.e.

$$T_i = T_0 + G_T dh.$$

The concentration at the interface, in the absence of surface energy and kinetic undercooling, must now follow the liquidus relation (2.14), i.e.

$$C_i = C_0 + dC_i = C_0 + dT_i/m = C_0 + G_T dh/m$$

This concentration perturbation should now be compared to the unperturbed concentration $C_B = C_0 + G_c dh$. As illustrated in figure 3, $C_i > C_B$ implies that the modulus

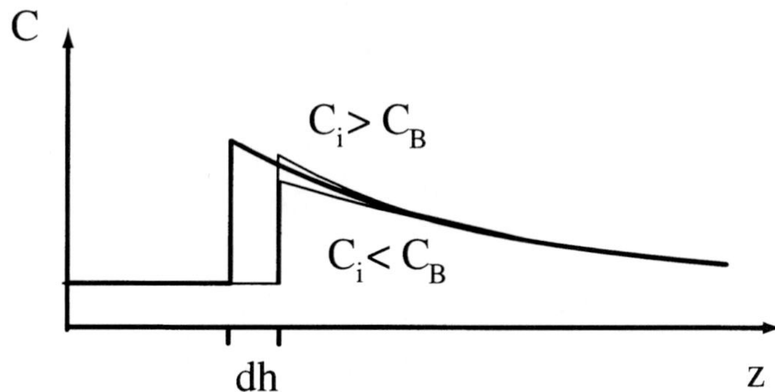

Figure 3. Sketch of the concentration profile ahead of a disturbance in interface speed.

of the concentration gradient ahead of the perturbation is increased. This increasingly negative gradient causes a larger diffusive transfer of solute away from the perturbation, and, as can be seen from equation (2.10), this implies an increased speed and thus the perturbation is amplified. Conversely, $C_i < C_B$ means that the perturbation will slow down and be damped. In terms of the relations above this gives

$$C_i = C_0 + G_T dh/m < C_B = C_0 + G_c dh,$$

which, remembering that m is typically negative, is equivalent to equation (2.35). With reference to how the base profiles in temperature and concentration deviate from the liquidus relation, the unstable situation $mG_c > G_T$ is called constitutional undercooling.

As an illustration, figure 4 shows contours of the growth rate σ for a Pb-Sn alloy which is solidified directionally at a speed $V = 30\mu m/s$, and a temperature gradient in the liquid of $G_T = 200K/cm$. The parameter values and the case is chosen according to Coriell et. al. (1980), except that we have here made the additional assumption of a frozen temperature, which may cause some numerical discrepancy. The horizontal axis shows the dimensional wavenumber of the disturbance, $a' = a/l$, the vertical axis is the bulk concentration of tin. The thicker contour corresponds to marginal stability, $\sigma = 0$. Contours above this curve correspond to positive σ values, contours below to negative. Thus the interface is unstable for concentrations above this thick contour.

Within the FTA employed here, the condition for constitutional undercooling, eq (2.35), together with eq (2.22), corresponds to a condition on the concentration

$$C_\infty < G_T D_l k_p/(-mV(1-k_p)) \tag{2.36}$$

This has the value 0.368 wt % for the parameter values in figure 4. Below this concentration, all disturbances are thus stable. This is also confirmed by the marginal stability curve in figure 4, which reaches a minimum value of 0.398, just above the value for constitutional undercooling.

In the limit of very large surface energy, and/or solidification speed, the interface will be stabilized again. In equation (2.13) it is seen that the parameter Γ gives a non-dimensional measure of the strength of the surface energy. For large Γ and M (or c_∞) it

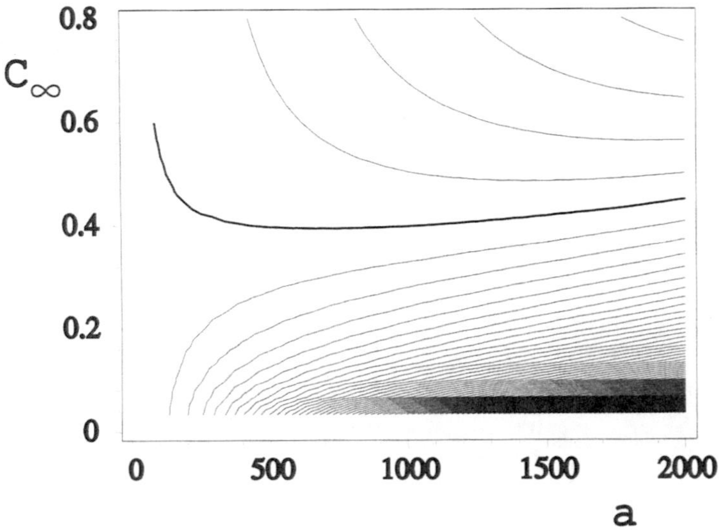

Figure 4. Contours of the growth rate σ as a function of C_∞ (wt %) and a' (cm^{-1}).

can be shown that all disturbances are stable if

$$\Gamma > \frac{M}{k_p} \tag{2.37}$$

In terms of concentration and solidification speed this condition becomes:

$$C_\infty < \frac{k_p^2}{1-k_p} \frac{T_m \sigma_0 V}{-m\rho_0 L_f D_l} \tag{2.38}$$

If we consider C_∞ as fixed, and increase V from zero, we see that the condition for constitutional undercooling tells us that the interface is stable for sufficiently low speeds. Increasing V further, positive growth rates become possible and the interface develops instabilities. Increasing V further still, the absolute stability condition (2.38) indicates that the interface will become stable once again, for a sufficiently high speed V.

2.4 Stability of a planar interface in the presence of natural convection

In the directional solidification problem that has been studied above, we have seen that solute is typically accumulated ahead of the front, as depicted in figures 2 and 5. The melt in this enriched region will have a different density, which in the presence of gravity may cause convection. Consider the situation in which the solidification proceeds vertically upwards, opposite to the direction of gravity. If the solute is lighter than the solvent, the enriched layer may be unstably stratified and there is potential for a convective instability in the melt ahead of the front. This convection will affect heat and mass transfer at the interface and thus local solidification rates and morphology. In this arrangement with melt overlying the growing solid, temperature increases upwards, which implies a stable stratification that would suppress convective instability. Clearly the competition between the destabilizing solutal buoyancy and the stabilizing thermal buoyancy must be important.

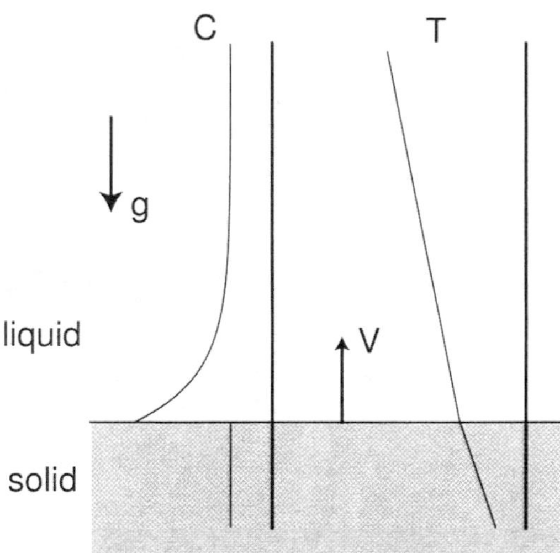

Figure 5. Sketch of concentration and temperature profiles in vertical directional solidification

This coupled phase change and convection is, within the Boussinesq approximation, governed by the equations above, namely conservation of heat and solute, equations (2.1-2.4), momentum balance (2.5) and incompressibility (2.6). As before the conditions at the solid-liquid interface are the conservation of heat and solute, eq (2.7) and (2.10), and the solidification kinetics in equation (2.13). In the present section we will, as before, disregard kinetic undercooling and solidification shrinkage, and assume that the diffusion coefficient in the solid is effectively zero.

This problem now involves diffusion and advection of both mass and heat, and the density is assumed to depend on both temperature and concentration. Such double diffusive convection problems, Turner (1985), have some unexpected features that we wish to illustrate by considering first the double diffusive version of the Benard problem, Nield (1967). Consider a horizontal layer of fluid of thickness l that is subject to mean gradients in both concentration and temperature. There is no phase change, and conditions at the lower and upper boundaries are zero shear stress, no normal velocity, and prescribed concentration and temperature. The relevant equations are thus equations (2.1, 2.3, 2.5, 2.6), with boundary conditions $\theta = c = 0$ at the lower boundary, and $\theta = c = 1$ at the upper. In the scaling we have thus taken the height of the layer equal to l, velocity $U = \nu/l$, and temperature and concentration differences according to $\Delta T = T_{top} - T_{bottom}$, $\Delta C = C_{top} - C_{bottom}$, the difference between the values prescribed at top and bottom.

The independent nondimensional quantities are then the thermal Rayleigh number

Ra, the solutal Rayleigh number Ra_c, and the Prandtl and Schmidt numbers, Pr and Sc. Notice that we now consider Ra, Ra_c to be signed quantities, as given by the sign of ΔT, β_T, etc, with a negative Rayleigh number indicating a stabilizing contribution to the vertical density gradient.

In the simple situation where the shear stress is set to zero at the upper and lower boundaries, the stability problem is solvable exactly, Nield (1967). The result is that the no motion state is stable if

$$Ra + Ra_c < Ra^* = \frac{27\pi^4}{4} \tag{2.39}$$

and

$$\frac{RaSc^2}{1+Sc} + \frac{Ra_c Pr^2}{1+Pr} < (Pr + Sc)\frac{27\pi^4}{4} \tag{2.40}$$

and unstable otherwise. In the present context we are interested in high Sc, small or moderately large Pr, and positive Ra_c and moderately negative Ra, i.e. destabilizing concentration gradient and stabilizing temperature gradient. Under these conditions it is the first, equation (2.39) that is limiting.

It should be noted at this point, that the gradient in the base state density is not necessarily a relevant indicator for stability. From the definitions of the Rayleigh numbers it is seen that the total density difference is proportional to

$$\Delta\rho = \rho(\beta_T \Delta T + \beta_c \Delta C) = \frac{\rho\nu^2}{gl^3}\left(\frac{Ra}{Pr} + \frac{Ra_c}{Sc}\right) \tag{2.41}$$

This does not coincide with the stability criteria (2.39, 2.40), and it is certainly possible to have instability even if the total density is decreasing upwards.

We return now to the full problem of a large melt domain and a solid growing upwards into the melt. This has been studied by, among others, Coriell et. al. (1980) and Hurle et.al. (1982). This amounts to carrying out a linear stability analysis in the style of that shown for the morphological stability above. The final result is obtained numerically. Figure 6 shows the stability boundary in terms of bulk concentration C_∞ vs V, with stable growth below the curves. This is based on the data of Coriell et. al. (1980) for a Pb-Sn system in a temperature gradient of $200K/cm$. The thick solid lines are the computations of Coriell et. al. (1980). The thick dashed lines are sketched to give a qualitative picture of the overall behavior of the system. Considering V fixed and increasing C_∞, we see that the convective instability precedes the morphological for low speeds, while the opposite holds at higher speeds. The two instabilities are also rather separated, which becomes evident when examining the wavelengths of the critical disturbances. At $V = 40\mu m/s$, just to the left of the junction between the convective and the morphological branches in figure 6, the convective instability gives a wavelength of $1.4mm$ (Coriell et. al. (1980)). At $V = 50\mu m/s$, the morphological mechanism gives a wavelength of $80\mu m$.

The morphological instability is bounded by the two asymptotic estimates discussed above, the constitutional undercooling in equations (2.35) or (2.36), (marked CU), and the absolute stability limit (AS), equation (2.37) or (2.38). This also illustrates the expected behavior that, keeping concentration fixed and increasing speed, the interface

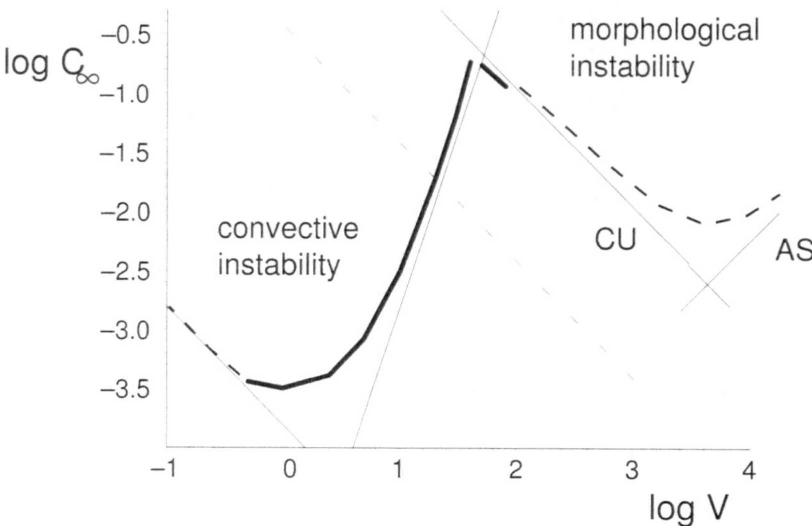

Figure 6. The stability limits in terms of critical $\log C_\infty$ (wt % Sn) as function of $\log V$ ($\mu m/s$) for a Pb - Sn system. The thick solid lines are based on the computations of Coriell et.al. (1980). The thick dashed lines are sketched. The various asymptotic stability boundaries are indicated as thin lines.

is morphologically stable for low speeds, becomes unstable at intermediate values of V, and is again stabilized at very high velocities. It should be mentioned however that at these high speeds different nonequilibrium effects, such as kinetic undercooling, may have to be taken into account.

The convective instability can also be bounded by simple estimates of stability. Returning to the double diffusive Benard problem discussed above, it would be natural to identify the thickness of the fluid layer with the thickness of the enriched solute layer ahead of the front, and use equation (2.39) to determine the onset of the convective instability. In fact, basing the scaling on the concentration difference over the enriched layer, $\Delta C = C_{top} - C_{bottom} = C_\infty - C_\infty/k_p = -C_\infty(1-k_p)/k_p$ (see figure 2), the temperature difference expected from the temperature gradient, $\Delta T = G_T l$, and the expected thickness of the layer $l = D_l/V$, we obtain the two Rayleigh numbers as $Ra = g\beta_T G_T D_l^4/(V^4\kappa\nu)$ and $Ra_c = -g\beta_c C_\infty D_l^2(1-k_p)/(V^3\nu k_p)$. Remembering that β_c is negative for a light solute, we see that Ra_c is positive. Introducing these into the stability criterion (2.39) and solving for C_∞, yields

$$C_\infty \quad > \quad \frac{k_p}{1-k_p}\frac{-\beta_T G_T D_l^2}{(-\beta_c)\kappa}\,\frac{1}{V} \quad + \quad Ra^* \frac{k_p}{1-k_p}\frac{\nu}{g(-\beta_c)D_l^2}\,V^3$$

Here β_T is negative for a fluid that has decreasing density with increasing temperature so that the first term is positive. The first term dominates for small enough V, and is shown as the line with a slope of -1 that bounds the convective instability from the left.

The second term, proportional to V^3, is the steep line to the right of the convective instability curve. This line also depends on the constant Ra^* in equation (2.39). It should be pointed out here that using a value for Ra^* that is obtained for a finite fluid layer, as the value $27\pi^4/4 = 657$ in equation (2.39), results in much too high values for the critical concentration. However, Hurle et.al. (1983) investigated solutal convective instability in the presence of an exponential concentration profile in a large fluid volume above a solid boundary, as the base profiles above, but without solidification and thermal buoyancy. This gives a much more accurate determination of the constant Ra^*. The value they obtained for Ra^* depends on Schmidt number and partition ratio, but it is bounded from below by the expression $2(1 + 1/Sc)$ which is obtained at $k_p = 0$. The line drawn to the right of the convective instability in figure 6 uses $Ra^* = 6.7$, which is entirely compatible with the data given by Hurle et.al. (1983) and gives a good fit. It can thus be concluded that the convective instability is indeed hydrodynamical in nature and that the morphological and the convective instabilities have a weak interaction.

Finally, to illustrate that the total density gradient is not the proper quantity for determining the stability properties, the line corresponding to constant density according to equation (2.41) is drawn as the thin dashed line with slope -1. Below this line the total density distribution is stable, but the convective instability sets in for much lower values.

The influence of convection on the stability of the front has been studied also in a number of different flow situations. The weak flow towards the front that is induced by a solidification shrinkage (i.e. when $\rho_l \neq \rho_s$) was studied by Caroli et.al. (1985) and Brattkus (1988). Brattkus showed the effects of different amount of solute rejection and Caroli et al. showed the effects of different interface speed. Delves (1971) studied solidification into a Blasius boundary layer, using the parallel-flow approximation and showed that the morphological stability was increased by a strong flow. Coriell et.al. (1984) numerically studied the linear stability of the interface in a plane Couette flow. Wheeler (1984) studied how a time dependent parallel flow affects the linear stability of the interface. Brattkus and Davis (1988a) and Brattkus and Davis (1988b) looked at the morphological instability of a solidifying rotating disc and a stagnation point flow. Forth and Wheeler (1989), and Hobbs and Metzener (1991) analyzed the linear stability in a parallel flow with an asymptotic suction profile. Merchant and Davis (1989) investigated a non-parallel flow produced by a temporally modulated stagnation-point flow. Schulze and Davis (1994, 1995) look at the situation with a linear combination of an asymptotic suction profile and a compressed Stokes layer.

A general result of these investigations is that disturbances with wave vector in the spanwise direction will be unaffected by the flow and disturbances in the flow direction can be both stabilized or destabilized. The reason for this alternating stabilization effect is that the flow does not only steepen the solute profile, which would make the interface more unstable, but the flow also distributes the solute laterally which can give a stabilization of the interface. Recently, Bühler and Davis (1998) showed that for moderately small ratios between the flow-field and the morphological disturbance wave number, the disturbed interface becomes localized between stagnation points of the flow and the pattern moves within a fixed envelope.

Some attempts to study effects of forced or large scale natural convection on the stability of plane interfaces in experiments on real crystals are reported by Carlberg (1995), Carlberg (1997), Tewari and Chopra (1992).

3 Dendrites

As we have seen above, in most practical situations solidification interfaces undergo a fingering instability. These often develop into what is called dendrites (from the greek *dendros*, tree) see figure 7. This picture shows the result of a representative simulation of isothermal solidification of a binary alloy. It shows a total of six dendrites that have grown in the six preferred growth directions from a small spherical nucleus. In many metals dendrites indeed resemble trees with a main stem and sidebranches, where the apparently anisotropic growth is due to the anisotropy of the growth kinetics. They may typically be of the order of micrometers up to fractions of a millimeter in size. Dendrites are maybe the most common microstructure that grows naturally during solidification of alloys and pure metals, see for instance the standard text by Kurz and Fisher (1992), Davis (2001), Huang and Glicksman (1981a), Huang and Glicksman (1981b) and Glicksman and Marsh (1993).

3.1 The Ivantsov solution for an isolated needle crystal

The study of fully developed dendrites start with the similarity solution obtained by Ivantsov (1947). This builds on the observation that the diffusion equation will allow a solution that is translating with constant speed, and where temperature is constant on paraboloids. We will first focus on growth in a pure quiescent melt. Ahead of the growing crystal the melt is assumed to have a constant temperature T_∞. The melt is undercooled, i.e. $T_\infty < T_m$, where T_m is the melting temperature. In order to obtain the solution we will assume that it grows at constant speed V, and that the solid-liquid interface has constant temperature. The last statement thus implies that curvature and kinetic effects are neglected in the Gibbs-Thomson relation, equation (2.13).

The mathematical problem thus reduces to solving for the temperature from equation (2.1) in a coordinate system that translates with a constant speed V, see figure 8. The temperature is calculated in parabolic coordinates and is obtained explicitly as function of the spatial coordinates, see Davis (2001) for an accessible account of the details. The most interesting result of this calculation is obtained when the solution for temperature thus obtained, is inserted into the flux condition, eq (2.7). This will place a restriction on the solution and thus on the relation between the tip speed V and the undercooling $\Delta T = T_m - T_\infty$, and via the solution for temperature, the geometry of the tip. This relation can be written as follows:

$$\Delta = P\, e^P\, E_1(P) \tag{3.1}$$

where E_1 is the first exponential integral

$$E_1(P) = \int_P^\infty \frac{1}{z} e^{-z} dz. \tag{3.2}$$

Here $\Delta = c_p(T_m - T_\infty)/L_f$ is the nondimensional undercooling and $P = RV/2\kappa$ is a growth Peclet number, with R denoting the radius of curvature at the tip, see figure 8.

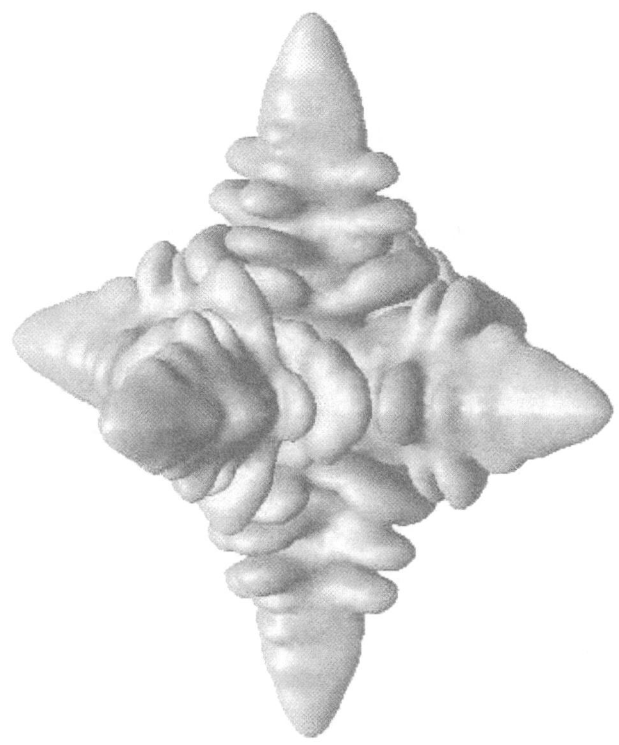

Figure 7. Example of a simulated dendrite (I. Loginova).

The expression (3.1) thus gives the growth Peclet number as a function of undercooling, but it does not specify the speed or the tip radius, only their product. This is a consequence of the fact that surface energy etc has been neglected, so that there is no natural lengthscale in the problem.

Some additional information is thus required in order to determine the tip speed and radius. We should remember that the Ivantsov solution only describes the diffusion ahead of the tip, and does not contain any information on the phase change kinetics at the solid-liquid interface. One way to add the missing information is to invoke the 'marginal stability hypotheses', Langer and Müller-Krumbhaar (1978), Langer (1980). The essential idea is to use the stability properties of the interface, similar to the Mullins-Sekerka stability studied in section 2.3. The hypothesis consists of assuming that the tip radius R is approximately equal to the wavelength of a marginally stable disturbance, so

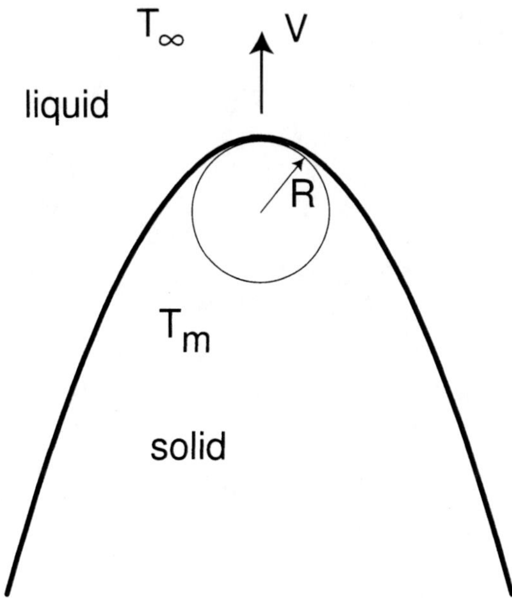

Figure 8. Sketch of a parabolic dendrite tip

that the following parameter is constant:

$$\sigma^* = (\frac{\lambda}{2\pi R})^2 \tag{3.3}$$

Here λ is the cut-off wavelength, i.e. a wavelength corresponding to a disturbance with growth rate zero, such that all shorter waves are damped. Crudely speaking, unstable disturbances with wavelengths longer than the tip radius will travel down the stem of the dendrite and may grow into side branches. Wavelengths shorter than the tip radius, had they been present, might grow and make the tip sharper, thus restoring the situation where $\lambda \approx R$.

The stability analysis gives the following expression for λ, for the thermal case

$$\lambda = 2\pi\sqrt{d_0 l}$$

where $l = \kappa/V$ is the diffusion length scale based on a growth velocity V, and $d_0 = c_p\sigma_0 T_m/L_f^2$ is a capillary length scale. Taken together the definition of the stability parameter σ^* may be written is

$$\sigma^* = \frac{d_0}{R}\frac{1}{P}, \tag{3.4}$$

where $P = RV/(2\kappa)$, the same growth Peclet number as defined above.

Equation (3.4) together with the Ivantsov solution (3.1) now determine the growth Peclet number and the tip radius. A value of $\sigma^* = 1/4\pi^2 \approx 0.025$ gives a reasonable fit to experiments.

More rigorous approaches include the microscopic solvability theory (MST), Meiron (1986), Kessler and Levine (1986), and interfacial wave theory (IWT), Xu (1997). Microscopic solvability derives conditions for the existence of steady state solutions. These would imply that surface energy anisotropy would be crucial for the existence of solutions; for isotropic non-zero surface energy there are no admissible solutions, but with finite anisotropy possible solutions emerge. In the interfacial wave theory, Xu (1997) carries out a rigorous stability theory on a needle crystal and finds timedependent growth solutions at very low levels of the anisotropy.

3.2 The phase field method

The hard part in numerically solving the system of equations, eq. (2.1-2.13), is the tracking of the solid/liquid interface in such way a that the boundary conditions on the solid/liquid interface are fulfilled accurately. This task gets very hard when the solid-liquid-domains change topology through fusion and fission of different solid domains. The phase-field method is a common and powerful method used to study the evolution of microstructures in solidification problems. The governing equations are derived from knowledge of the thermodynamic potentials of the system, together with the assumption of a diffuse solid/liquid interface. Even though there exists numerous different phase-field formulations, for pure melt, the numerical properties are rather similar. Numerically, the interface is given a finite width δ and the front is tracked implicitly by an extra variable, ϕ (order parameter), that is governed by an extra field equation. Also, the heat equation is modified to include a source term for the release of latent heat in the interface region. The drawback is the extra field equation, an interface region with large spatial gradients in ϕ and numerical problems for low undercoolings.

Kobayashi (1991) pioneered the use of the phase field method in this context. See also Karma and Rappel (1996b). Wang et.al. (1993), Fried and Gurtin (1996), Sekerka and Bi (2000), Anderson et.al. (2000), Penrose and Fife (1990) and Davis (2001), ch 11, Provatas et.al. (1999b), for more information on the foundation of the method. The method has been applied to several different systems, which are not covered here, where we in the following will focus on thermal solidification of a pure material. Warren and Boettinger (1995) shows an application of the method to a binary alloy solidifying isothermally. Loginova et.al. (2001) have made simulations of a nonisothermal binary alloy.

We will follow Karma and Rappel (1996a) and Karma and Rappel (1996b) and formulate a phase field method that is used to track the solid/liquid interface. The variable ϕ is used as the phase-field variable (or the order parameter) and it varies between -1 in the melt to 1 in the solid. In this method the interface kinetic and interface energy term have the same β dependency, thus $\zeta(\beta) = \eta(\beta)$ in equation (2.7). A common solid-lattice structure as BCC of Nickel gives a fourfold symmetry in the anisotropy and therefore η is assumed to vary as $\eta(\beta) = 1+\epsilon cos(4(\beta-\beta_0))$, where ϵ is the magnitude of the anisotropy, as written in eq (2.12). In the simulations it is assumed that the lattice of the nucleus has an arbitrary orientation and this orientation is controlled by the parameter β_0. This results in that without any flow, one of the main-branches would eventually grow in the direction given by $\beta = \beta_0$.

In the following we will restrict ourselves to the case of pure materials and thermal solidification, and thus disregard any concentration effects (c=0 in the equations above). We will invoke the same Boussinesq approximations as above, and also assume that all relevant properties are equal in the solid and liquid phases. In this case the equation for the order parameter ϕ that distinguishes solid and liquid is:

$$\tau\eta^2\frac{\partial\phi}{\partial t} = [\phi - \Delta\lambda\theta\,(1-\phi^2)](1-\phi^2) + \nabla\cdot(\eta^2\nabla\phi) - \frac{\partial}{\partial x}\left(\eta\frac{\partial\eta}{\partial\beta}\frac{\partial\phi}{\partial y}\right) + \frac{\partial}{\partial y}\left(\eta\frac{\partial\eta}{\partial\beta}\frac{\partial\phi}{\partial x}\right), \quad (3.5)$$

where τ and λ are constants which are chosen according to Karma and Rappel (1996b),Karma and Rappel (1996a), to impose the modified Gibbs-Thomson condition. In the equations in this section we have chosen a diffusive velocity scale, i.e., $U = \kappa/l$.

When used in equation (3.5), the angle β is calculated using the spatial derivatives of ϕ, according to:

$$\beta = \arctan\left(\frac{\partial\phi}{\partial y} \Big/ \frac{\partial\phi}{\partial x}\right).$$

Since the interface is now treated as a diffuse region, the conditions at the liquid-solid interface, eq (2.7,2.13) must instead appear in the equations, which must be formulated to be valid in both the solid, liquid and interfacial regions. In addition to the phase-field equation (3.5), the equation for conservation of heat, eq (2.1), is modified to

$$\frac{\partial\theta}{\partial t} + \frac{30g(\phi)}{\Delta_c}\frac{\partial\phi}{\partial t} + \bar{u}\cdot\nabla\theta = \nabla^2\theta, \quad (3.6)$$

where the ϕ-term corresponds to release of latent heat and $g(\phi) = \phi^2(1-\phi)^2$.

The momentum equation for melt flow, eq (2.5), must also be modified to be applicable over both phases. Here this is done by allowing the viscosity to depend on the phase field variable ϕ:

$$\frac{1}{Pr}(\frac{\partial\bar{u}}{\partial t} + \bar{u}\cdot\nabla\bar{u}) = -\nabla p + \nabla\cdot(f_\nu[\nabla\bar{u} + (\nabla\bar{u})^*] - Ra\theta\bar{e}_g), \quad (3.7)$$

where Pr and Ra are the Prandtl and Rayleigh numbers as before, and * means the transpose of the tensor. f_ν is a function of the phase-field variable ϕ and it introduces an increase of the viscosity on solidification. We have used the following definition of f_ν:

$$f_\nu = \begin{cases} 1 & \phi < -0.6 \\ 1 + 10\,(\phi + .6)^2 & \text{otherwise} \end{cases} \quad (3.8)$$

Inside the solid, where ϕ is close to one, the velocity is set to zero. The dependency of viscosity on ϕ in the interface gives a smooth increase of drag there. The simulated results are rather insensitive to a variation of f_ν and $f_\nu \equiv 1$ gives only a small decrease in accuracy.

Beckermann et. al. (1999a) suggest another way to immobilize the material in the solid phase. They use a modification of the Navier-Stokes equation by adding a Darcy term, i.e. considering the interface region as being porous, as a very thin mushy layer.

Still neglecting solidification shrinkage, i.e. assuming $\rho_l = \rho_s$, the volume of the material must be conserved during the process, i.e.

$$\nabla\cdot\bar{u} = 0, \quad (3.9)$$

as before.

Equations (3.5, 3.6, 3.7, 3.9) constitute a set of equations for the phase, the temperature and the flow and pressure. Below some cases where this has been solved will be discussed.

3.3 Convection effects on the growth of fully developed dendrites

Convection in the melt is often an important effect in solidification and may be desirable, or at least difficult to avoid, in practical processes. One driving force for convection may be gravitation in the presence of density variations within the melt. In many practical materials processes strong forced convection is maintained in the melt, for instance to achieve mixing by using stirrers of different kinds. As another example, large temperature gradients over a free surface may cause Marangoni convection, which is known to be the primary driving force for melt convection in welding. A much weaker flow that may still be important is the Stefan wind that appears when a material solidifies in the presence of solidification shrinkage.

The existence of convection in solidification processes sometimes plays a fundamental role in the phase transition. Convective heat and mass transfer alters the concentration and the thermal gradients in the vicinity of the solid/liquid interface, and thus the growth rate and the size of the crystal.

Forced convection

The case of forced convection with a direction perpendicular to the growth direction of the tip has been treated with different approximations of the fluid flow, such as Oseen, Euler, Stokes and matched Stokes and Oseen flows. In these theories the flow is assumed not to modify the shape of the tip. The isotherms (or solute isopycnals) turn out to be parabolic, with the exception of the matched solution. By an approach inspired by microsolvability theory Bouissou and Pelce (1989b) succeeded to estimate how the stability parameter is affected by an Oseen type fluid flow. Lee et.al. (1992) solved the Navier-Stokes equation numerically by considering a local region very close to the tip of an isothermal dendrite. They concluded that Oseen viscous flow and Stokes flow approximations overpredict the growth Peclet number, P, when the flow Peclet number is greater than 2.

Experiments (Lee et.al. (1993), Bouissou et.al. (1989a)) have shown that for a large range of fluid flows the shape of the tip does not change, only the stability parameter σ^* is affected. The change is significant only for fluid flows that are ten times larger than the tip velocity. Many experiments Bouissou et.al. (1989a); Lee (1991), indicate that the operating point of the tip of a single dendrite is rather independent of the component of melt velocity that is perpendicular to the growth direction. This changes when the flow direction makes more than 60 degrees angle with the growth direction. The perpendicular component of the fluid flow velocity however has a large effect on the sidebranches. Tönhardt and Amberg (1998) have done 2D simulations of a dendrite that grows in a shear flow. They showed that growth of the side-branches were promoted and restrained on the upstream and downstream side of the tip, respectively. Moreover, a split of the tip of the dendrite could occur for strong flows.

Experimentally it has been shown that a forced flow above an array of dendrites will result in a tilted growth of the array Fredriksson et.al. (1986), Murakami et.al. (1983) and Murakami et.al. (1984). Fredriksson et.al. (1986) investigated the effect of stirring the melt (Al-Cu) during the solidification. They concluded that the tilting increased in the columnar zone as function of increased stirring rate to reach a maximum. It was also observed that the copper concentration increased the tilting.

Dantzig and Chao (1986) investigated this numerically by studying an array of needle crystals of an assumed simple shape in a shear flow. They investigated the steady state shape of the array in a translating coordinate system and succeeded, for most parameters, to compute the tilting angle of the array. In this work the surface energy was assumed isotropic and no interface kinetics was used. The maximum computed tilting angle was 30 degrees. The 3D simulations by Dantzig and Chao (1986) revealed the qualitative difference between 2D and 3D. In 3D the fluid may flow horizontally around a vertical main-branch. Hence, this gives an effective transfer of heat or solute from the upstream to the downstream side of the main-branch. In contrast, in a 2D simulation all transfer must go vertically around the tip of the main-branch and therefore the transfer is less effective.

Esaka et.al. (1996) have also studied the tilting angle as function of carbon content in an iron-carbon system, and compared calculation with experiments where a melt flowed down an inclined plate. They used two prescribed 2D needle dendrites and an empirical relation for the primary spacing. The tilting angle was assumed to be the same as the direction of the steepest concentration gradient and they observed an angle increasing with the carbon content at low concentrations (0.1 %mass), reaching about 20 degrees at around 1 %mass.

Convective effects on fully developed dendrites have not been studied using first principle simulations until quite recently. Sekerka and Bi (2000) and Anderson et.al. (2000) have derived phasefield formulations from thermodynamics including convection. The growth of dendrites in a uniform forced flow has been studied by Tong et.al. (2000), Beckermann et. al. (1999a), and Diepers et.al. (1999b) and Lan and Hsu (2001). These simulations are all two dimensional, but fully three dimensional simulations of dendritic growth in a uniform forced flow has been done by Al-Rawahi and Tryggvason (2000), Jeong et.al. (2001). The growth of a dendrite into a shear flow, Tönhardt and Amberg (1998, 2000a), and natural convection effects, Tönhardt and Amberg (2000b), will be discussed in detail below.

Natural convection

During the last decades there has been a continuing interest in natural convection effects on the morphology and growth of individual dendrites. One reason for this is that natural convection is frequently suspected to influence experimental results that are primarily intended for studying the diffusion limited growth, Glicksman et.al. (1994), Glicksman et.al. (1986), Glicksman et.al. (1995), Kallungal et.al. (1977) and Huang and Barduhn (1985). A recent review is available in Lee et.al. (1996). For pure Succinonitrile (SCN) there are quantitative measurements of the velocities and radii of the tips of the main stems (Glicksman et.al. (1994), Glicksman et.al. (1986), Glicksman et.al. (1995), Lee

(1991), Koss et.al. (1996)). Experimental and theoretical work has shown, rather unexpectedly, that the effects of natural convection increases at low undercooling, due to the increasing size of the dendrites, Ananth and Gill (1988). One indication of this is that when terrestrial measurements are compared to theories that disregard convection, such as the classical Ivantsov theory discussed above, the agreement is good only for a small interval of undercooling. For undercooling above 4-5 K there seems to be interface kinetic effects. For growth below this undercooling, but above 2 K, the Ivantsov solution agrees well with experiments. Moreover, the Ivantsov solution in combination with a constant value of the stability parameter, gives a theoretical tip velocity and tip radius that are almost the same as measured in terrestrial experiments. Below 2 K the terrestrial growth again deviates from this theoretical operation point of the tip. Under terrestrial conditions it has also been observed that the dendritic growth depends on the direction of the gravitational field. The side-branches close to the tip of the dendrite and the operation point of the tip change depending on the growth direction relative to gravity. Also, it has been found that the operating point of the tip of a dendrite is different in terrestrial and micro-gravity experiments. It has also been shown that the volume-change at the solid/liquid interface which results in a Stefan wind has only a minor effect on the operating point of the tip Pines et.al. (1996), McFadden and Coriell (1986).

Gill and coworkers have investigated theoretically the effects of forced and natural convection on dendrite growth in a series of papers, Lee (1991), Ananth and Gill (1988), Ananth and Gill (1997), Lee et.al. (1992), Lee et.al. (1993), using an approximation that is valid close to the dendrite tip. Ananth and Gill (1988) use this to solve the Navier-Stokes equations with thermal buoyancy for a shape preserving paraboloidal dendrite tip, and compute for instance the growth Peclet number as a function of undercooling. They show, as was also observed experimentally, that natural convection becomes important for sufficiently small undercooling, and that the radius of the dendrite tip is the characteristic length scale for the thermal convection. Canright and Davis (1991) use a different approximation, which assumes small buoyancy, but is valid in a larger region around the tip. They study among other things the influence of the Prandtl number. Sekerka et.al. (1995) introduce the natural convection fluid flow by replacing the dendrite array by an isothermal sphere that encloses the dendrites and then treating the case of the sphere surrounded by cold melt. The stagnant thermal boundary layer around the sphere is then used to modify the Nusselt number for the tip of a dendrite.

Simulation of dendritic growth into a shear flow

We study the evolution of a small nucleus in a pure undercooled melt, where the melt flows due to an applied shear stress far away from the nucleus. This boundary condition was chosen because the local flow around the crystal will then resemble that which would exist on a solid wall in the presence of a large scale exterior flow provided that the separation of length scales is sufficient. The computational domain is rectangular with the lower boundary representing a solid wall, see fig. 9. The melt flows from left to right, entering at the left vertical boundary and leaving to the right. On the top boundary a constant shear is applied, which produces a linear velocity profile (except for the influence of the growing dendrite).

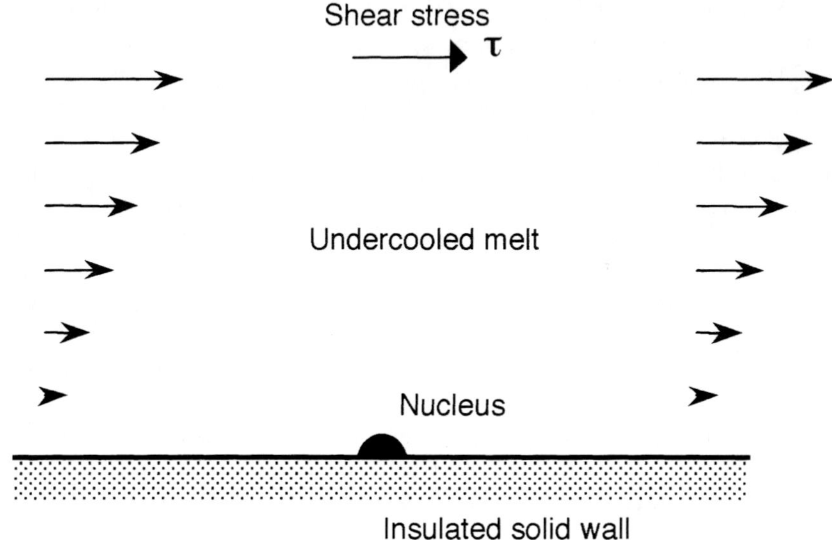

Figure 9. Sketch of a nucleus growing from a solid wall into a shear flow.

The assumed constant value of the shear stress τ at the top gives a dimensional velocity gradient τ/μ. With the diffusion velocity taken as reference velocity the nondimensional velocity gradient is then $(\tau l/\mu)(l/\kappa) = Pe$, i.e. a Peclet number based on the velocity at a distance l from the wall. This is the parameter that will be used to quantify the strength of the flow. In the simulations of shear flow below, the length scale l was chosen to be $2.1 \cdot 10^{-4} cm$. This value of l corresponds to approximately ten times the marginal stability wavelength when a dendrite grows in an undercooled melt of pure Nickel and the undercooling is $\Delta_c = 0.5$. To reduce the influence of the artificial boundaries the computational domain is of the size 20W×10W here. We have used a value for the anisotropy parameter ϵ of 0.0045 for the simulations with $\Delta_c = 0.5$.

In the convection free case a dendrite with a bcc-lattice will grow faster in the four 100-lattice-directions. To begin with we will assume that these are (arbitrarily) oriented in the directions normal and parallel to the wall. In the expression for anisotropy, eq. (2.12) this means that we have a constant value of $\beta_0 = 0$. With convection, the parts of the crystal positioned to the left and right of the tip of the vertical main stem are referred to as the upstream and downstream part, respectively.

Fig.10 shows the result of a simulation with Pe=0, the different contours represent the solid/liquid interface position at different times (t=0,...,0.9). The distance to the artificial outer boundaries does not seem to affect the simulation, since the tips have a constant velocity after an initial transition. A calculation of the growth Peclet number, $P = VR/2\kappa$, result in that $P = 0.185$ which is in good agreement with the two dimensional version of the Ivantsov similarity solution. The sidebranches closest to the initial nucleus are rather large, this is due to that the shape of the initial nucleus was a half circle with a small disturbance. The disturbance survives for a long time and grows to form the first secondary sidebranches. Hence, this shows how sensitive the growth of the sidebranches is to disturbances at the interface. All the tips of the main stems evolve with almost exactly the same radius, R, and velocity, V, and also the sidebranch evolution is almost identical on the different main stems. Another observation is that on the vertical main stem the sidebranches on the opposite side of the tip seem to evolve symmetrically, at least close to the tip, and this occurs with a well defined wavelength. It is not clear why this symmetry exists, but it might be due to that the disturbance that triggers the sidebranches is generated at the very tip of the vertical main stem and then travels symmetrically down the tip.

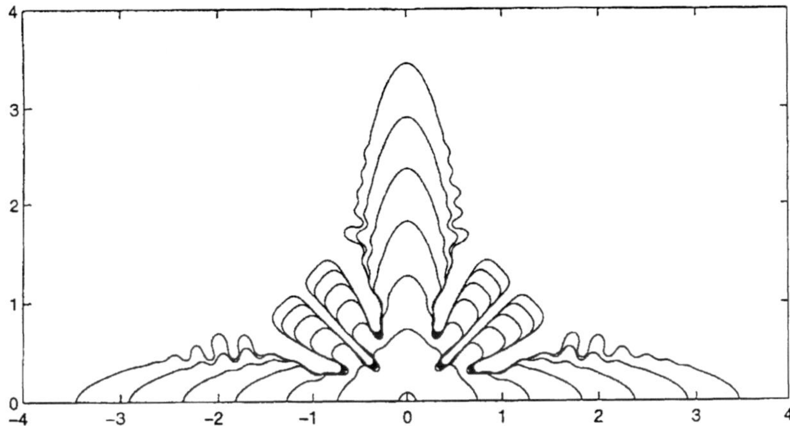

Figure 10. The evolution of a nucleus in the absence of convection. The contours represent the position of the liquid/solid interface at different times. The time difference between consecutive contours is 0.15, with the innermost and outermost contours at $t = 0$ and 0.9, respectively. From Tönhardt and Amberg (1998).

Fig. 11a shows the solid/liquid interface and the isotherms for Pr=0.03 and Pe=1 at t=0.9. The small asymmetry in the thermal field reveals some convective influence. This Peclet number corresponds to the moderate Reynolds number, $Re = Pe/Pr$, of 33 and therefore it is not surprising that the flow follows the interface without any creation of separated vortices downstream of the main stem. From the upstream position cool melt is transported towards the crystal, causing a steeper temperature gradient on the upstream side of the crystal. At this time, the sidebranches on the upstream side of the vertical main stem are almost of the same size. On the downstream side the secondary branches are prevented from growing, because warm fluid is transported around the tip and pushed along the solid/liquid interface. This flow also hinders growth of secondary branches on the downstream horizontal main stem. Conversely, growth of secondary branches is promoted on the upstream horizontal main stem.

Fig. 11b shows the case of Pr=0.03 and Pe=2.5 at t=0.9. The morphology of this

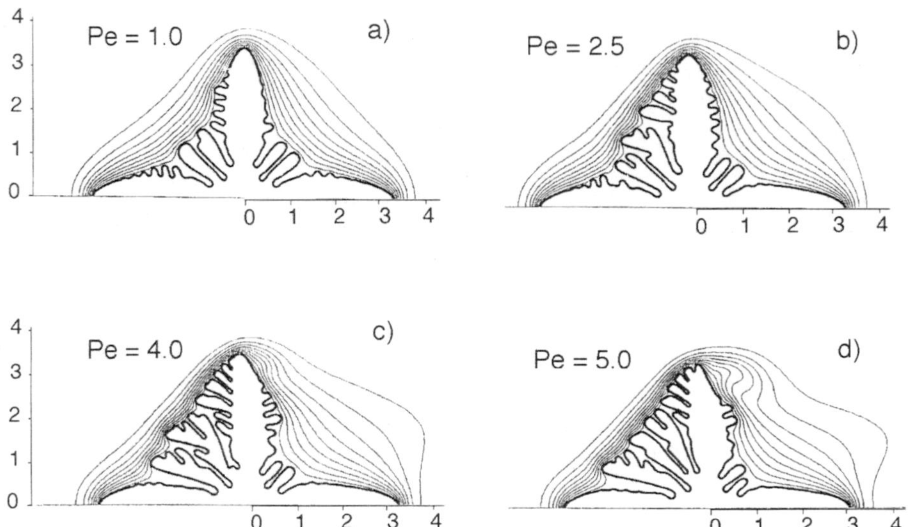

Figure 11. The solid liquid interfaces and the temperature field in the melt for a sequence of increasing Peclet numbers. The parameters are $\Delta = 0.5$, $Pr = 0.03$, and the dendrites are displayed at time $t = 0.9$. The values of the Peclet number are 1 in a), 2.5 in b), 4 in c) and 5 in d).

crystal is quite different from the one in fig. 11a. As the isotherms reveal, the Reynolds number is high enough to create a small separated vortex in the bulk. The vertical main stem starts to tilt towards the upstream direction. On the upstream side of this main stem the growth of the sidebranches is rapid and it also starts very close to the tip. At the tip there is a compressed boundary layer which increases the local Peclet number. The compressed isotherms promote the growth of small disturbances that otherwise would hardly be visible. On the upstream side of the vertical main stem most of the sidebranches are large. In between some of the large sidebranches there are some small sidebranches. These small sidebranches have almost stopped growing, because they are shielded from the cold melt and the large sidebranches are growing nearly in the horizontal preferred crystal direction. On the downstream side the sidebranches have also been promoted, due to the small vortex, and they are of almost the same size. The two horizontal main stems have only small sidebranches. On the downstream side the fluid flows more or less along the solid/liquid interface. This flow along the horizontal main stem results in that the sidebranch growth is inhibited. The three main stems are growing at the same velocity, despite the very different sidebranch structure.

When Pr=0.03 and the Peclet number is increased to 5 then at t=0.9 this results in fig. 11d. The vertical main stem has become more tilted and the isotherms reveal a large advective effect. The separated vortex behind the main stem transports cold bulk melt that hits the downstream side of the main stem at a distance about one third of the dendrite length from the tip. This causes the two distinct sidebranches and the flat side of the tip. The local Peclet number at the tip has become large enough that the asymmetrical compression of the isotherms overcome the anisotropy of the solid/liquid interface and the tip splits. The downstream horizontal main steam looks like a needle crystal and this is due to the slow fluid motion at this interface. Upstream sidebranches look a bit isotropic, but their envelope is a straight line. If the growth of the side-branches was isotropic then the envelope would have the form of a circle segment. The individual tips of the sidebranches grow not radially, they rather grow in the direction of the preferred horizontal lattice direction. The sidebranch closest to the solid wall has got tertiary sidebranches and ,judging from the isotherms, they will grow slowly in the future. At this stage, the main stems grow almost with the same velocity, except for the downstream horizontal main stem that grows a bit slower. In fig. 12 streamlines are plotted for this simulation. The flow field shows a characteristic moderate Reynolds number flow over an obstacle, with a weak recirculating vortex behind the tip of the main stem. This separated vortex is also evident from the isotherms.

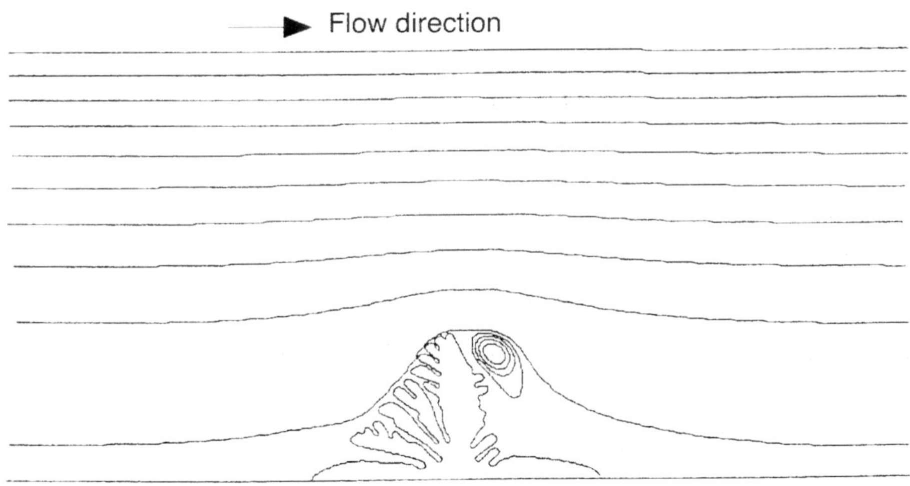

Figure 12. Streamlines around the dendrite for the case displayed in figure 11d.

We have so far used a flow Peclet number, Pe, based on l and U as the characteristic length and fluid velocity, respectively. The most interesting effects occur at the tip of the vertical main stem. Therefore it is more appropriate to use a local Peclet number, Pe^*, at the tip when discussing the physics of the growth. We use $Pe^* = U_\infty H/\kappa$, where H is the length of the dendrite from the tip to the solid wall and U_∞ is the free stream flow velocity at the vertical distance H above the wall i.e. $U_\infty = \tau H/\mu$. In the same way, we use a local growth Peclet number $P^* = V_n H/\kappa$, where V_n is the actual growth velocity at the tip. Since the growth is controlled by the local heat transfer, a local Nusselt number is also of interest. In analogy with forced convection heat transfer we use $Nu^* = Pe^{*1/2}$ and $Nu^* = Pr^{-1/6}\, Pe^{*1/2}$ when $Pr = 0.03$ and $Pr = 23.0$, respectively.

The growth is controlled by local heat transfer from the tip. There are different contributions to heat transfer, which may be summarized as follows:
1) quasi-steady growth. The thermal field has not reached its steady state, but is quasi-steady due to a slow growth and weak convection. This may appear typically in 2D in the case of very low undercooling.
2) diffusion controlled growth. The convection is small, but the growth is rapid enough to cause a formation of a thermal boundary layer around the tip.
3) flow controlled growth. The thermal field is determined by a sufficiently rapid convection past the tip.
In terms of the flow Peclet number, Pe^*, and growth Peclet number, P^*, we can determine the type of growth according to:

$$max(1, P^*, Pe^*) = \begin{cases} 1 & = \text{quasi static diffusion controlled growth} \\ P^* & = \text{diffusion controlled growth} \\ Pe^* & = \text{flow controlled growth} \end{cases} \quad (3.10)$$

In intermediate cases when Pe^* and P^* are of order one, the statement is of course not precise.

The dimensionless numbers of interest for the different simulations have been summarized in table 3.3, including also several other flow cases that have not been described in detail here, see Tönhardt and Amberg (1998).

Fig.11a shows a simulation with a slow convection and according to table 3.3, $Pe^* \approx P^* \approx 13$. The isotherms are a bit asymmetric and there is a thermal boundary layer at the tip. Close to the tip the isotherms are almost symmetric and the growth is mainly diffusion controlled with small convective effects. In this respect this case is an intermediate case which is also predicted by eq. (3.10). When the fluid velocity is increased the result is shown in fig.11b-d. For both the simulations with nominal $Pe = 2.5$ and 5, $Pe^* > P^* \approx 13$ and in both simulations the isotherms are very asymmetric at the tip. This is thus examples of increasingly flow controlled growth, in agreement with eq. (3.10).

Influence of the orientation of the nucleus.

As was discussed above it has been observed experimentally that a flow past a growing array of dendrites that is perpendicular to the overall direction of growth, may result in a tilting of the dendrites, Fredriksson et.al. (1986), Esaka et.al. (1996). The major reason for this is the asymmetrical heat and mass transfer at the tip. If the anisotropy of the kinetics is sufficiently weak, the asymmetric cooling may cause the tip to grow in a direction that deviates substantially from the preferred direction, without actually altering this. Another possibility is that, as the grain grows, lattice faults are generated so that the crystal lattice orientation actually changes and turns toward the one that is promoted by the asymmetrical cooling.

figure	Δ	Pr	Pe	P^*	Pe^*	Nu^*	H/l
10	0.5	0.03	0	13	0	0	3.5
11a	0.5	0.03	1.0	13	12.5	3.5	3.5
11b	0.5	0.03	2.5	13	30	5.4	3.5
11d	0.5	0.03	5.0	13	60	7.7	3.25
9 in TA98	0.5	23	5.0	15	80	5.3	4.1
10 in TA98	0.5	23	10.0	15	160	7.6	4.25
11 in TA98	0.5	23	15.0	15	240	9.3	4.6
12 in TA98	0.1	0.03	0	0.2	0	0	10.5
13 in TA98	0.1	0.03	2.5	0.35	12.5	3.5	15
14 in TA98	0.1	0.03	5.0	0.35	25	5.0	16

Table 1. The approximative values of the local growth Peclet number P^*, the local flow Peclet number Pe^*, the dimensionless height and the local Nusselt number Nu^* for the different figures. The bottom 6 entries refer to figures in Tonhardt and Amberg (1998) (TA98), where more details are given.

However, in the simulations that we have carried out, we have for reasonable values of the anisotropy always found that the growth direction of the main stem remains quite close to the preferred direction, as in figure 11, where the main stem grows close to the preferred direction perpendicular to the wall. At least in these simulations, the anisotropy is much too strong to allow a tilting of the dendrites away from the growth direction of the magnitude that is observed in experiments. Our simulations do not include the possibility of creation of crystal faults that will change the preferred growth direction, though.

Still another possibility that would allow a large tilting of dendrites in a shear flow is to consider how the competition between initially randomly oriented nuclei on a wall with a shear flow would evolve as time proceeds. One could ask which orientation is optimal, in the sense that a nucleus oriented this way may outgrow differently oriented nuclei. Our approach is to look at an isolated nuclei and study its evolution. It is thus assumed that nuclei are placed sufficiently far apart, and interaction between neighboring dendrites is not considered here. Still, if the growth is significantly faster for a certain orientation, this could be indicative of the one that dominates after an initial period. We will thus take the inclination obtained for an isolated dendrite with optimal growth rate as an indication of that which might be encountered experimentally. In particular we will note that, in the presence of reasonable crystal anisotropy, the growth rate of the dendrite is quite sensitive to the orientation of the nucleus, and the main stem of the dendrite deviates very little from the preferred direction even in the presence of rather strong flow.

In this section, all simulations are for the dimensionless undercooling Δ equal to 0.1, a Prandtl number, Pr, of 23.1, and an anisotropy parameter $\epsilon = 0.015$, approximately corresponding to reasonable values for SCN. The influence of variations in β_0 and Pe has been investigated. The orientation of the nucleus, β_0, has been varied between -45 and 45 degrees, typically by a step of 7.5 degrees. Note that the β_0 values -45 and 45 degrees represent the same case, due to the four-fold symmetry in the anisotropy of the surface energy and the boundary conditions at the solid wall. The reference length l is here taken as $1.33 \cdot 10^4\ d_0 = 34 \mu m$, with $d_0 = c_p \sigma_0 T_m / (\rho_0 L_f^2)$ and data for SCN. The computational domain size extends from $-520l$ to $780l$ in the horizontal and from 0 to $520l$ in the vertical direction.

Figure 13 shows the result for ϵ=0.015 and $Pe = 1$. The figure shows the solid/liquid interface with and without convection for four different values of β_0, at time=20. Figure 13a, for β_0=-45 (or 45) degrees, shows the longest branch that was obtained, for this time. The flow has compressed the isotherms at the upstream tip and this results in the increased growth velocity. The nondimensional tip radius and velocity is about 0.082 and 0.31, at this time, and the velocity is increasing. The Reynolds number, $Re = Pe/Pr$, based on the free-stream velocity at the tip of the upstream branch and the vertical distance from the wall to the tip is less than 1.

Between the upstream branch and the solid wall there is a weak clockwise vortex which has a minor influence on the heat flux at the interface. About 3 radii below the very tip of the vertical branch on the upstream side, the main flow begins to follow the solid/liquid interface and remains attached all around the tip. At the downstream branch the flow has a moderate effect on the growth velocity and the fluid field is almost parallel to the isotherms.

As β_0 is increased to $\beta_0 = 0$, some small secondary-branches appear on the vertical branch as shown in figure 13b, with the corresponding temperature field in 13e. The vertical branch is the one that has had the fastest growth. It has grown less than the upstream branch in figure 13a, but it has nevertheless reached a larger vertical distance from the wall. Due to the low Re, the main flow follows the interface all around the vertical tip from about 4-5 tip radii below the vertical tip on the upstream side.

At the downstream branch there is a vortex that transports fluid towards the tip, but this is warm fluid that has been heated by the release of latent heat from the upstream and vertical branches. Therefore, the growth velocity is reduced for this branch.

On the upstream and vertical branches some instabilities have appeared. Apparently the convection has promoted this instability. The growth velocity for these secondary branches decreases with time due to the shielding effect of the (main-) branches.

In figure 13c, the flow follows the interface around the tip of the upstream vertical branch from about 5 and to about 8 tip radii below the tip on the upstream and downstream side, respectively. The flow increases the growth of both the upstream vertical and the (wall) upstream branch.

At the tip of the downstream branch there are two weak clockwise vortices, one is located above the vertical side of the branch while the other is located a bit downstream of the tip. These vortices cause warm fluid to flow towards the tip, and the growth of the branch is decreased.

Figure 13d is for $\beta_0 = 30$ degrees. This value of β_0 results in the branch that has grown the farthest vertically, at this time. The main flow follows the interface around the upstream vertical tip. The attachment point for the main flow is about 3 tip radii below the tip on the upstream side and the main flow follows the interface until a bit beyond the tip of the downstream branch, where a small clockwise vortex appears. The resulting isotherms are shown in figure 13f.

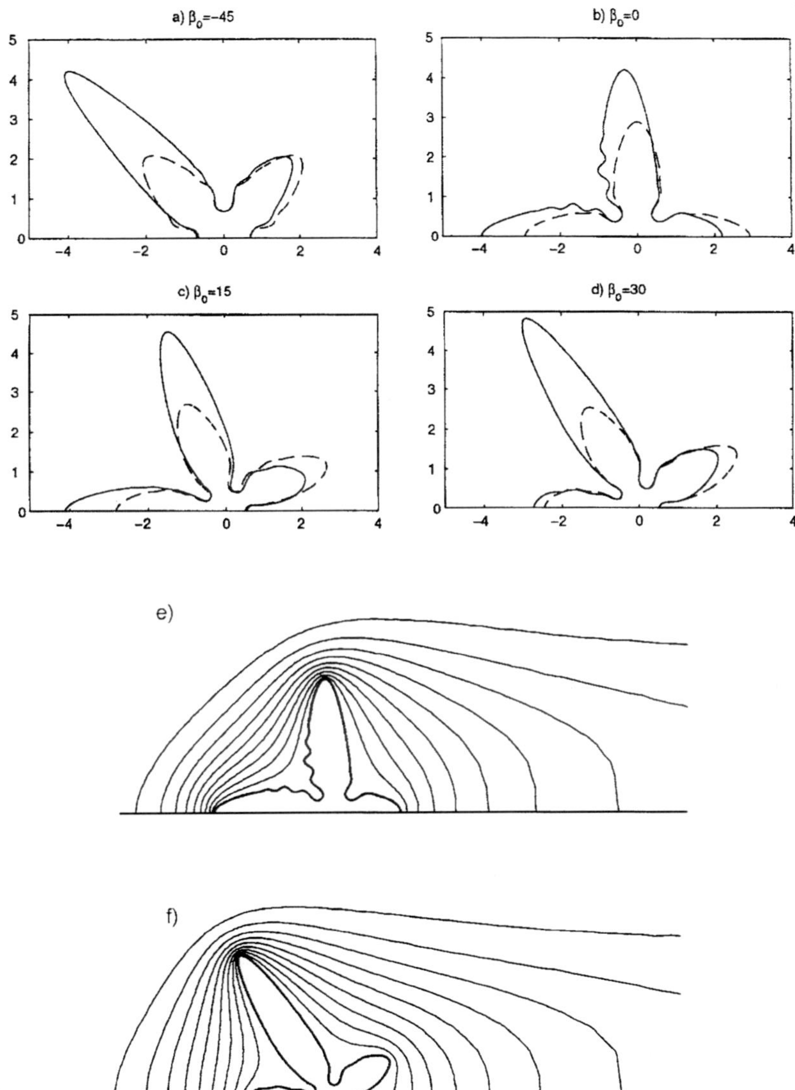

Figure 13. The solid/liquid interface for $\epsilon = 0.015$, $\Delta = 0.1$, $Pr = 23.1$, $Pe = 1$ at time=20. The dashed line shows the interface position without convection, $Pe = 0$. In a) $\beta_0 = -30.0$ b) $\beta_0 = 0.0$ c) $\beta_0 = 15$ and d) $\beta_0 = 30$. e) The temperature field for the case $\beta_0 = 0$. f) The temperature field for the case $\beta_0 = 30$. The temperature levels are $(-0.1, -0.2, ..., -0.9)$. The vertical main branch in fig. d) has reached the largest vertical distance at this time.

Figure 14 shows the vertical distance as a function of β_0 for $\epsilon = 0.015$ and $Pe = 1$. Each curve is normalized by the maximum vertical distance between the tip and the wall that is reached for all β_0 values at that time. To avoid overlap between the curves, they have been separated with an offset (of 0.15) in the vertical direction of the figure. Each curve represents a different time and from the bottom to the top the times are 5, 10, 15 and 20. Again there is a distinct increase in the optimal β_0 value with time. At time=10 the optimal β_0 is 22.5 degrees, and at the final time 20 it has increased to 30 degrees.

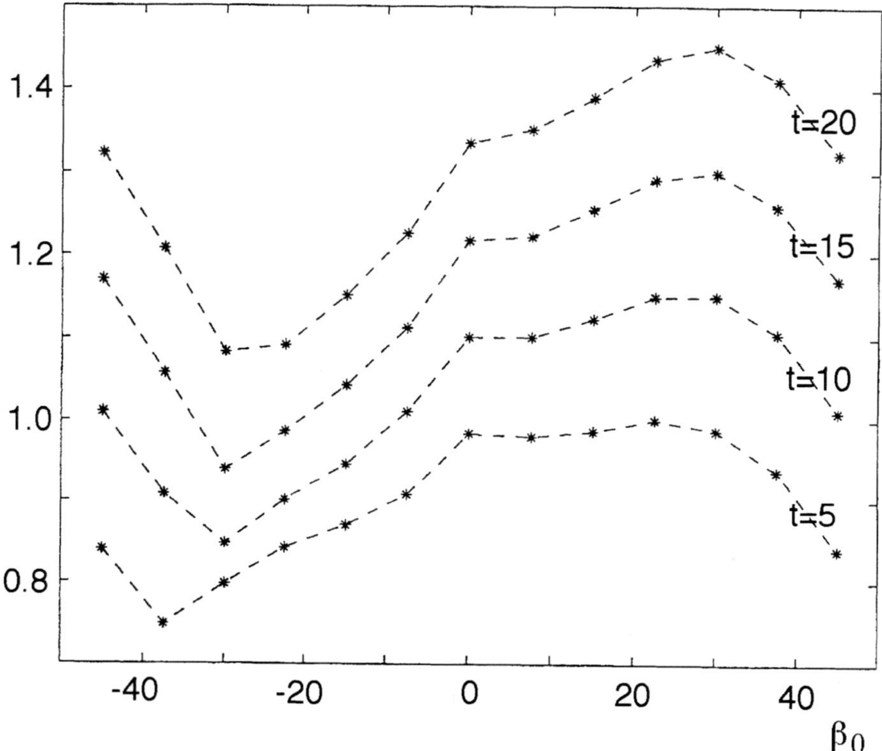

Figure 14. The normalized vertical position as function of β_0 for $\epsilon = 0.015$, $\Delta = 0.1$, $Pr = 23.1$ and $Pe = 1$. Each curve represents a time and from bottom curve to the top the time is 5, 10, 15, 20. With the exception for the result with β_0=0, the curves are smooth. At time=20 , the maximum and minimum vertical distances are achieved for $\beta_0 = 30.0$ and $\beta_0 = -30.$, respectively.

We conclude that the anisotropy in the cases we have studied seems strong enough to effectively lock the dendrite branch to grow along its preferred growth direction. The role of the convection is instead to alter the growth rate along this direction. We have also found that this growth rate may vary significantly with the orientation of the nucleus.

We conjecture that a possible mechanism for tilt is that there exists an optimal nucleus orientation that will give a maximum growth rate for nuclei having that orientation and that dendrites that have grown from these nuclei may eventually dominate a fully developed cluster of dendrites.

Our conclusions are of course limited to the restricted range of cases that we have been able to study. It is certainly possible that for other materials and parameter combinations the flow may cause the growth direction to deviate significantly from the preferred one. There may also of course be other phenomena, for instance an actual turning of the preferred growth direction, via lattice defects that are induced at the solidification interface. We would still like to draw some attention to the orientation selection mechanism, since it seems to have been overlooked in the past, and since it is dominant for the cases we have studied.

Natural convection around an individual dendrite

In this section phase field simulations of the natural convection that will arise around a nucleus growing in an undercooled melt will be described, Tönhardt and Amberg (2000b). As the crystal grows and latent heat is released, the melt around the crystal is heated, which causes natural convection. As described above, this is suspected to contaminate ground based experiments that are intended to study the diffusion controlled growth. We will consider a small nucleus placed in the center of a large rectangular closed container in two dimensions.

Five different values of the dimensionless undercooling, Δ, 0.08, 0.04, 0.02, 0.01 and 0.005, will be discussed. With material properties for pure succinonitrile This corresponds to values of the undercooling between 1.92 K and 0.12 K, which is in the range of experiments.

The reference length l was chosen in order to reflect the tip lengthscales, consequently it is taken differently depending on the undercooling. For the five values of undercooling from $\Delta = 0.08$ down to 0.005, the corresponding values for the reference length are $l = 140.8d_0$, $500d_0$, $1000d_0$, $1666.66d_0$ and $3333.33d_0$, where d_0 is the capillary length $d_0 = c_p\sigma_0 T_m/L_f^2$ as before. In order to see the natural convection effects, the undercooling had to be chosen in this very low range, and this is the reason for the large scale separation in the problem that is evident from the large ratio between reference lengths and capillary lengths. The formal validity of the phase field model may be questionable at these extremely low values of the undercooling, but nevertheless we have found reasonable agreement in calibration runs as long as the tip radii are well resolved.

Figure 15 shows the temperature field in the entire container for Δ=0.005, at nondimensional time $3 \cdot 10^6$. The overall picture is that of a plume rising from a concentrated heat source at the crystal. In figure 15 the plume has just reached the top of the container and starts to build up a stratification. This plume shows typical features that would be expected from a plume above a two dimensional heat source. As expected at the Prandtl number 23.1 used here, the velocity field in the plume (not shown) has a wider profile than the temperature. The fluid in the plume continues to accelerate above the crystal, giving a flow velocity which is maximal on the centerline far above the crystal.

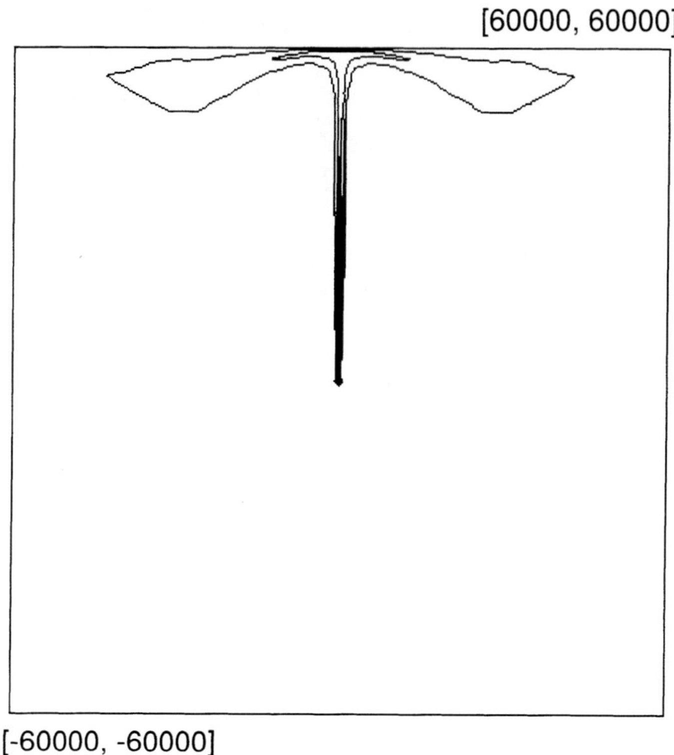

Figure 15. The temperature field in the entire domain at the latest stage, $t = 3 \cdot 10^6$, for the $\Delta = 0.005$ case.

Figure 16 shows the thermal field around the crystal for $\Delta = 0.02$, at nondimensional time $1.64 \cdot 10^6$. The nominal Rayleigh number Ra based on the reference length l has the value $2.24 \cdot 10^{-7}$. A more representative Rayleigh number is obtained by basing it on the dendrite arm length H (nondimensional) instead. For this case this gives, using the length of the downward growing arm $H = 500$, $Ra_H = H^3 \cdot Ra = 27.8$. The temperature has returned to the undisturbed initial value at a distance from the crystal which is approximately the overall crystal size. This is consistent with the small value of Ra_H, but the isotherms also reveal clear convective effects. Above the crystal the beginning of the thermal plume is visible. The velocity field reaches much further from the crystal than the temperature disturbance, as is to be expected in any high Prandtl number natural convection flow.

The isotherms reveal that the heat flux is increased at the tip of the downward growing branch, while it is decreased at the upward growing branch. The growth Peclet number is defined as $P = VR/2$, where V and R are the nondimensional tip growth velocity and radius, respectively. The value of the growth Peclet number is around 0.00172 for the tip

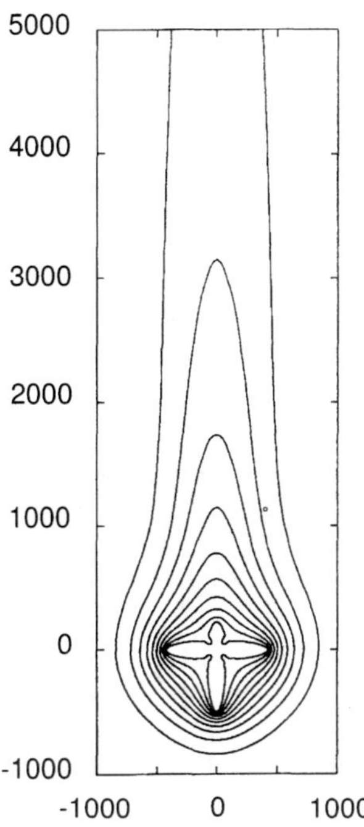

Figure 16. The temperature field around a growing crystal, close up around the crystal at $1.64 \cdot 10^6$ for $\Delta = 0.02$.

of the downward growing branch. The convective effects that can be seen in the shape of the isotherms are thus not the consequence of the translation of the tip, but instead of the fluid motion. This observation is also supported by the fact that the flow velocity below the crystal is larger than the tip growth velocity (a factor 6 higher at a distance 100 below the downward growing tip).

The time history for the growth velocity for the downward growing branch is interesting. In all cases without convection the growth does not reach a steady state. This is expected at these low undercoolings, since in the presence of very slow growth, the convective effects due to the motion of the solidification front are largely absent. The thermal field will then evolve in time as pure diffusion from a localized heat source. In two dimensions, as is well known, the far field is logarithmic and a steady state is possible only after the diffusion has reached the outer boundary. Provatas et.al. (1999a) investigate the scaling of this time dependent growth at undercoolings down to $\Delta = 0.05$. Both with and without convection there is an initial transient when the initial condition for temperature is equilibrated over a length scale comparable to the size of the initial nucleus.

However, with convection the growth velocity soon reaches a quasi-steady state. With the natural convection flow the diffusion of heat is balanced by the convective removal of heat, and the disturbance of the temperature field is confined within a region which for the higher Rayleigh numbers starts to resemble a convective boundary layer of thickness $H \cdot Ra_H^{-1/4}$.

As an example figure 17 shows the time history of the dendrite tip speeds for the $\Delta = 0.04$ case. The speed is shown for the downward, horizontal and upward growing tips, with the downward being the fastest. With convection the time history of the downward tip growth is typically that, after an initial rapid growth, the tip speed drops to an almost constant value. This may typically increase slightly at later times. This is due to the fact that the size of the crystal is continuously increasing and this causes a continued increase of the natural convection, but this is a minor effect as the growth velocity of the tip is increasing very slowly. The horizontally growing tip shows a history resembling the downward one, approaching a quasi-steady state, but at a somewhat lower speed. The upward growing branch is shielded from the convective flow by the crystal and is not enhanced by the convection. The tip speed decreases monotonically and does not become steady.

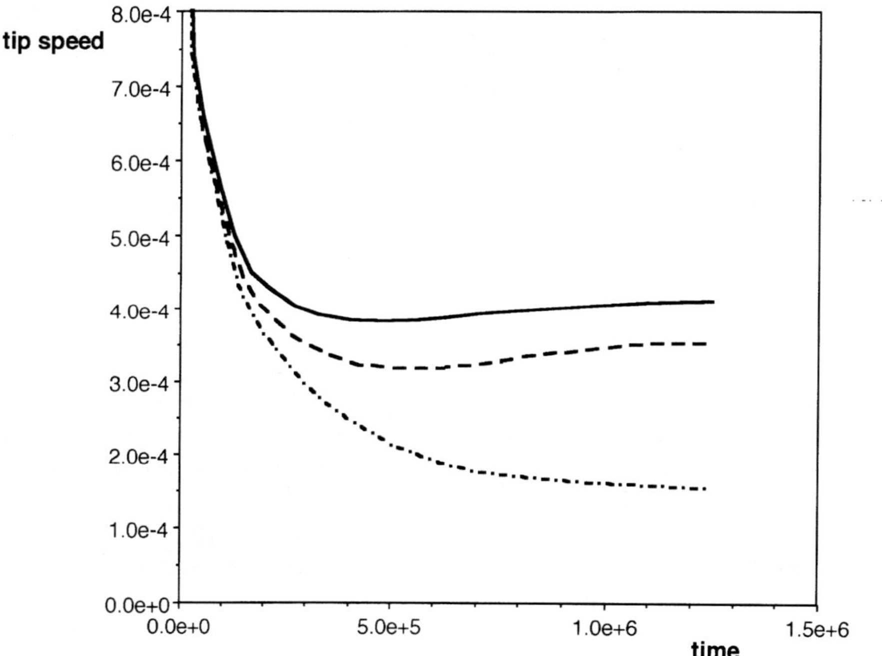

Figure 17. The history of the tip speeds for the $\Delta = 0.04$ case. The downward growing tip is shown as the solid curve, the horizontal is the dashed, and the upward is the dash-dot curve.

The slow increase of growth velocity with crystal size is understandable as a consequence of an increasing flow speed around the tip, as the crystal grows in size. With a larger crystal the velocity scale of the convective motion increases. The tip will thus experience an increased convective flow which will give an enhanced heat transfer. Note that, if the tip had been so large that there had been well resolved boundary layers around it, the flow would be self-similar and the local heat transfer at the tip would be independent of the overall crystal size. Here however the tip region is very small, and indeed there is no self similar boundary layer region at all. A reasonable conceptual model for the heat transfer at the tip would rather be a small object in a forced flow, where the flow is driven by the convection on the dendrite size scale.

Figure 18 shows the growth Peclet numbers P for the simulations as a function of the undercooling. In the graph, the dash-dotted and the dashed lines are the growth Peclet numbers according to an Ivantsov solution without convection in 2D and 3D, respectively. The squares are experimental results for SCN in terrestrial environment by Glicksman et.al. (1995). The most striking feature is that the growth Peclet number increases with the undercooling in very much the same way for our simulations and the experiments, despite the fact that the simulations are 2D.

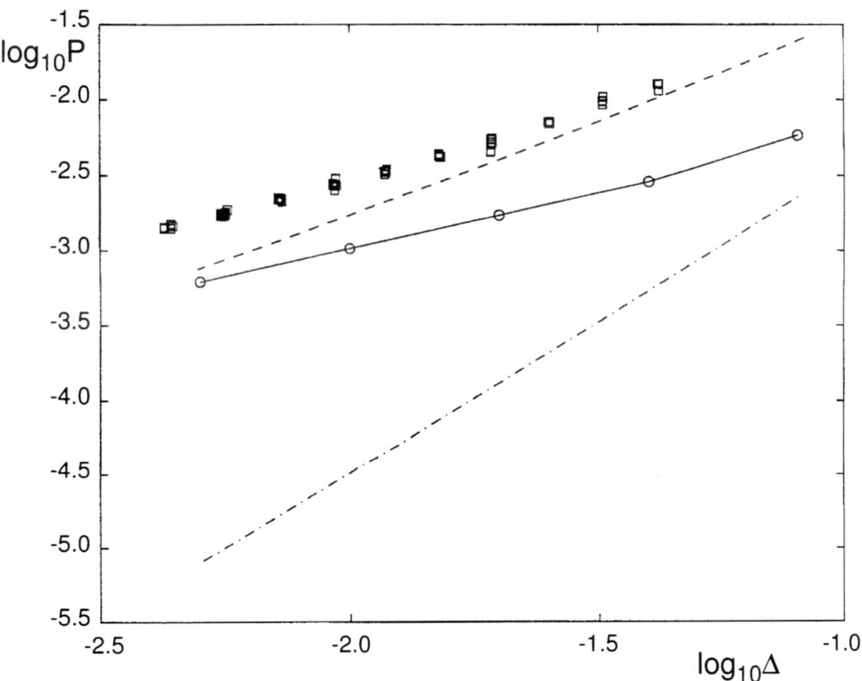

Figure 18. Growth Peclet number vs nondimensional undercooling. The solid line shows the present simulations; dashed line shows a 3D Ivantsov; squares are experimental results by Glicksman et.al. (1995); dash-dot shows a 2D Ivantsov. From Tönhardt and Amberg (2000b).

The large difference between the 2D and 3D Ivantsov curves in figure 18 is related to the logarithmic far field singularity in a steady 2D diffusion problem which is absent in 3D.

However, with convection, a finite length scale for the thermal field is established and the heat transfer becomes independent of the state far from the crystal, also in the 2D case. There is a quantitative difference between the simulation results and experimental results, but that is a constant factor, the slope of the curves agree quite well. This indicates that the essential features of the convective effects are similar in the experiments and the 2D simulation.

Figure 19 shows the relation between a Nusselt number for the downward growing branch and the Rayleigh number. The Nusselt number is based on the tip radius and the release of latent heat, which should be the relevant scales governing the heat transfer around the tip. This gives an expression as $Nu_R = VR/\Delta$, where V and R are the nondimensional tip growth speed and radius, as before. Using the argument outlined in connection to the discussion of the time dependence of the growth above, we expect the heat transfer around the tip to be governed by the flow caused by the overall natural convection. This would give $Nu_R \approx Pe_R^{1/2}$, where $Pe_R = UR$ is a Peclet number based on the flow velocity U (nondimensional) and the tip radius. This flow velocity is assumed to be driven by the overall convection, i.e. $U \approx Ra_H^{1/2}/H$. This gives a relation between Nu_R and Ra_H according to $Nu_R(H/R)^{1/2} \approx Ra_H^{1/4}$.

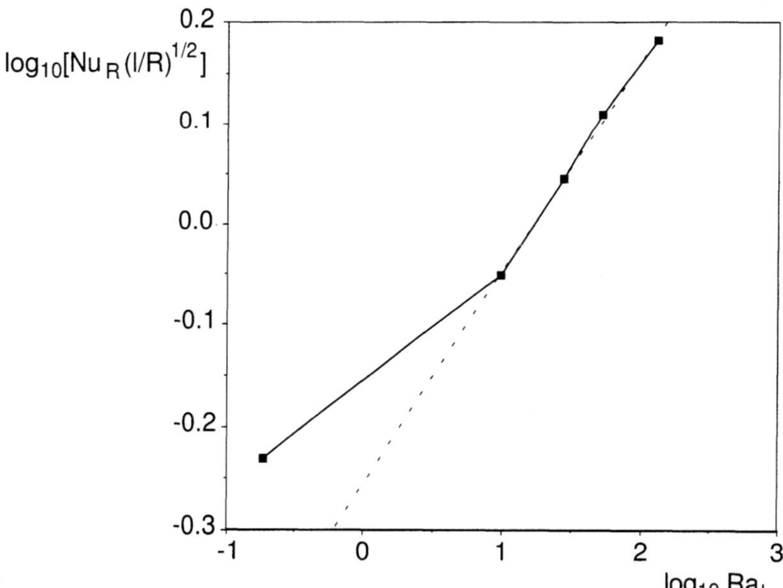

Figure 19. Relation between tip Nusselt number and overall Rayleigh number. Solid line and squares show the present simulations, the dashed line is a fit to the data with $\Delta \leq 0.04$, giving the relation $Nu_R(H/R)^{1/2} = 0.555 \cdot Ra_H^{1/4.79}$. From Tönhardt and Amberg (2000b).

This relation has been tested in figure 19. It is seen that $log_{10}[Nu_R(H/R)^{1/2}]$ varies linearly with $log_{10}(Ra_H)$ as expected, except for the lowest value of Ra_H corresponding to $\Delta = 0.08$, where we do not expect convection to be important. With this choice of parameters it is also clear that the switchover from convectively to diffusively controlled growth happens for $Ra_H \approx 1$. A fit of a power law to the points where convection dominates gives the result that $Nu_R(H/R)^{1/2} = 0.555 \cdot Ra_H^{1/4.79}$, in reasonable agreement with the expected law. It is an attractive feature of this relation that it involves both the tip radius and the overall length scale, but, in view of the limited parameter range that was covered, we do not wish to overemphasize the significance of this.

4 Chimney formation in mushy zones

As these dendrites continue to grow, they may after some time occupy a macroscopic region and can then be treated as a porous material made up of solid crystals bathed in residual melt. For instance in a casting, where a molten alloy is solidified by cooling the walls of the mold, dendrites usually grow from the walls into the melt. This effectively porous region is referred to as a 'mush'. The speed at which this mush advances is in principle determined by the conditions at the edge of the mush, i.e. the local heat and mass transfer at the tips of the dendrites that constitute the mush edge, i.e. the interaction between dendritic growth and flow that is addressed above.

Another important class of phenomena relate to the mass transfer inside the mush, coupled to the phase change. Due to the thermodynamics of phase change, the melt in the mush is typically enriched in alloying elements. This may induce density gradients and thus convective flow, Huppert (1990), which may have diverse consequences, depending on the particular motion.

Here we will be interested in an idealized model problem, which we might call Benard convection in a solidifying mush: Consider directional solidification from below, such that a planar solidification front is unstable and that a dendritic mushy layer of a finite extent is formed. Figure 20a sketches the situation found in experiments such as those by Tait et.al. (1992) and Chen and Chen (1991), in which a mush propagates upwards into a deep bath of a superheated mixture. In this case the mush is bounded at the top by a free interface and freezes at a rate that is time-dependent, and limited by thermal diffusion within the mush, Huppert (1990). In such experiments so called chimneys often appear, see Chen and Chen (1991), Copley et.al. (1970), Tait et.al. (1992), Tait and Jaupart (1992) , Sample and Hellawell (1984), Sarazin and Hellawell (1988), Worster (1997). These constitute completely molten channels through the mushy layer that are typically oriented in the direction of gravity. When finally frozen, the fossil record of these chimneys often manifests itself by localized regions of different chemical composition in the sample known as "freckles".

The origin of freckles has been speculated by many investigators to be related to a buoyancy-driven instability in the mush by the following mechanism: during freezing from below, solute is rejected from the solid dendrites and is enriched in the interdendritic liquid. If the rejected constituent is lighter, as is frequently the case, the cold, solute rich liquid near the bottom is often lighter than the warmer liquid above. The liquid is thus unstably stratified and thermal-solutal convection may occur. This fundamental stability

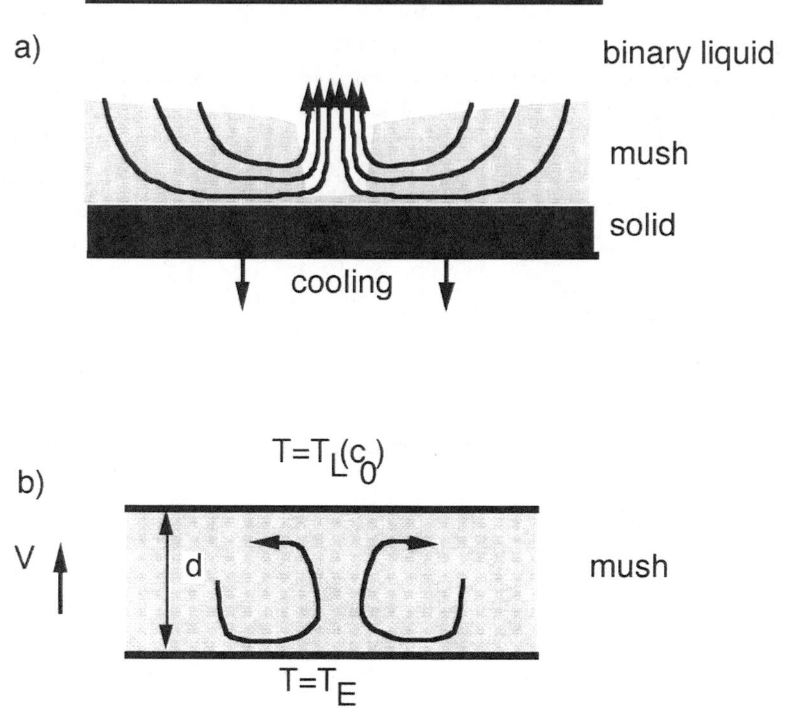

Figure 20. Sketch of chimney formation in directional solidification of a mushy zone, a). b) shows the simplified configuration.

problem has generated a number of articles during the last decade, Worster (1997).

Once convection has started an important coupling between the convection and the porosity of the mush becomes possible. The mush will attain a temperature, composition and solid fraction that is dictated by a combination of the local thermodynamic state and the phase diagram, leading to the possibility of either remelting or enhanced freezing of the solid. If the details are such that remelting occurs in regions of upflow, the local porosity will increase there, leading to a reduction in the viscous resistance to flow, and a positive feedback mechanism for further focusing of the upflow.

The effects on the final solid of such convective motions go by many different names in materials science. In the directional casting of single crystal turbine blades, the fossil traces of a convecting plume in the mush is called a 'freckle', as mentioned above, and is highly detrimental. In large scale castings the enriched melt in the mushy region may slowly convect through the porous mush to the top or the bottom of the cast, as the case may be, causing accumulation of light elements at the top and heavy at the bottom of the cast. In the course of this convection, related instabilities may cause 'A-segregates' or 'V-segregates'. In the continuous casting of steel there are many important phenomena related to the use of strong magnetic fields to control the melt flow, Davidson (1999).

Our main interest here is to study instabilities within the mush, and we therefore adopt a simplified model geometry, Amberg and Homsy (1993), where we assume that the mush is solidified directionally at a prescribed velocity V, see figure 20b. In a coordinate system translating with the solidification velocity V, the mush is assumed to be contained in a layer of thickness d between a completely solidified lower boundary held at the eutectic temperature, and a top interface. Liquid of composition C_0 is introduced through the upper boundary to replace the solids that are withdrawn at the bottom. The lower boundary is at the eutectic temperature, so that the mixture is completely solid at $z < 0$. The upper boundary is kept at the liquidus temperature of the mixture, and the entire space is assumed to be filled with a mush. The main motive for studying this geometry is to simplify the stability problem and enable analytical progress, and to decouple the instability mechanism that is inherent to the interior of the mush, from the instability in the liquid layer above the mush. As shown by Worster (1992), the two instabilities are well separated in wavelength and have typically a weak interaction. We also note that, at least in principle, also this simplified situation would be possible to realize experimentally, utilizing a porous upper plug that is kept at a temperature near the liquidus.

4.1 Stability analysis

The basic equations that govern the flow in the mush are derived by considering conservation of heat, solute, momentum etc over control volumes that are large compared to dendrites, but still small compared to macroscopic lengths, Hills et.al. (1983). Here we will formulate the equations for a simple special case of a Boussinesq fluid, equal thermal diffusivities in the solid and liquid, no solidification shrinkage, and zero mass diffusion in the solid. Also a crucial assumption is that the system is near the eutectic point in the phase diagram, Fowler (1985), see the sketch in figure 21. This is reflected by the choice of nondimensional temperature and concentration as

$$\theta = \frac{T - T_L(C_0)}{T_L(C_0) - T_E}$$

$$\eta = \frac{C - C_0}{C_0 - C_E}.$$

Here C is the (dimensional) concentration in the liquid, C_0 is the concentration of the liquid entering from above, and C_E is the eutectic concentration. T_L is the liquidus temperature as a function of the concentration, as in equation (2.14), except that here the reference temperature is taken as the eutectic temperature T_E, i.e.

$$T_L = T_E + m\,(C - C_E). \tag{4.1}$$

This equation can be seen as derived from equation (2.13) by averaging over many dendrites. The contribution from the curvature term in equation (2.13) is then neglected, as is the kinetic term, since the growth considered here is rather slow.

The nondimensional form of equation (4.1) is simply

$$\eta = \theta. \tag{4.2}$$

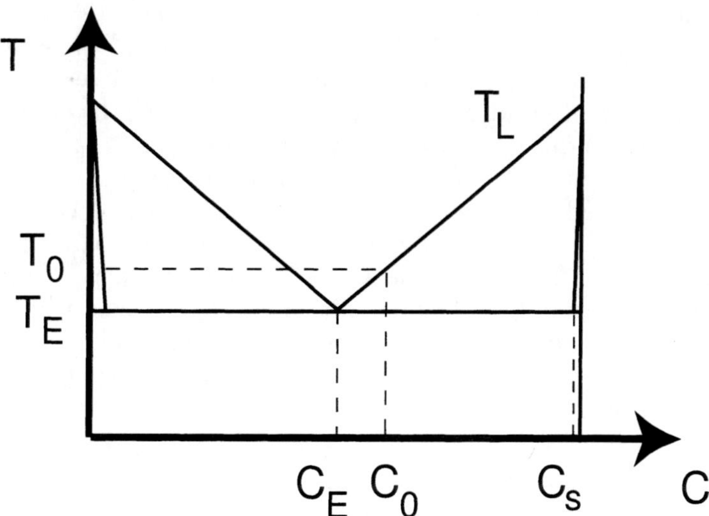

Figure 21. Sketch of a phasediagram.

The near-eutectic approximation will make use of the assumption that the temperature T_0 and composition C_0 of the liquid entering above the mush, which are assumed to be on the phasediagram as indicated in figure 21, are close to T_E and C_E respectively.

The conservation of heat and solute, equations (2.1-2.4), take the following nondimensional form, as averages over control volumes containing solid and liquid:

$$-\frac{\partial \theta}{\partial z} + \nabla \cdot (\bar{u}\theta) = \nabla^2 \theta - Ste\,\frac{\partial \phi}{\partial z}, \tag{4.3}$$

$$-\frac{\partial}{\partial z}(\tilde{C}\phi + (1-\phi)\eta) + \nabla \cdot (\bar{u}\eta) = \frac{1}{Le}\nabla^2 \eta, \tag{4.4}$$

Here velocities have been scaled with the growth velocity V, and lengths with the thermal diffusion lengthscale $l = \kappa/V$. Le is the Lewis number $Le = \kappa/D_l = Sc/Pr$, which is typically very large, so that the right hand term in equation (4.4) is negligible. $Ste = L_f/c_p(T_L(C_0) - T_E))$ is the Stefan number, essentially equal to $1/\Delta$ defined above. $\tilde{C} = (C_s - C_0)/(C_0 - C_E)$, is an important parameter in the problem, actually the nondimensional concentration in the solid. ϕ is the local solid volume fraction in the mush, one of the primary unknowns in the problem. The equations are formulated as steady, since we wish to find conditions that allow steady nontrivial solutions.

The melt flow in the mush is assumed to follow Darcys law for flow through a porous material. In this situation it takes the form

$$0 = -\nabla p - K(\phi)\bar{u} - Ra_m \theta \bar{e}_z, \tag{4.5}$$

together with the continuity equation

$$\nabla \cdot \bar{u} = 0. \tag{4.6}$$

The velocity $\bar{u}$ is here the Darcy velocity, defined as the volume flux across a control surface, divided by the area of the surface. $K(\phi)$ denotes the dependence of the permeability in Darcys law on the solid fraction.

The parameter Ra_m is an effective Rayleigh number in the mush defined as

$$Ra_m = \frac{g(T_L(C_0) - T_E)(\beta_c/m - \beta_T)\Pi(0)}{\nu V} \tag{4.7}$$

Since the temperature and concentration are related via the phase diagram (4.1) in the mush, the density of liquid in the mush is

$$\rho - \rho_0 = \rho_0(\beta_c(C - C_E) - \beta_T(T - T_E)) = \rho_0(\beta_c/m - \beta_T)(T - T_E).$$

Thus the combination $\beta_c/m - \beta_T$ is the effective expansion coefficient in the mush.

$\Pi(\phi) = \Pi(0)/K(\phi)$ is the permeability as a function of solid fraction. $\Pi(0)$ is the value of the permeability at zero solid fraction, which is assumed finite, and is used as a reference value. The functional dependency of Π on ϕ is crucial for the appearance of chimneys. Here we use the following expression

$$K(\phi) = \Pi(0)/\Pi(\phi) = 1/(1 - \phi)^3 \tag{4.8}$$

Equations (4.2,4.3, 4.4) can be seen as determining the three unknowns θ, η, ϕ, while the flow quantities $\bar{u}, p$ are determined by equations (4.5,4.6).

The boundary conditions are, using the simplified model described above, that

$$\bar{u} \cdot \bar{e}_z = 0, \quad \theta = -1, \quad \text{at} \quad z = 0.$$

$$\bar{u} \cdot \bar{e}_z = 0, \quad \theta = 0, \quad \phi = 0, \quad \text{at} \quad z = \delta.$$

Here δ is the nondimensional thickness of the mushy layer, $\delta = Vd/\kappa$, with d denoting the dimensional thickness of the layer.

These equations were studied in the asymptotic limit of $\delta \to 0$, while $\tilde{C} = O(1/\delta)$ and $Ste = O(1)$. The first of these amounts to assuming that the mush is thin compared to the thermal diffusion layer, or that growth is slow. The second is the assumption of solidification close to the eutectic, see the definition of $\tilde{C}$ above and figure 21.

The solution is obtained by splitting the unknowns in a base state and a perturbation, according to

$$\theta = \theta_B(z') + \epsilon\hat{\theta}(x', y', z')$$
$$\bar{u} = 0 + \frac{\epsilon}{\delta} R\hat{\bar{u}}(x', y', z')$$

$$p = Rp_B(z') + \epsilon\hat{p}(x', y', z')$$
$$\phi = \delta\phi_B(z') + \epsilon\hat{\phi}(x', y', z') \tag{4.9}$$

Here lengths have been rescaled according to $(x, y, z) = \delta \cdot (x', y', z')$, and the Rayleigh number R_m has been replaced by R according to $R^2 = \delta R_m$.

The solution proceeds by first determining the base state. In this limit of small δ and large $\tilde{C}$ this becomes quite simple,

$$\theta_B(z') = -(1 - z') + O(\delta),$$

$$\phi = \delta\phi_B(z') = (1 - z')\frac{1}{\tilde{C}} = O(\delta),$$

i.e. linear profiles in ϕ and θ, a small solid fraction, and stagnant melt , see figure 22.

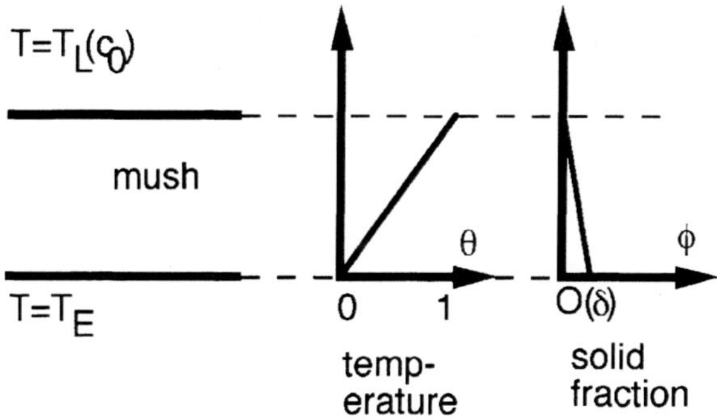

Figure 22. Sketch of the base state profiles for temperature and solid fraction.

In order to study possible steady solutions, the perturbations above are expanded in powers of the disturbance amplitude ϵ, introduced into the equations, which are solved for each order separately.

In order to study the linear stability of the base state, only linear terms in ϵ are retained. The result of this procedure is, keeping only first order in δ,

$$R = 2\pi + \frac{\pi(K_1 - 2Ste)}{2\tilde{C}} + O(\delta) \tag{4.10}$$

Here K_1 is the first coefficient in a power series expansion of $K(\phi) = 1 + K_1\phi + K_2\phi^2 + O(\phi^3)$. Equation (4.8) gives a value for K_1 as 3. This expression gives the value of R, and thus of Ra_m, that will allow a nontrivial steady disturbance. With the present boundary conditions, the most unstable perturbation is a sinusoidal variation of temperature and solid fraction along the layer, together with an alternating up- and downflow, such that upflow corresponds to a low temperature and a low solid fraction. The flow field can be described as two-dimensional rolls, i.e. parallel two dimensional vortices that rotate in alternating directions. Each roll has a width approximately equal to the thickness d of the mush, so that the wavelength of the disturbance is $2d$.

The first term is the classical stability limit for convection in a homogeneous porous medium. The second term shows a positive contribution for increasing K_1. This is understandable, since a larger value of K_1 means a flow resistance that increases more rapidly with increasing solid fraction.

An increasing Ste implies a decreasing R, i.e. a destabilizing effect. In the absence of latent heat release, i.e. $Ste = 0$, fluid that rises will carry a lower concentration upwards, and will tend to be colder than its surroundings, even if temperature fluctuations are expected to be weak due to the large thermal diffusivity. If latent heat is now added, the enhanced melting in the upflow will cause adsorption of latent heat and some additional cooling of the rising fluid. A larger Ste will thus cause a larger temperature disturbance and thus a less stable base state.

The influence of the near-eutectic parameter $\tilde{C}$ is less clear. In equation (4.10) it is seen that the effect on R of changing $\tilde{C}$ depends on the relative magnitudes of K_1 and Ste; For the case of a small K_1 an increasing $\tilde{C}$ means a more stable situation.

Worster (1992) studied the linear stability of the full problem of a mushy layer with a large overlying body of liquid, growing upwards at a constant speed. The treatment was fully numerical and was thus not restricted to the near asymptotic limit used above. His analysis gives a much more complex picture, that also accounts for the influence of parameters on the base state solid fraction, etc. The qualitative features are however compatible with those discussed above. In addition to the mushy layer instability above, Worster identified a boundary layer mode that was associated with an unstably stratified layer in the liquid just above the mush. Worster also made a comparison with the experiments of Tait et.al. (1992) and found a reasonable quantitative agreement.

The linear stability discussed so far only describes infinitesimal disturbances, i.e. far from the fully developed chimneys that are the ultimate object of this study. As a first step towards these, weakly nonlinear solutions were derived by extending the expansions of the perturbations to higher orders in the amplitude ϵ. The essential strategy in such a weakly nonlinear theory is to let the nonlinear terms in the equations determine the amplitude of essentially the linear modes that were obtained in the linear stability problem. In so doing, it is necessary to consider all relevant superpositions of the linear modes. The first natural candidate that need to be discussed is two-dimensional rolls.

The essential result can be expressed as

$$R = 2\pi + \frac{\pi(K_1 - 2Ste)}{2\tilde{C}} + \delta^2 R_{2\delta} + \epsilon^2 R_{2\epsilon} + O(\epsilon^3) \tag{4.11}$$

where

$$R_{2\epsilon} = \frac{1}{\tilde{C}_s^2}(-g_1\tilde{C}_s K_1 - g_2 K_1^2 + g_3 K_2 + g_4 \tilde{C}_s^2) \tag{4.12}$$

and

$$\tilde{C}_s = \tilde{C}\delta = O(1) \tag{4.13}$$

$$g_1 = 15.503, g_2 = 82.683, g_3 = 85.267, g_4 = 7.7516 \tag{4.14}$$

Here ϵ is the amplitude of the disturbance in temperature, solid fraction and velocity, as indicated in equation (4.9). Notice that the amplitude may be negative, and that a change of sign in this two-dimensional case where the disturbances vary sinusoidally in one direction, corresponds to a translation of a half wavelength. Thus, if ϵ gives a steady roll solution, then $-\epsilon$ does too. Due to this symmetry there is no linear term in ϵ, i.e. $R_{1\epsilon} = 0$. $R_{2\delta}$ is the second order correction in δ, but it is independent of the disturbance amplitude, and will thus only affect the linear stability.

Given a value of R from experiment or otherwise, the amplitude is solved from (4.11) to give

$$\epsilon = \pm\sqrt{\frac{R - R_{lin}}{R_{2\epsilon}}}$$

where R_{lin} is the sum of the first three terms on the right hand side of (4.11), that determines the linear stability. The sign of $R_{2\epsilon}$ thus determines the character of the bifurcation as supercritical for a positive $R_{2\epsilon}$, or subcritical for a negative $R_{2\epsilon}$.

In the present case we see that a permeability that is independent of solid fraction, i.e. $K_1 = K_2 = 0$, gives a positive $R_{2\epsilon}$ and thus a supercritical bifurcation. On the other hand, a permeability that decreases with increasing solid fraction, positive and sufficiently large K_1, gives a subcritical bifurcation. In this case it is thus possible to have a convecting solution at a value of R that is less than that predicted from linear theory. This is reasonable since a solution with convection will have a lower solid fraction in an upflow region, and thus a lower flow resistance there. If the dependency of permeability on solid fraction is strong enough, the motion may thus be maintained at values of R below the linear stability threshold.

The other relevant possibility is to have three-dimensional hexagons, i.e. a solution composed of three superimposed rolls which have an orientation 120 degrees apart in the horizontal plane of the mushy layer. Calculating the superimposed modes, temperature and solid fraction perturbations will show hexagonal patterns in horizontal planes. The expansion in ϵ is now different and instead of eq (4.11), we obtain

$$R = 2\pi + \delta\frac{\pi(K_1 - 2Ste)}{2\tilde{C}_s} - \epsilon\frac{3\pi^2}{2}\frac{K_1}{\tilde{C}_s} + O(\epsilon^2) \tag{4.15}$$

For three-dimensional hexagons the symmetry in amplitude is destroyed, and we obtain a nonzero linear term in ϵ. The expansion is then truncated at this level. The bifurcation is now transcritical, with one supercritical and one subcritical branch. The sign of ϵ now has a different meaning than in the two-dimensional case. For $\epsilon > 0$ the vertical velocity is upward at the center of hexagonal cells, and downward at the cell boundaries. For $\epsilon < 0$ it is instead upward at cell boundaries, and downward at cell centers. For the solution in eq (4.15), the term linear in ϵ is negative, and the subcritical branch exists for positive ϵ, i.e. hexagons with upflow at the centers.

The solution would be unstable on the subcritical branch close to the bifurcation point, but presumably this branch turns forward into a stable branch at higher amplitudes, if not sooner this would happen when the upflow regions are completely molten, and fully developed chimneys have appeared. This however happens at higher amplitudes where the weakly nonlinear expansion does not hold.

Anderson and Worster (1995) studied another limit, where they assumed that K_1 is small (proportional to δ) instead of $O(1)$. This means that the turning point where the subcritical branch turns forward into a stable solution branch appears for much smaller ϵ, where weakly nonlinear theory applies. A sketch is shown in figure 23. For these much smaller K_1, it turns out that the coefficient of ϵ in the expansion of R can change sign. Furthermore the quadratic terms are retained so that the turning point on the subcritical branch is captured. The sketch in figure 23 for simplicity includes only the three-dimensional hexagonal solutions. Anderson and Worster also discuss the two-dimensional roll solutions, which are unstable up to a certain value of R.

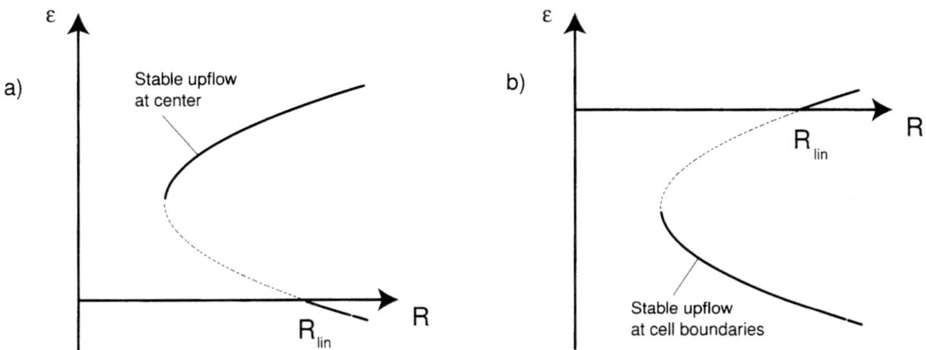

Figure 23. Sketch of bifurcated solution. a) $R_{1\epsilon} < 0$, stable hexagons with upflow at cell centers, b) $R_{1\epsilon} > 0$, stable hexagons with upflow at cell boundaries.

For nonzero K_1 it is possible, depending on other values of the parameters, to have the subcritical branch exist for positive ϵ, see figure 23a, i.e. hexagons with upflow at the center. This is thus the situation that would be expected to lead eventually to chimneys.

For sufficiently small K_1, the subcritical branch appears for negative ϵ, as depicted in figure 23b. This thus describes a stable situation with downflow at cell centers, and upflow along the boundaries of hexagonal cells. It is interesting to note that Tait et.al. (1992),Tait and Jaupart (1992) found experimentally precisely this structure: In experiments with very slow solidification they obtained instead of chimneys, completely molten trenches that would be rather straight and and meet in junctions of three. Even if the pattern was not a very regular one it clearly shows the possibility of hexagons with downflow at the center.

So far we have assumed that bifurcated solutions would be steady. However, Anderson and Worster (1996) revisited the linear stability and found that the problem allows for a Hopf bifurcation to a an oscillatory state in certain parameter ranges. A necessary condition for oscillatory convection is that the parameter group $Ste/(\delta(Ste + \tilde{C})^2)$ is order one or larger (thus ruling out oscillatory convection in the particular asymptotic limit above). These oscillatory solutions take the form of temperature disturbances and convection rolls that are essentially oriented vertically and quite similar in spatial structure to steady disturbances. However, as solidification proceeds, the temperature, solid fraction and velocity patterns translate horizontally. This produces a pattern in the solid fraction where the regions of low solids content are not vertical but are inclined in the direction of the translation of the mode.

4.2 The fully developed chimneys

The fully developed chimney, for instance as sketched in figure 20a, is a strongly non-linear phenomenon which is difficult to treat analytically. Fully numerical treatments have been tried however, using models of different degrees of complexity. The essence of such models is still that given in equations (4.2-4.6), Hills et.al. (1983), i.e. conservation equations for temperature and concentration; The phase diagram relation is imposed in the mush, which gives an equation from which the solid fraction can be determined; The convective melt motion is determined from a momentum balance that effectively reduces to the Darcy law in the mush and the Navier-Stokes equations in the completely molten region. Such numerical codes have been developed and tested and several applications to chimney formation have been presented, Neilson and Incropera (1993), Huang et.al. (1996), Poirier and Heinrich (1994), Schneider and Beckermann (1995a), Schneider and Beckermann (1995b). While such simulations may incorporate quite comprehensive material models and phase diagrams, and realistic geometry, the problem of chimney formation is very tough computationally, due to the large span in lengthscale between the details around a chimney, and the size of the mush, or the overall geometry. The results are generally quite reasonable qualitatively, but due to the difficulties in obtaining a truly resolved solution, the quantitative results should be treated with caution.

To study the detailed structure of one fully developed chimney, Schulze and Worster (1998) solved a somewhat simplified numerical model in two dimensions. The flow through the mush was computed numerically and was matched to a semi-analytic solution for the flow in the chimney. The chimney was assumed to have a uniform width, which was determined as part of the solution. In this way the problem with scale separation discussed above is at least partially circumvented. In the course of this work Schulze and Worster (1998) also derived scaling laws for a strongly convecting, thin mushy layer, i.e. $R_m \sim \tilde{C} >> 1$, $Da << 1$, while ${R_m}^{4/9} Da^{1/3} << 1$, where the Darcy number Da is given by $Da = \Pi(0)V^2/\kappa^2$. In this limit they obtained a scaling of the chimney width as $Da^{1/3} {Ra_m}^{-2/9} (\kappa/V)$.

The chimney formation problem has been studied with a number of additional features. Chiareli and Worster (1995) studied the instability related to the coupling between phase change and a forced convection through the mush that is driven by solidification shrinkage. Feltham and Worster (1999) studied the influence of a forced flow in the melt region outside the mush, parallel to the mushy layer. It was found that the interaction between the melt flow and the mush-liquid edge, could cause an instability. In experiments the solidification typically proceeds from the cold bottom or top of a tank. The solidification speed is then not constant but rather governed by diffusion, giving a speed roughly $V \sim 1/\sqrt{t}$. The stability of the mush in this more realistic situation has been addressed by Emms and Fowler (1994), and Hwang and Choi (2000). It has been hypothesized in the past that rotation and magnetic fields would be efficient in suppressing chimneys; a recent stability study of rotation effects is given by Guba (2001), and magnetic effects by Riahi (2000).

5 Acknowledgment

This work was partially supported by the Swedish Research Council for Engineering Science (TFR) and the Swedish Research Council (VR).

References

Al-Rawahi, N. and Tryggvason, G., "Effect of Melt Flow on Dendritic Solidification". Talk presented at 53rd annual meeting of the Division of Fluid Dynamics of the American Physical Society, Washington DC, Nov 19-21 (2000).

Amberg, G. and Homsy, G.M. 'Nonlinear Analysis of Buoyant Convection in Binary Solidification with Application to Channel Formation', *J. Fluid Mech.*, vol 252, pp 79-98, (1993).

Ananth, R., and Gill, W.N., *J. Crystal Growth* vol 91, p.587, (1988).

Ananth, R., and Gill, W.N., *J. Crystal Growth*, vol 179, p.263, (1997).

Anderson, D.M., McFadden, G.B. and Wheeler, A.A., *Physica D*, vol 135, p.175-194, (2000).

Anderson, D.M. and Worster, M.G., *J. Fluid Mech.*, vol 302, p.307-331, (1995).

Anderson, D.M. and Worster, M.G., *J. Fluid Mech.*, vol 307, p.245-267, (1996).

Brattkus, K. Ph.D. Thesis ,1988, Directional solidification of dilute binary alloys, Nothwestern University

Beckermann, C., Diepers, H.J., Steinbach, I., Karma, A., and Tong, X., "Modeling Melt Convection in Phase-Field Simulations of Solidification", *J. Computational Physics*, Vol. 154, pp. 468-496, (1999a).

Boettinger, W.J., Coriell, S.R., Greer, A.L., Karma, A., Kurz, W., Rappaz, M. and Trivedi, R., "Solidification Microstructures: Recent developments, future directions", *Acta Materialia*, vol 48, pp43-70, (2000).

Bouissou, P. Perrin, B. and Tabeling, P., *Phys. Rev. A*, vol 40, p. 509, (1989a):

Bouissou, P. and Pelce, P. *Phys. Rev. A*, vol 40, p. 6673, (1989b).

Bühler L. and Davis, S.H. *J. Crystal Growth*, vol 186, pp. 629-647, (1998).

Brattkus, K. and Davis, S.H.. emphJ. Crystal Growth, vol 87, p. 385, (1988a).

Brattkus, K. and Davis, S.H.. emphJ. Crystal Growth, vol 89, p. 423, (1988b).

Canright, D., and Davis, S.H., *Journal of Crystal Growth*, vol 114, pp. 153-185, (1991).

Carlberg, T., *Microgravity Q*, vol 5, pp135-145, (1995).

Carlberg, T., *Acta Astronautica*, vol 40, pp407-414, (1997).

Chen, F. and Chen, C.F., *Journal of Fluid Mech.*, vol 227, pp. 567-586, (1991).

Copley, S.M., Giamei, A.F., Johnson, S.M. and Hornbecker, M.F. 1970 'The Origin of Freckles in Unidirectionally Solidified Castings', *Met. Trans.*, vol 1, p 2193-2205.

Caroli, B. Caroli, C. Misbah, C. and Roulet, B. *Journal de Physique C*, vol 46(3), p. 401, (1985).

Coriell, S.R., Cordes, M.R., Boettinger, W.J. and Sekerka, R.F. 'Convective and interfacial instabilities during unidirectional solidification of a binary alloy', *J. Crystal Growth*, vol 49, pp 13-28, (1980).

S.R. Coriell, G.B. McFadden, R.F. Boisvert and R.F. Sekerka, *J. Crystal Growth*, vol 69, p. 15, (1984).

Chiareli, A.O.P. and Worster, M.G., *J. Fluid Mech.*, vol 297, p. 293-305, (1995).

Dantzig, J.A. and Chao, L.S. , *Proc. 10th U.S. National Congress of Applied Mechanics*, J. Lamb, ed., ASME, pp. 249-255, (1986).

Davidson, P.A., "Magnetohydrodynamics in Materials Processing", *Annu. Rev. Fluid Mech*, vol 31, pp 273-300, (1999).

Davis, S.H., "Hydrodynamic interactions in directional solidification", *J. Fluid Mech.*, vol 212, pp241-262, (1990).

Davis, S. H. (2001). *Theory of Solidification*, Cambridge University Press.

Delves, R.T. *J. Crystal Growth*, vol 8, p. 13, (1971).

Diepers, H.J., Beckermann, C., and Steinbach, I., "Simulation of Convection and Ripening in a Binary Alloy Mush Using the Phase-Field Method", *Acta Materialia*, Vol. 47, pp. 3663-3678, (1999b).

Emms, P.W. and Fowler, A.C., *J. Fluid Mech.*, vol 262, pp. 111-139, (1994).

Esaka, H., Suter F. and Ogibayashi, S., *ISIJ International*, vol 10(36), pp. 1264-1272, (1996).

Feltham, D.L. and Worster, M.G., *J. Fluid Mech.* vol 391, p. 337-357, (1999).

Forth, S.A. and Wheeler, A.A., *J. Fluid Mech.* vol 202, p. 339 (1989).

Fowler, A.C., *IMA J. Appl. Math.* vol 35, p. 159-174 (1985).

Fredriksson, H., El Mahallawy, N., Taha, M., Liu Xiang and Wänglöw, G. *Scandinavian Journal of Metallurgy*, vol 15, p. 127, (1986).

Fried, E., Gurtin, M.E., "A phase-field theory for solidification based on a general anisotropic sharp-interface theory with interfacial energy and entropy", *Phys. D*, 91, pp. 143-181, (1996).

Glicksman, M.E. and Marsh, S.P., "The Dendrite", in: D.T.J. Hurle (Ed.), *Handb. Cryst. Growth*, vol.1, ch.13, Elsevier, Amsterdam, pp. 1077-1122, (1993).

Glicksman, M.E., Koss, M.B., and Winsa, E.A., *Phys. Rev. Lett.*, vol 73, p. 573, (1994).

Glicksman, M.E., Coriell, S.R., and McFadden, G.B., *Ann. Rev. Fluid Mech.*, vol 18, p. 307 (1986).

Glicksman, M.E., Koss, M.B., Bushnell, L.T., LaCombe, J.C. and Winsa, E.A. , *ISIJ International*, vol 35(6), p. 1216, (1995).

Guba, P., *J. Fluid Mech.*, vol 437, p. 337-365, (2001).

Hills, R.N., Loper, D.E. and Roberts, P.H., *Q. J. Mech. appl. Math.* vol 36, p. 505-539 (1983).

Hobbs, A.K. and Metzener, P. *J. Crystal Growth* vol 112, p. 539 (1991).

Hwang, I.G., and Choi, C.K., *J. Crystal Growth* vol 220, p. 326-335 (2000).

Huang, J.S., and Barduhn, A.J., *AIChE J.* vol 31, p. 747 (1985).

Huang, H.W., Heinrich, J.C. and Poirier, D.R., *Modelling Simul. Mater. Sci. Eng.* vol 4, p. 245-259 (1996).

Huang, S.-C. and Glicksman, M.E., "Fundamentals of dendritic solidification. I. Steady-state tip growth". *Acta Met.*, vol. 29 no.5, pp. 701-716, (1981a).

Huang, S.-C. and Glicksman, M.E., "Fundamentals of dendritic solidification. II. Development of sidebranch structures". *Acta Met.*, vol. 29 no.5, pp.717-734, (1981b).

Huppert, H., "The fluid mechanics of solidification", *J. Fluid Mech.*, vol 212, pp209-240, (1990).

Hurle, D.T.J., Jakeman, E. and Wheeler, A.A. "Effect of solutal convection on the morphological stability of a binary alloy", *J. Crystal Growth*, vol 58, pp163-179, (1982).

Hurle, D.T.J., Jakeman, E. and Wheeler, A.A. "Hydrodynamics stability of the melt during solidification of a binary alloy", *Phys. Fluids*, vol 26, pp624-626, (1983).

Ivantsov, G.P., *Dokl. Akad. Nauk SSSR*, vol 58, p567, (1947).

Jeong, J.H., Goldenfeld, N. and Dantzig, J.A., *Physical Rev. E*, 64, 041602, (2001).

Kallungal, J.P. and Barduhn, A.J., *AIChE J.*, vol 23, p. 294, (1977).

Karma, A. and Rappel, W.-J., *Physical Review Letters*, vol 77, pp. 4050-4053, (1996a).

Karma, A. and Rappel, W.-J., "Phase-field method for computationally efficient modeling of solidification with arbitrary interface kinetics", *Physical Review E*, vol 53, pp R3017-R3020, (1996b).

Kessler, D.A. and Levine, H., *Phys. Rev. Lett.*, vol 57, pp3069-3072, (1986).

Kobayashi, R., *Bull. Jpn. Soc. Ind. Appl. Math.*, vol 1, p 22-, (1991).

Koss, M.B., Bushnell, L.T., LaCombe, J.C., and Glicksman, M.E., *Chem. Eng. Comm.*, Vol. 152-153, p. 351, (1996)

Kurz, W. and Fisher, D.J., *Fundamentals of Solidification*, Trans Tech Publications (1992)

Lan, C.W. and Hsu, C.M., *Journal of Crystal Growth*, submitted (2001).

Langer, J.S. and Müller-Krumbhaar, H., *Acta Metall. Mater.*. vol 26 pp.1681-1688, (1978).

Langer, J.S., "Instabilities and pattern formation in crystal growth". *Reviews of Modern Physics*. vol 52 pp.1-28, (1980).

Lee, Y.-W., Ananth, R. and Gill, W.N., *Chem. Eng. Comm.* vol 116, p. 193, (1992).

Lee, Y.-W. Ananth, R. and Gill, W.N., *J. Crystal Growth*, vol 132, p. 226, (1993).

Lee, Y.-W., Ph.D. Thesis, Rensselaer Polytechnic Institute, Troy, New York, (1991).

Lee, Y.-W., Smith, R.N., Glicksman, M.E., and Koss, M.B.,'Effects of buoyancy on the growth of dendritic crystals', in *Annual Review of Heat Transfer*, C.-L. Tien (Ed.), vol 7, pp 59-139, (1996).

Loginova, I., Amberg, G. and Ågren, J., *Acta Materialia*, vol 49, p. 573-581, (2001).

McFadden, G.B. and Coriell, S.R., *J. Crystal Growth*, vol 74, p. 507, (1986).

Meiron, D.I., *Phys. Rev. A*, vol 33, pp2704-2715, (1986).

Merchant, G.J. and Davis, S.H., *J. Crystal Growth*, vol 96, p. 737, (1989).

Mullins, W.W. and Sekerka, R.F., "Morphological Stability of a Particle Growing by Diffusion or Heat Flow", *Journal of Applied Physics*, Vol. 34, pp. 323-329, (1963).

Mullins, W.W. and Sekerka, R.F., "Stability of a Planar Interface During Solidification of a Dilute Binary Alloy", *Journal of Applied Physics*, Vol. 35, pp. 444-451, (1964).

Murakami, K., Fujiyama, T., Koike, A., and Okamoto, T. , *Acta Met.* vol 31(9), p. 1425, (1983).

Murakami, K., Aihara, H. and Okamoto, T.,, *Acta Met.*, vol 32(6), p. 933 (1984).

Nield, D.A., "The thermohaline Rayleigh-Jeffreys problem", *Journal of Fluid Mechanics*, Vol. 29, pp. 545-558, (1967).

Neilson, D.G. and Incropera, F.P., *Int. J. Heat and Mass Transfer*, Vol. 36, pp. 489-505, (1993).

Penrose, O. and Fife, P.C., *Physica D*, vol 43, p. 44-62, (1990).

Pines, V., Chait, A., and Zlatkowski, M., *J. Crystal Growth*, vol 167 p. 383, (1996).

Poirier, D.R. and Heinrich, J.C., *Materials Characterization* vol 32, p. 287-298 (1994).

Provatas, N., Goldenfeld, N., Dantzig, J., LaCombe, J.C., Lupulescu, A., Koss, M. B., Glicksman, M. E. and Almgren, R., *Phys. Rev. Lett.*, vol 82, p4496-, (1999a).

Provatas, N., Goldenfeld, N., Dantzig, J., "Adaptive Mesh Refinement Computation of Solidification Microstructures using Dynamic Data Structures", *J. Comp. Phys.*, vol 148, pp 265-290 (1999b).

Riahi, D.N., *J. Crystal Growth*, vol 216, pp501-511, (2000).

Sample, A.K. and Hellawell, A., 'The Mechanisms of Formation and Prevention of Channel Segregation during Alloy Solidification', *Met. Trans. A*, vol 15, pp2163-2173, (1984).

Sarazin, J.R. and Hellawell, A. 'Channel Formation in Pb-Sn, Pb-Sb, and Pb-Sn-Sb Alloy Ingots and Comparison with the System NH4Cl-H2O' *Met. Trans. A*, vol 19, pp1861-1871, (1988).

Schneider, M.C., and Beckermann, C., "A numerical study of the combined effects of microsegregation, mushy zone permeability and flow, caused by volume contraction and thermosolutal convection, on macrosegregation and eutectic formation in binary alloy solidification", *Int. J. Heat Mass Transfer*, vol 38, pp 3455-3473, (1995a).

Schneider, M.C., and Beckermann, C., "Formation of Macrosgregation by multicomponent thermosolutal convection during the solidification of steel", *Metallurgical and Materials Transactions A*, vol 26, pp 2373-2388, (1995b).

Schulze, T.P. and Davis, S.H. *J. Crystal Growth*, vol 143, p. 317, (1994).

Schulze, T.P. and Davis, S.H. *J. Crystal Growth*, vol 149, p. 253, (1995).

Schulze, T.P. and Worster, M.G. *J. Fluid Mech.*, vol 356, p. 199-220, (1998).

Sekerka, R.F., Coriell, S.R., and McFadden, G.B., *J. Crystal Growth*, vol 154, p.370, (1995).

Sekerka, R.F. and Bi, Z., "Phase field model of multicomponent alloy solidification with hydrodynamics", *Proc of 'Interfaces for the Twenty-First Century', Monterey, August 1999.*

Tait, S., Jahrling, K. and Jaupart, C., *Nature*, Vol. 359, pp.406-408, (1992).

Tait, S. and Jaupart, C., *J. Geophysical Res.*, vol 97(B5), pp6735-6756, (1992).

Tewari, S.N. and Chopra, M.A., *J. Crystal Growth*, vol 118, pp183-, (1992).

Tong, X., Beckermann, C., and Karma, A., "Velocity and Shape Selection of Dendritic Crystals in a Forced Flow", *Physical Review E*, Vol. 61, pp R49-R52, (2000).

Tönhardt, R. and Amberg, G., "Phasefield Simulation of Dendritic growth in a Shear Flow", *J. Crystal Growth*, vol 194, pp 406-425, (1998).

Tönhardt, R, and Amberg, G., Dendritic growth of randomly oriented nuclei in a shear flow, *Journal of Crystal Growth*, vol 213, pp 161-187, (2000a).

Tönhardt, R, and Amberg, G., Simulation of natural convection effects on SCN crystals, *Physical Review E*, vol 62, pp828-836, (2000b).

Turner, J.S., "Multicomponent convection", *Ann. Rev. Fluid Mech.*, vol 17, pp11-44, (1985).

Wang, S.-L., Sekerka, R.F., Wheeler, A.A., Murray, B.T., Coriell, S.R. and Braun, R.J., *Physica D*, vol 69, pp 189-, (1993).

Warren, J.A., and Boettinger, W.J., *Acta Metall. Mater.*, vol 43, pp 689-703, (1995).

Wheeler, A.A. *J. Crystal Growth*, vol 67, p. 8, (1984).

Worster, M.G., "Convection in mushy layers", *Ann. Rev. Fluid Mech.*, vol 29, pp91-122, (1997).

Worster, M.G., "Instabilities of the liquid and mushy regions during solidification of alloys", *Journal of Fluid Mechanics*, vol 237, pp649-669, (1992).

Xu, J.-J., "Interfacial wave theory of pattern formation", Springer Verlag, New York, (1997).

Microscopic-Macroscopic Modelling of Transport Phenomena during Solidification in Heterogeneous Systems

Piotr Furmański

Institute of Heat Engineering, Warsaw University of Technology, Warsaw, Poland

Abstract. Microscopic-macroscopic approach to modelling of solidification in heterogeneous systems (the mushy zone of binary alloys, in manufacture of metal-matrix composites and in porous media) is addressed in the paper. Microscopic phenomena accompanying solidification and microscopic equations are presented. Averaging procedures (volume and ensemble averaging techniques) and their limitations are discussed. Macroscopic equations are derived and macroscopic phenomena described. Conditions for existence of local thermal and chemical equilibrium during solidification on the macroscopic scale are shown. Some examples of equilibrium and non-equilibrium solidification based on macroscopic approach are presented.

1 Introduction

Solidification problems play an important role in nature and material processing. Some naturally occurring examples are freezing of ice in lakes and rivers or formation of igneous rocks in geological systems. When undercooled water droplets strike a cold object they freeze and form an ice deposit. Familiar examples of this phenomenon are rime ice formation during winter fogs, growth of hailstones, icing of object like masts, power lines, trees, automatic meteorological stations. Ice accretion on structures may lead to financial losses or even constitute a risk to life in some cases as icing of an aircraft. Arctic or cold environments provide a variety of unique freezing phenomena in water, air, earth and biological fields. Engineering in cold regions has received interest due to petroleum exploration, construction of roads, airfields, pipelines and other structures. The thermal interaction of theses engineering systems with the environment may result in unexpected failures.

Solidification processes in porous media saturated with solutions are important for several engineering practices. In civil engineering the ground-freezing technique is used to support ground in coalmines under water saturated soil layers or in tunnels located near seashore. In thermal energy storage systems the latent heat of fusion of water saturated soil is used. Salt solutions, as a phase change materials (pcm), may also be applied in these systems (Okada et al., 1998). They allow attaining a desired phase change temperature. A high thermal conductivity of granular material that is mixed with the solution helps to enhance the heat transfer rate during loading or unloading of the storage system. Instead of salt solutions, the impure are used for solar energy storage, in passive thermal control of standard electronic modules in avionics applications and in spacecraft thermal management.

Building materials also belong to liquid saturated porous media. When water is capillary absorbed in the pores and freezing occurs the material surface can break up. This phenomenon leads to deterioration of stones. Use of protective agents can then help in preserving the cultural heritage of the country (Santoli et al., 1998).

Freezing can slow down or stop some biological reactions for preservation. The cryo-preservation technique is used for food stuff or biological tissues (skin, human cornea and organs) which are often considered as materials composed of solid particles and solution. Control of the undercooling and solidification is then very important. At low cooling rates, ice formation in living cells is restricted to the extracellular region and water is removed from the cell osmotically to avoid intracellular freezing. In this case the cell may be damaged by the effect of concentrated solution. At high cooling rates, probability of the intracellular ice nucleation increases and the cell may also be damaged. The problem then arises how to reduce injuries in the tissues during freezing, for example, by use of cryo-protective agents.

Analysis of solidification processes can also be of interest in nuclear engineering when solidification of a molten pool, namely the corium, in nuclear vessel can occur during hypothetical accident.

Perhaps the greatest field of application of solidification can be found in material formation and processing. Different techniques are used for carrying out solidification of pure metals and alloys, glass forming, production of metal-matrix composites or obtaining pure crystals of semiconductors from melts and solutions. Metal and alloy solidification processes are associated with such technologies as casting (die, continuous, form, ingot, precision, investment), micro-droplet soldering, spray deposition, directional solidification techniques used in single crystal growth (like Bridgeman or Czochralski methods), semi-solid forming in manufacture of composites in situ from eutectic alloys, etc.

The quality of the single crystal grown from the melt strongly depends on growth morphology and on macrosegregation in the solid caused by convection effects. A low gravity environment (i.e., microgravity with residual accelerations of the order of hundreds μg) produces conditions in which convection is eliminated or at least decreased to a level at which crystal growth is diffusion controlled (Timchenko et al., 1998). Some other problems with formation of ideal single crystals are, however, faced there. They are connected with Marangoni effects. Unlike natural convection, which is driven by density differences generated by either temperature or concentration gradients in bulk of the melt, the Marangoni convection is driven by surface tension forces brought about by temperature or concentration gradients along the external or internal free surfaces (called thermo-capillary and soluto-capillary convection, respectively). Solidification experiments carried out in space are often compromised by the evolution of unwanted bubbles or voids in the melt. These bubbles or voids form the internal free surfaces in the melt. On Earth, the gravity forces and the natural convection are dominant, especially in apparatuses with large volume-to-surface ratio. They quickly allow removing voids from the melt. However, in the microgravity environment of the orbiting spacecraft surface forces and the Marangoni convection become very important. It is estimated that in these conditions the thermo-capillary convection can be at least three orders of magnitude larger than the natural, buoyancy-driven one caused by residual gravity. As a result, the bubbles generated Marangoni convection will alter the flow, temperature and concentration fields in the melt. These effects can even extend to the solidification front, change the interfacial temperature and concentration gradients and lead to the unexpected growth conditions. Although the bubbles are highly undesirable, there is at present practically no effective means of preventing their formation or eliminating their adverse effects during microgravity solidification experiments carried out in space (Naumann, 1995).

Metal-matrix composites consist of a combination of two or more distinct components. The metallic component is continuous and the other discontinuous components are treated as reinforcing agents. The discontinuous component may occur in the form of fibres, micro-balloons or platelets. The metal-matrix composites offer enhanced material properties over traditional homogeneous materials. Their production techniques embrace die-casting, stir-casting, chemical vapour deposition or pressure infiltration (sometimes called the squeeze casting). For instance during the first stages of the pressure infiltration process the molten metal completely infiltrates the fibrous prepregs or other preforms (Khan and Tong, 1998). In the second stage the liquid metal is allowed to solidify to form the composite.

Solidification of metals can also be observed during secondary manufacturing processes as welding, soldering, brazing, cladding and sintering.

1.1 Motivation for Mathematical Modelling of Solidification

In case of solidification of metals there exists a relation between microstructure and bulk properties of the material. For example its mechanical strength is directly connected to the grain size, dendrite arm spacing (the lamellar or fibrous spacing in case of in situ composites), the number and size distribution of inclusions and the crystallographic texture of micro-structural matrices, i.e., directionality and anisotropy of the solid material. In manufacture high precision and high quality products are awaited, homogeneous without species segregation (resulting from uncontrolled constituents transport in the material), with desirable microstructure, without shrinkage formation, residual thermal stresses or pore and void formation due to improper filling of the casting. The structural perfection and the microstructure of the casting depend on solidification parameters: growth rate, temperature gradients at the solid/liquid interface, thermal undercooling and there exists close interaction between these parameters and heat transfer processes in the solid and the melt. Convection effects play here important role. For example quality of the single crystals depends not on the growth morphology but also on the species macrosegregation caused by convection effects. The convective plumes rising from narrow chimneys in the two-phase region (mushy zone) of alloys may lead to formation of freckles or other defects in the resultant solid (Viskanta, 1988). Voids formed by entrapped gases at the melt or at the mould-metal interface are carried up, due to buoyancy effects, to the free surface of the melt. Detachment and movement of the bubbles is significantly hindered in the microgravity experiments leading to formation of large voids in the crystal.

The solidification processes are accompanied by a release of thermal energy in form of the latent heat. The major problem in obtaining desirable microstructure during solidification is extraction of this heat in a controlled manner. It is not an easy task to perform as cooling rates during solidification can range from 10^{-5} to 10^{10} K/s and the solidification systems can extend from a few micrometers to several meters.

In case of other solidification processes, thermal behaviour of the latent heat of fusion storage systems that are periodically loaded and unloaded is of interest. Efficiency of these systems is determined by the amount of heat that can be extracted or delivered to the system in the prescribed time. Carrying out proper freezing of the biological tissues, not leading to their destruction, is necessary and controlled by the cooling rate. Ice accretion of aircraft may occur in a very short time and may lead to rapid change in aerodynamic characteristics of the flying objects. Time of switching on de-icing devices may then be important. In nuclear engineering solidification of the molten curium may lead to thermal explosion.

Solidification phenomena are very complex. They involve nucleation of the new phase, undercooling, phase transformation kinetics often subject to crystallographic constraints in solid phase, moving liquid-solid interface (i.e., evolution of microstructure during solidification), equilibrium and stability of the liquid/solid interface (Viskanta, 1988). Also such phenomena as liberation of thermal energy during solidification, remelting in some parts of the solidifying system, heat and species diffusion are present. Natural convection in the liquid caused by thermal and solutal buoyancy effects, Marangoni convection driven by surface tension variation with temperature and species concentration is often observed during solidification. Convection induced by shrinkage or forced convection, occurring during mould filling, or residual flows after mould filling, strip casting or squeeze casting at production of the metal-matrix composites or during ice accretion on aircraft elements occur in many cases of solidification processes. During solidification the fracture or break-off of the small parts of a crystal due to stresses in the solid phase ensuing liquid may occur. Deformation of the solid skeleton due to temperature differences or constituent inhomogeneities can also be present. Liberation of gases in the melt or at the mould/casting interface caused by variation in solubility of some species in the melt exerts impact on convection in the melt in the microgravity environment. These gases together with the shrinkage of the casting and/or the thermal expansion of the mould participate in formation of gaps at the mould/casting interface. Heat flow in these gaps occurs predominantly by conduction, in lesser extent by convection and at high temperatures by thermal radiation. As heat transfer processes in gases are less intensive than in liquids or solids, formation of this gap causes additional resistance to heat flow between the solidifying metal and its surroundings (Wang and Qiu, 2002). In order to reduce this resistance avoiding excessive thermal stresses in the mould and the cast is desirable.

Radiative transport of energy during solidification is also present in semitransparent materials like glass, some semiconductors and ceramics. The volumetric processes of absorption, emission and scattering of radiation occurring in bulk of the material lead to different behaviour of these semitransparent materials from the opaque ones. Liquid/solid interface is also the place where optical properties of the material change, radiative non-equilibrium exist and where enhanced radiation reflection is present. Most of the phenomena mentioned above are interrelated and lead to very complex interactions inside solidifying material.

In most cases of solidification processes there exist serious problems in carrying out proper experimental studies. To these problems contribute opaqueness of most materials (e.g., metals), high temperatures at which experiments should be carried out (due to high melting points of metals or ceramics), uncertainties involved in measurements of temperature, concentration, flow and microstructure in the solidifying materials. These uncertainties may also be connected to properties of the phases and coefficients describing transport phenomena at the liquid/solid interface or other experimental conditions. The above mentioned problems cause that not many reliable, experimental data are available for proper analysis and validation of theoretical models.

Presence of many physical phenomena, presence of multi-constituent systems, complex microstructure, mutual interaction of extreme number of factors during solidification and difficulties in carrying out experimental studies are the main factors explaining why an attempt of mathematical modelling of the solidification process was undertaken. The subject matter is very broad and multi-disciplinary field. It encompasses thermodynamics, material science, fluid and solid mechanics, heat and mass transfer, phase transformation kinetics and many other related disciplines. Impressive progress have been made in investigation of solidification phenomenon in separate disciplines and now one of the problems is how to put separate results obtained in these fields together to obtain better, more realistic and general mathematical description of the process. This will give both better insights in the involved, detailed phenomena. This approach should enable to work out better numerical codes for application in analysis of solidification processes occurring both in nature and manufacturing. However, the mathematical modelling of the process is faced with difficulties (Viskanta, 1988). The phenomena associated with solidification occur on different length scales ranging from 10^{-9} meters for nucleation, interface structure and atomic attachment kinetics (*atomic scale*); through single crystal-melt interactions, i.e., capillarity, interface instabilities, local interface non-equilibrium conditions and movement (*interface scale*). Further, local transport processes in microstructural parts of the system lead to different types of microstructural elements like dendrites or equiaxed crystals, spacing of dendrite arms or lamella and fibres in eutectic alloys, coarsening (*grain scale*). Finally, microstructure patterns (porosity variation), macroscopic inhomogeneities in constituents distribution and transport processes (*system scale*) associated evolution of the latent heat, cooling rates and volume change of system as a whole (from 10^{-2} to several meters) occur.

The topics discussed further in the text will be limited to solidification of multi-component systems (mainly alloys). However, many of the ideas presented here may also be applied in theoretical description of other related cases of solidification phenomena. At first more discussion of solidification phenomena occurring in multi-component, multiphase systems will be given. Then the equations describing transport phenomena in the so-called microscale will be presented. The solidification systems often appear to have complex multi-phase form. For example, two-phase mushy zone appears in case of pure metals or alloys while three-phase media can be met in case solidifying granular media or metal matrix composites. During solidification structure with distinct phase boundaries and different properties of the phases is formed. These systems may be then treated as heterogeneous media. The region close to the mould-casting interface may also be considered as heterogeneous as thermal conditions for heat removal from the melt are varying along the interface. For analysis of transport phenomena in the heterogeneous systems, the so-called, macro modelling is often used. The micro- versus macro modelling will be discussed in detail and motivation for macro modelling will be presented. The main body of the presentation will be devoted to macro modelling of the solidification process. Discussion of the averaging techniques used in this approach, macroscopic closure problems (constitutive relations), and estimation of macroscopic properties and coefficients describing mould-casting thermal interaction will be given. Examples of application of the macroscopic approach to equilibrium and non-equilibrium solidification will also be shown. Finally, unresolved problems in macro modelling will be presented and directions for further developments in this approach suggested.

2 Physical Phenomena Accompanying Solidification in Pure and Multi-component Materials

In pure materials solidification occurs at the distinct temperature T_m which is known as equilibrium temperature of freezing (melting). This temperature is solely dependent on pressure and is kept constant if solidification proceeds at a slow rate, no curvature of the liquid-solid interface is present and no significant difference in specific heats between phases exists. For phase transformation occurring at the equilibrium temperature the heat liberated per unit mass of solid phase, i.e. the latent heat of fusion L_f per unit volume, is also only dependent on pressure.

The multi-component systems (like alloys) have some special features that distinguish their solidification from pure materials. Their equilibrium freezing temperature is dependent both on pressure and concentration of the dissolved species. As concentration may vary during the process of material cooling so phase transformation is taking place over range of temperatures and liquid and solid phases can coexist in equilibrium over this range of temperatures (see Fig.1). For a given initial concentration C of the binary mixture nucleation of small solid grain occurs at a temperature slightly below the liquidus line. As the equilibrium temperature is lowered the concentration of the solid and liquid on each side of the interface is continuously varying till the eutectic temperature is attained. This is a direct indication of another future of the multi-component media that the chemical species have different solubility in the liquid and solid phases. Ratio of solubility of the particular species in the solid and liquid phases is known as the partition factor κ_p. If the equilibrium is present, κ_p is proportional to the ratio of species concentration in the solid C_s and liquid phase C_l and it is smaller than unity for the left branch of the equilibrium diagram (see location of the solidus line in Figure 1). Solubility of species varies not only with temperature but also with curvature of the solid-liquid interface.

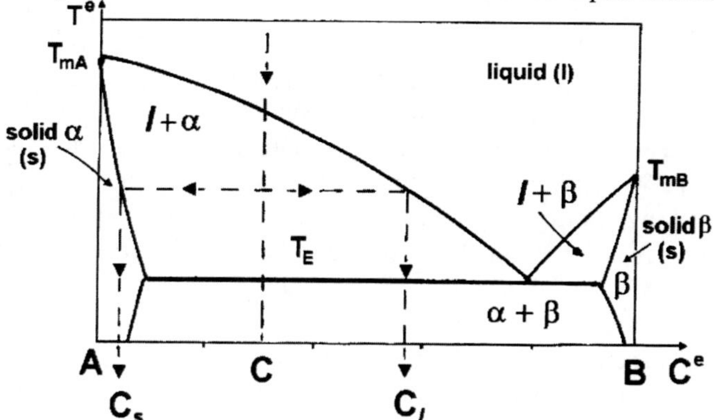

Figure 1. A typical equilibrium phase diagram for binary mixture.

Latent heat of fusion, in the considered case, is related not only to pressure but also to composition of the mixture. The third feature of the multi-component media is that concentration of the dissolved species can vary in space and time due to species diffusion and advection caused by liquid motion.

2.1 Nucleation

The first stage in formation of the new (solid) phase from the old (liquid) one is known as nucleation. Two forms of nucleation can be met: *homogeneous* and *heterogeneous* one. *The homogeneous nucleation* occurs in bulk of the undercooled liquid. It is the starting point for the activated nuclei to grow into separate grains in the bulk undercooled melt. This kind of solidification process is called an equiaxed crystal growth. For semitransparent materials, significant melt undercooling may develop prior to crystalline nucleation because of volume cooling by internal radiation. If strong catalytic agents are distributed in the melt, nucleation may take place with little undercooling, resulting in equiaxed solidification in the practically isothermal zone. *The heterogeneous nucleation* takes place on surfaces that catalyse solidification, such as inclusions or other impurities in the material being solidified, the walls of the container (in which liquid is being held) or the surfaces of the casting which are cooled from the outside. It is the starting point for directional (columnar) solidification.

From the thermodynamic point of view the driving force for solidification is the difference in Gibbs free energy ΔG between the liquid and the solid. Two effects contribute to variation in ΔG. One is associated with formation of a new surface while the other with phase transformation. The difference in Gibbs free energy between phases varies with a radius of nucleus of the solid phase. The radius at which ΔG attains maximum value is known as *the critical radius* r^* of the nucleus of solid in the liquid (see Fig. 2). For subcritical radius of the nucleus the new phase will disappear so for a viable nucleus, i.e. nucleus to grow, its size should be equal or greater than radius r^*.

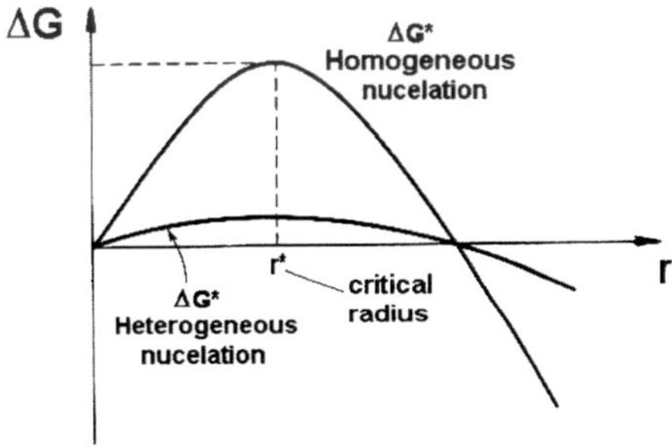

Figure 2. Variation of ΔG versus nucleus radius for homogeneous and heterogeneous nucleation.

For pure material the critical radius can be calculated from the expression

$$r^* = -2\sigma / \Delta G_v \tag{2.1}$$

where σ denotes surface tension (surface energy between solid and liquid) while ΔG_v is the volumetric Gibbs free energy. The latter is equal to

$$\Delta G_v = \frac{L_f (T - T_m)}{T_m} \tag{2.2}$$

where L_f is the latent heat of solidification per unit volume of the substance. As the undercooling of the liquid ($T - T_m$) increases nuclei of smaller value of r^* become viable.

The critical Gibbs free energy (i.e., corresponding to the critical radius) for the heterogeneous nucleation depends on the nucleus volume $V^*(\theta)$, where θ is the wetting angle, and can be expressed as

$$\Delta G^* = -\frac{1}{2} V^*(\theta)\,\Delta G_v \tag{2.3}$$

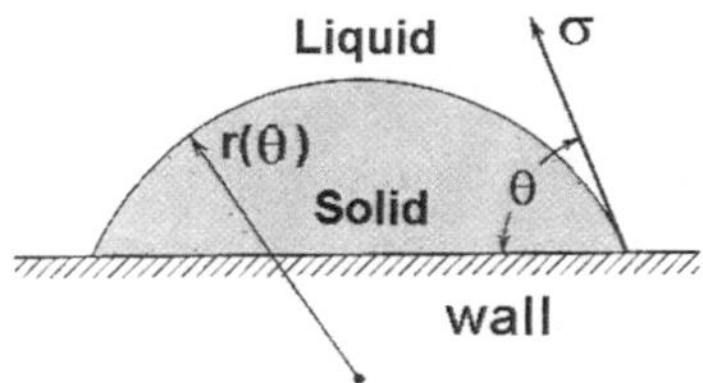

Figure 3. Shape of a nucleus on the catalysing surface.

Presence of the catalysing agent, in form of the external surface, changes shape of the nucleus when the latter is wetting the surface (see Fig. 3). The associated reduction in the critical Gibbs free energy of formation of the nucleus (see Fig.2) makes nucleation more probable.

The rate of nucleation is determined by the concentration N of nuclei of the critical size and the rate they are activated to grow. It is approximately given by the following relation (Garde et al., 1993)

$$\frac{dN}{dt} = f(T)(N_0 - N)\exp\left[-\frac{\Delta G^*}{kT}\right] \tag{2.4}$$

where $f(T)$ is the frequency factor, N_o - the initial concentration of nuclei of the critical size and k stands for Boltzmann constant. As ΔG^* is proportional to undercooling so the concentration of the activated nuclei grows with ($T - T_m$). The nucleation rate attains the maximum value at certain $T < T_m$ and than decreases with undercooling. This leads to the asymptotic concentration of the activated nuclei when ($T - T_m$) is increasing.

When cooling occurs through walls of the container (mould) solid may nucleate on the surface (heterogeneous nucleation). On the microscopic scale the surface of a mould is not completely smooth but is rough. The roughness of the surface can be imagined as a great number of small asperities protruding from the surface profile. When liquid melt first approaches the mould surface, contact occurs at the peaks of asperities. Rapid cooling at these peaks causes solidification of the liquid to nucleate from these sites. At the same time, surface tension of the liquid and subsequent rapid growth of solidification from the nucleation sites prevents the solidifying material from wetting the valleys between asperities on the surface profile.

2.2 Growth of Solid Phase and Formation of the Mushy Zone

During solidification the solid/liquid interface is moving as the amount of solid increases. The velocity of the interface w_i is controlled by heat generation at the interface due to phase transformation and heat removal from the interface by conduction, convection or radiation. For the case of the last two modes absent, the interface velocity is given by the expression

$$w_i = \left[\lambda_s \left(\frac{\partial T}{\partial n} \right)_s - \lambda_l \left(\frac{\partial T}{\partial n} \right)_l \right] / L_f$$

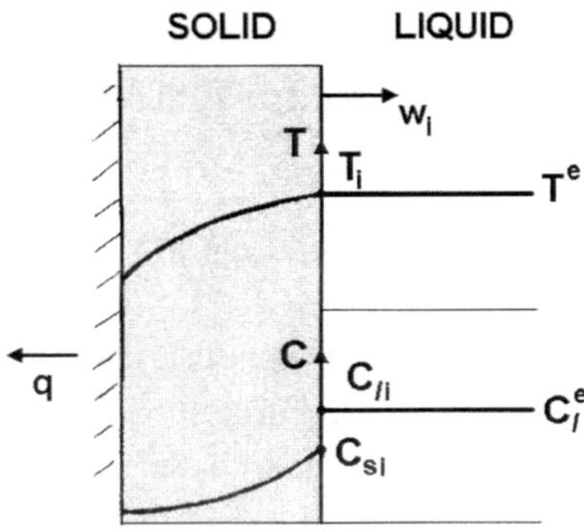

Figure 4. Shape of the solid/liquid interface for slow rate of solidification.

For low values of the interface velocity the solidification front tends to be planar and corresponds to minimum surface energy configuration (Fig. 4). Thermodynamic conditions at the solid/liquid interface correspond to the equilibrium ones, i.e., temperature is equal to the equilibrium temperature $T_i = T^e$ (e.g., the melting temperature T_m for the pure materials) and dissolved species concentrations are equal $C_{li} = C_l^e$, $C_{si} = C_s^e$ on the liquid and solid side of the interface (for multi-component mixtures), respectively. In case of the multi-component mixture the equilibrium temperature T^e and the equilibrium concentrations C_l^e, C_s^e are related to each other by the phase diagram (see Fig. 1).

For increasing value of the interface velocity (rate of solid growth), and thus greater departure from equilibrium, the interface breaks down, first into the cellular (Fig.5a) and then into a dendritic structure (Langer, 1989 – Fig.5b). The two-phase region characterised by presence of such irregular interfaces is often called the *mushy zone.* In pure or nearly pure materials, departure from equilibrium conditions and the break of the solidification front are associated with undercooling of the liquid ahead of the solidification front. The interface temperature T_i is lower than the equilibrium temperature T^e (corresponding to liquidus temperature for flat interface) and may be written as (Aziz, 1996)

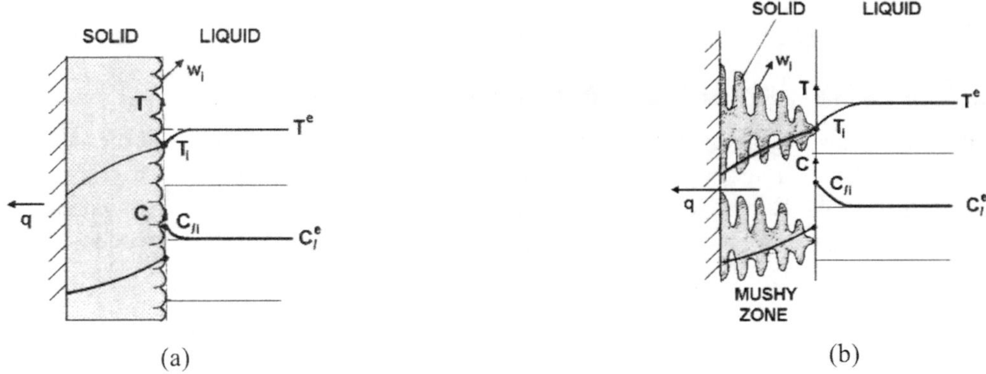

Figure 5. Shape of the solid/liquid interface for slow rate of solidification:
(a) columnar cellular, (b) columnar dendritic.

$$T_i = T^e - \frac{2\,\sigma(\vartheta_o, s)}{\Delta S_{ls}} \kappa - \mu_k(\vartheta_o, \gamma)\, w_i \tag{2.5}$$

where σ denotes coefficient of surface tension that depends on the angle ϑ_o that the crystallographic axis makes with normal to the surface (see Figure 6a), s is the strength of the anisotropy of the solid crystal and ΔS_{ls} a difference in the entropy of liquid and solid phases per unit volume of the solid phase for composition corresponding to the solid phase. Symbol κ stands here for curvature of the surface while μ_k is the kinetic coefficient. The second and the third terms on the r.h.s. of eq. (2.5) correspond to the curvature undercooling and the kinetic undercooling of the interface, respectively. The group of quantities standing in front of curvature in the second term is known as the first Gibbs-Thomson coefficient.

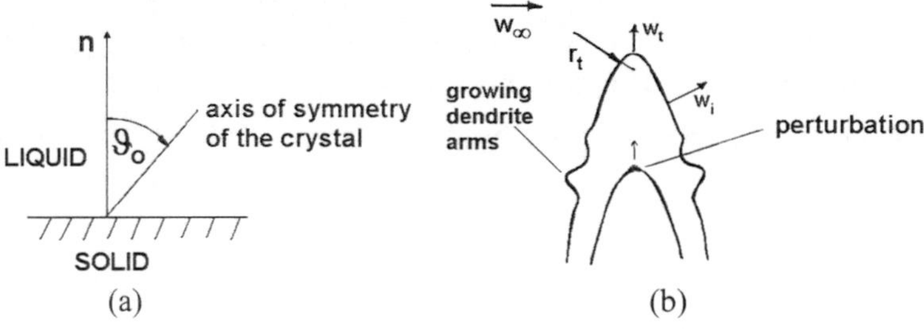

Figure 6. Relation between crystallographic axis of the growing crystal and normal to the solid/liquid interface (a) and formation of the side arms (branches) close to tip of the growing dendrite (b).

During solidification of a binary mixture, solute enrichment or depletion of the remaining liquid phase results from differences in solute solubility in the solid and liquid phases. The break down of the solidification front occurs more readily when solute elements are present, either as impurities or as deliberate alloying components. In the multi-component mixtures species (solute) concentration gradients, in most cases, control evolution of the microstructure. On the liquid side of the interface most of the solute is rejected laterally in direction perpendicular to direction of heat flow, i.e., between the dendrite arms or eutectic lamellae. For solute solubility lesser in the solid than in the liquid a solute concentration gradient ahead of the solidification front exists in the liquid. Species concentration is thus higher at the interface than in bulk of the liquid (see, for example, Fig. 7). The difference between the concentration in these two locations is usually referred to as the *solutal (or constitutional) undercooling*. The term undercooling stems from the practice to convert the two concentrations to temperatures via the liquidus line of the phase diagram (see Fig. 8). Equilibrium solute concentration at the interface on the liquid side is additionally increased due to curvature of the interface and also contributes to the constitutional undercooling. For the binary mixture the liquid concentration C_{li} at the curved interface is greater than for the flat one and can be estimated from the Thompson-Freundlich equation (Mortensen and Flemings, 1996)

$$C_{li} = C_l^e + \frac{\sigma(\vartheta_o, \gamma)\, T^e}{L_f\, m_L} \kappa \tag{2.6}$$

where m_L is the slope of the liquidus line.

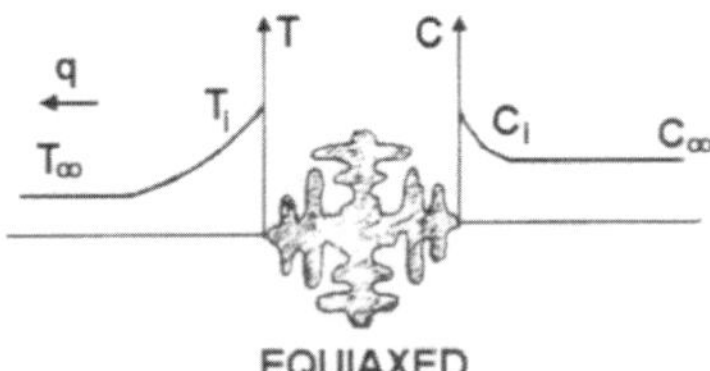

Figure 7. Temperature and solute distribution close to dendritic structure of the equiaxed crystal.

As the temperature is lowered, the concentration in the solid at the interface continuously changes along the solidus line (see Fig.1). As drop in temperature is different in different part of the solidifying mixture a microscopic concentration profile develops in the solid. This effect is termed the *microsegragation*. This profile can change during solidification due to species diffusion, however, owing to the very small mass, diffusivities of the dissolved species in the solid the micro-segregation usually persists in the solidified alloy.

Anisotropy of surface tension, thermal and constitutional undercooling and instabilities at the growing tip of the cell can lead to side branching. A cell with side branches resembles a tree and so is termed "dendrite". In nature and industrial processes the normal growth mode is dendritic (Viskanta, 1988). The dendrite thus exists in the solidifying material because it is the most efficient morphology for diffusion of solute and the dissipation of heat in order to reduce undercooling in the melt. The velocity of the interface at tip of the dendrite w_t (see Fig.6b), for no convection present, is related to the radius of curvature of the dendrite tip (Bouissou and Pécle, 1989):

$$w_t r_t^2 = \text{const} \tag{2.7}$$

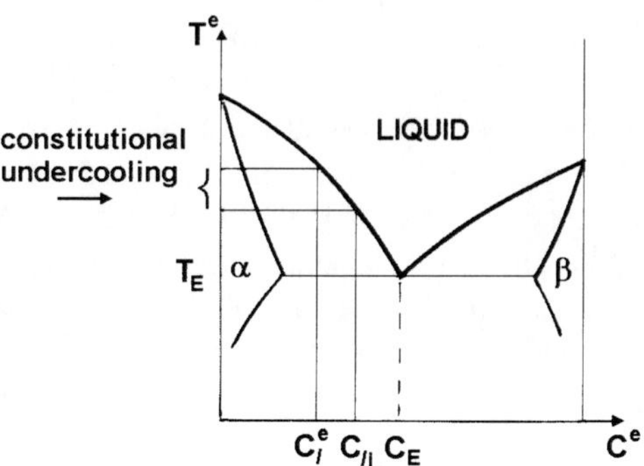

Figure 8. Definition of constitutional undercooling of the liquid phase at the interface.

While the growth direction of cells is determined by the maximum thermal gradient, the growth direction of dendrites deviates from it. It is a compromise between the direction of the maximum thermal gradient and of the crystallographic easy growth direction. The primary dendrite spacing is related to velocity of the interface, degree of undercooling, partition factor, diffusion coefficient of the liquid, mean concentration of the solute and temperature gradient on the liquid side. The secondary arm spacing is one of the most important length scales of dendritic microstructure since it determines the periodicity of solute segregation profile in the solidified material. For pure semitransparent materials (e.g., ceramics, plastics), radiative supercooling generates a large negative gradient ahead of the interface that may lead to an unstable interface and the possibility of developing thermal cells and dendrites as described above.

When the columnar mushy zone is formed heat may be quickly removed by conduction from the interface through the solid phase towards the wall, as thermal conductivity of the solid phase is greater than the liquid phase. For high cooling rates and thus high velocity of the interface non-equilibrium conditions at the interface develop and the interface temperature becomes lower than the equilibrium temperature (see Fig. 5). The melt ahead of the columnar dendrite tips becomes undercooled allowing equiaxed crystals to be formed (see Fig.7). As mentioned earlier the equiaxed crystals can be formed from nuclei that are floating in the melt. They may also be formed due to detachment of the broken dendrite arms and their movement from the dendrites due to convection currents. The equiaxed crystals are completely surrounded by the melt and so their interfacial temperature should be greater than the liquid nearby in order to remove the latent and sensible heat (Fig. 7). In opposition to columnar crystal structure where heat is removed directionally from the interface towards the cooling wall in case of the equiaxed crystals heat removal is multidirectional.

At very high growth velocity w_i the perturbation wavelengths become smaller. When the perturbations are small, surface tension plays an important role, comparable in magnitude to that of the solute field and can again stabilise the plane front of solidification.

In *eutectic alloys* systems, of initial solute concentration equal to C_E, two solid phases (α and β) grow simultaneously from the liquid phase (see Fig. 7). Columnar eutectic with lamellar or rod morphology or

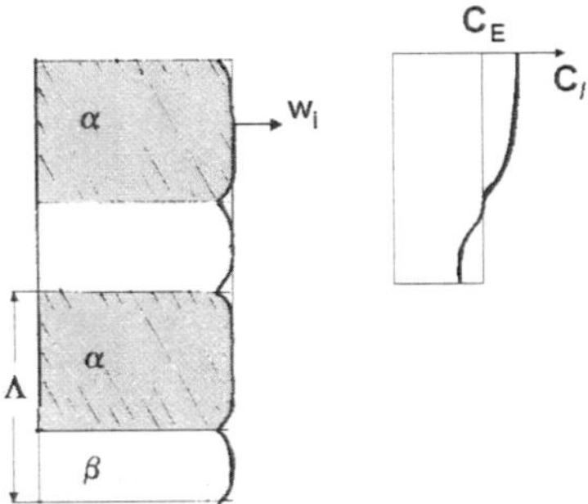

Figure 9. Solute concentration distribution in front of the growing eutectic crystal having regular lamellar structure.

regular equiaxed eutectic solid structure may be formed. The microstructure exhibited by the eutectic solid is also varied from regular to irregular, i.e., showing no regularity of distribution of the two phases. The regular eutectic growth was modelled by Hunt and Lu, 1996. For the maximum growth rate of solid, for a small constant undercooling, with no convection in the melt present, the eutectic spacing Λ for the steady state of growth is given by the expression

$$w_i \, \Lambda^2 = const. \tag{2.8}$$

which parallels similar expression for the relation between the radius of dendrite tip and its velocity (see eq. (2.7)). Ahead of the solidification front, during eutectic solidification, species concentration gradient is formed along the interface due to different solubility of the dissolved species in each of the solid phases α and β (see Fig. 9).

2.3 Role of Liquid Convection in the Solid Phase Growth and the Constituents Distribution

Fluid flow in the melt plays an important role in solid growth rate, formation of the mushy zone microstructure and segregation of the dissolved species. Convection in the liquid may be caused by: (a) external forces (*forced convection*) resulting from imposed pressure gradients, electric, magnetic forces or movement of external surfaces surrounding the melt, (b) buoyancy forces (*natural convection*) caused in presence of gravity by density differences in the liquid resulting from temperature or concentration gradients (thermo-solutal or double-diffusive convection), (c) thermo- and soluto-capillary forces in absence of gravity caused by variation in surface tension along the interface resulting from temperature and solute concentration surface gradients (*Marangoni convection*), (d) *solidification shrinkage* induced flows or (e) with fluid flow induced by the growth of the solid phase that is heavier or lighter than the liquid phase and by movement of solid grains (Viskanta, 1988). Fluid movement can also be associated with residual flows occurring just after mould filling.

Fluid velocity and direction effects velocity of the dendrite tip (Schrage, 1999; Amberg, 2002). They change orientation of the dendritic microstructure toward direction of the flow; cause increased growth rate in this direction and bending of the separate dendrites. In alloys there is a direct relation between direction of fluid flow due to buoyancy forces and type of the solute that is rejected by the solid into liquid due different solubility of the phases. Rejection of the solute that is heavier (or lighter) than the solvent causes differences in density close to the interface and can augment or hamper convection driven by temperature differences in the liquid. In case of cooling the melt from below and growth of the mushy zone from the bottom of the container upwards completely molten channels through the mushy zone, known as chimneys, can be formed. They are typically oriented in the direction of gravity, have different chemical composition than the surrounding mush and are the origin of freckles in the final product (Amberg, 2002).

Solid grains (equiaxed crystals) formed in the liquid or parts of the broken dendrite arms can also move around the melt either due to solid/liquid density difference or floating with the liquid. The solid crystals may be transported in the melt from locations of lower to higher temperature where they can partially remelt.

Movement of the liquid and, particularly, of the solid grains is associated with species advection. It can induce significant compositional inhomogeneities in the whole system known as the *macro-segregation*.

2.4 Variation in Microstructure of the System during Solidification

During solidification microstructure changes occur in the whole system. Mixed columnar – equiaxed structures occur. The modelling of the *columnar to equiaxed transition* is considered to be one of the major areas of the contemporary research. Free-floating grain structures are often associated with impinging of one grain on another and impinging of grains with the columnar solid structure. Sedimentation of floating solid grains at the bottom part of the mould may lead to some structural defects in the solidified material. The number of the equiaxed grains floating in the melt is being increased with progress of solidification. This contributes to increased interaction of grains and formation of solid skeleton. The solid volume fraction of the

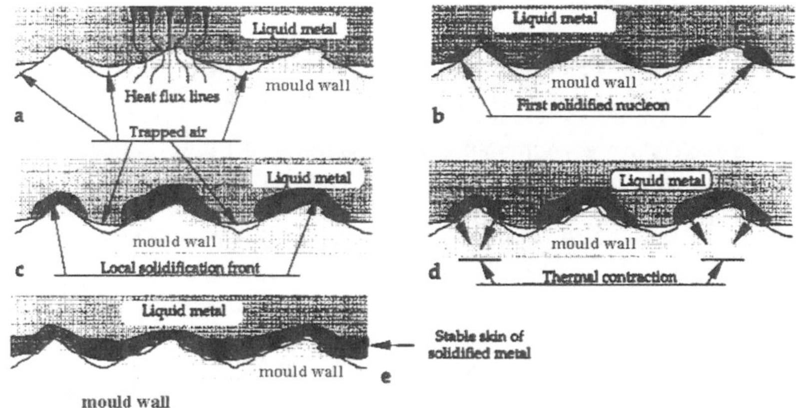

Figure 10. Evolution of the solidification process at the alloy-mould interface (after Lolou et al., 1999).

equiaxed grains at which the latter process occurs is known as the *dendritic coherency point.* It can be determined in torque experiments (Arnberg et al., 1993). The cooling rate, equiaxed grain sizes and the concentration of the dissolved species influences the dendritic coherency point. It is estimated to vary from 0.1 up to 0.5.

2.5 Processes at the Mould-Casting Interface

The surface of a container encompassing the solidifying material or the mould in case of castings is not smooth but rough. The roughness is characterised by presence of many asperities protruding from the surface. In the beginning of the solidification process the melt that spreads over the surface of the mould is in contact with the wall just through peeks of asperities (Fig.10a). The heat is removed from the melt mainly by conduction occurring at the direct contact areas. The air or the gaseous constituents that were initially dissolved in the liquid are rejected from it upon solidification due to their lesser solubility in the solid than in the liquid. These gases are trapped in the micro-cavities existing between asperities. In the short time the equilibrium is established between surface tension forces and pressure of these residual gases. The nucleation begins at the points where temperature is the lowest, i.e., at places where contact between molten material, wall and trapped gases occurs (see Fig.10b). Around each asperity the nucleation process proceeds and a solidified ring is formed around its top. As the rough surface is characterised by great number of asperities so many such solidification fronts appear. At this stage of the process an almost uniform distribution of the crystal growth orientations is present. The ring grows in time; its inner diameter decreases to zero and finally the cap of the solidified material (see Fig.10c) covers the whole top of the asperity. As solid has usually greater density than the liquid the metal contracts and the mould material expands forming greater interface gap between the casting and the mould (see Fig.10d). Subsequently the local solidification fronts merge and form a unique solidification front. The continuous solid crust is formed that separates liquid from the wall (see Fig.10e). As the thickness of the crust increases the shrinkage phenomenon becomes dominant. Due to shrinkage the number the size of the contact zones and number of contacts is reduced gradually while the gaseous gap grows in size. For sufficiently thick solid crust the interface gap tends to be stable and does not increase anymore (Lolou et al., 1999; Browne and O'Mahoney, 2001).

3 Micro-modelling. Equations Describing Transport Phenomena in Solidifying Medium at the Micro-scale

There are in general two different groups of methods that were worked out for mathematical modelling of the complicated solidification phenomena. The first of them is called micro modelling. In the micro modelling each of the phases is considered separately as continuum and all details of developing interface are taken into account. Each phase (either solid or liquid) is treated as homogeneous medium. It is assumed that the constitutive relations between stress and strain or rate of deformation, heat flux and temperature gradient, solute flux and concentration gradient, as Fourier or Fick laws, are valid. In case of the liquid phase Newtonian behaviour is assumed. Continuity of temperature across the interface is adopted but not always its equilibrium value T^e, for the flat interface, is used. Instead relations taking into account influence of the curvature or the interface velocity on the liquid/solid interface temperature, like eq. (2.5), are adopted. The dissolved species concentrations are assumed to be discontinuous at the interface obeying the equilibrium relation coming from the phase diagram (Fig. 1).

The set of equations used in the micro modelling consists of the conservation equations for mass, momentum, energy and jth species mass fraction in each kth phase of the mixture

$$\partial_t(\rho_k) + \nabla \cdot (\boldsymbol{w}_k \rho_k) = 0 \tag{3.1}$$

$$\partial_t(\rho_k \boldsymbol{w}_k) = -\nabla \cdot (\boldsymbol{w}_k \rho_k \boldsymbol{w}_k - \boldsymbol{\sigma}_k) + \rho_k(T, C_k)\boldsymbol{f}_k \tag{3.2}$$

$$\partial_t(\rho_k h_k) = -\nabla \cdot (\boldsymbol{w}_k \rho_k h_k + \boldsymbol{q}_k) \tag{3.3}$$

$$\partial_t(\rho_{jk} C_{jk}) = -\nabla \cdot (\boldsymbol{w}_{jk} \rho_k C_{jk} + \boldsymbol{j}_{jk}) \tag{3.4}$$

where ∂_t denotes time derivative of the respective quantity. In the above equations ρ_k, $\boldsymbol{w}_k$ and h_k are density, velocity and specific enthalpy of kth phase, respectively, while C_{jk} denotes mass concentration (mass fraction) of jth species in kth phase. Symbols $\boldsymbol{\sigma}_k$ and $\boldsymbol{q}_k$ stand for stress tensor and heat flux vector in kth phase while $\boldsymbol{j}_{jk}$ denotes mass flux of jth species in kth phase.

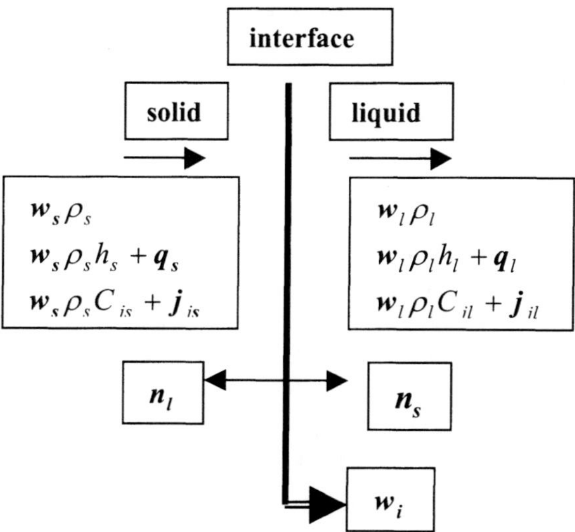

Figure 11. Schematic of balances of mass, energy and jth species at the liquid/solid interface.

In modelling of the solidification phenomena it is usually assumed that liquid and solid phases are incompressible, i.e., $\rho_k = const$, but it is convenient to retain the full form of the mass conservation equation, eq. (3.1). Separate constituents, which do not undergo phase transformations, as fibres in metal/matrix composites or grains of in the heterogeneous storage systems, may be included in these conservation equations by increasing number of phases. It is also very often assumed that the solid phase of the phase changing material (as well as other solid phases present in the system) is rigid. In case of the freely floating equiaxed grains additional equation of the moment of momentum should also be included in set of the conservation equations. The last term on the right hand side of eq. (3.2) corresponds to the body force that is exerted on kth phase. As natural convection is caused by buoyancy forces related to density variation in kth phase, dependence of the density on temperature and species concentration was retained in this term.

The mass, momentum, energy and species balances should also be satisfied at the liquid/solid interface which is moving with velocity $\boldsymbol{w}_i$ (see Figures 11 and 12). The respective equations are given below

$$\rho_s(\boldsymbol{w}_s - \boldsymbol{w}_i)\cdot\boldsymbol{n}_s + \rho_l(\boldsymbol{w}_l - \boldsymbol{w}_i)\cdot\boldsymbol{n}_l = 0 \tag{3.5}$$

$$[\rho_s\mathbf{w}_s(\mathbf{w}_s - \mathbf{w}_i) - \boldsymbol{\sigma}_s]\cdot\mathbf{n}_s + [\rho_l\mathbf{w}_l(\mathbf{w}_l - \mathbf{w}_i) - \boldsymbol{\sigma}_l]\cdot\mathbf{n}_l = -2\kappa\sigma\mathbf{n}_s - \nabla_t\sigma \tag{3.6}$$

$$[\rho_s h_s(\boldsymbol{w}_s - \boldsymbol{w}_i) + \boldsymbol{q}_s]\cdot\boldsymbol{n}_s + [\rho_l h_l(\boldsymbol{w}_l - \boldsymbol{w}_i) + \boldsymbol{q}_l]\cdot\boldsymbol{n}_l = 0 \tag{3.7}$$

$$[\rho_s C_{js}(\boldsymbol{w}_s - \boldsymbol{w}_i) + \boldsymbol{j}_{js}]\cdot\boldsymbol{n}_s + [\rho_l C_{jl}(\boldsymbol{w}_l - \boldsymbol{w}_i) + \boldsymbol{j}_{jl}]\cdot\boldsymbol{n}_l = 0 \tag{3.8}$$

In eq. (3.7) the energy contribution due to interface stretching was assumed as negligible. In the momentum balance, eq. (3.6), discontinuity of normal and tangential stresses was included (see Fig. 12). The last term on the r.h.s. of this equation corresponds to variation of surface tension along the interface (Marangoni effect).

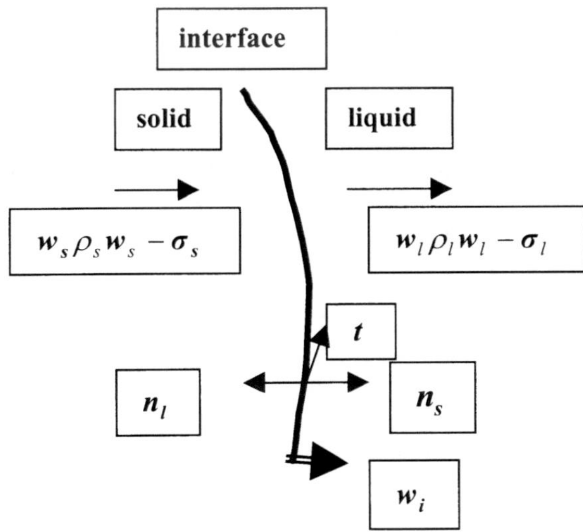

Figure 12. Schematic of momentum balance at the liquid/solid interface.

Interface mass balance, eq. (3.5) can be used to eliminate liquid velocity from the other interface balances. Then the following relations are obtained:

$$(\boldsymbol{\sigma}_s - \boldsymbol{\sigma}_l]\cdot \boldsymbol{n}_s = \rho_s(\boldsymbol{w}_s - \boldsymbol{w}_l)(\boldsymbol{w}_s - \boldsymbol{w}_i)\cdot \boldsymbol{n}_s - 2\kappa\,\sigma\,\boldsymbol{n}_s - \nabla_t \sigma \tag{3.9}$$

$$(\boldsymbol{q}_s - \boldsymbol{q}_l)\cdot \boldsymbol{n}_s = \rho_s(h_l - h_s)(\boldsymbol{w}_s - \boldsymbol{w}_i)\cdot \boldsymbol{n}_s \tag{3.10}$$

$$(\boldsymbol{j}_{js} - \boldsymbol{j}_{jl})\cdot \boldsymbol{n}_s = \rho_s(C_{js} - C_{jl})(\boldsymbol{w}_s - \boldsymbol{w}_i)\cdot \boldsymbol{n}_s \tag{3.11}$$

where velocity of the solid phase is very often assumed to be equal to zero. In case of other, internal interfaces inside the material, such as fibre-metal matrix in composites or solid, foreign grains –phase changing material in the thermal storage systems, the right hand sides of the above equations are also often assumed to be zero.

In addition to these conservation equations other conditions should be satisfied at the liquid/solid interface. Continuity of temperature and velocity at the interface, i.e., $T_l = T_s$, $\boldsymbol{w}_l = \boldsymbol{w}_s$ should be held.

Due to different solubility of species in the solid and liquid phases the mass concentration of species is discontinuous across the interface. As it is not convenient to deal with discontinuity, either analytically or numerically, a modified concentration $\widetilde{C}_{jk}$ was introduced. It is defined by the expression

$$\widetilde{C}_{jk} = M_{jk} C_{jl} \tag{3.12}$$

and obeys condition of continuity at the interface. The solubility M_{jk} of jth species in kth phase was defined as

$$M_{jk} = \begin{cases} 1 \text{ for } \boldsymbol{x} \in \boldsymbol{V}_s \\ \kappa_{pj} \text{ for } \boldsymbol{x} \in \boldsymbol{V}_l \end{cases} \tag{3.13}$$

This kind of substitution allows eqs (3.4), (3.8) and (3.11) to be rewritten in another form

$$\partial_t(\widetilde{\rho}_{jk}\widetilde{C}_{jk}) = -\nabla\cdot[\boldsymbol{j}_{jk} + \boldsymbol{w}_{jk}\widetilde{\rho}_{jk}\widetilde{C}_{jk}] \tag{3.14}$$

$$[\widetilde{\rho}_{js}\widetilde{C}_{js}(\boldsymbol{w}_s - \boldsymbol{w}_i) + \boldsymbol{j}_{js}]\cdot \boldsymbol{n}_s + [\widetilde{\rho}_{jl}\widetilde{C}_{jl}(\boldsymbol{w}_l - \boldsymbol{w}_i) + \boldsymbol{j}_{jl}]\cdot \boldsymbol{n}_l = 0 \tag{3.15}$$

$$(\boldsymbol{j}_{js} - \boldsymbol{j}_{jl})\cdot \boldsymbol{n}_s = \widetilde{\rho}_{js}\widetilde{C}_{js}(M_{js}/M_{jl} - 1)(\boldsymbol{w}_s - \boldsymbol{w}_i)\cdot \boldsymbol{n}_s \tag{3.17}$$

where the modified density is defined as

$$\widetilde{\rho}_{jk} = \rho_k / M_{jk} \tag{3.18}$$

The conservation relations should be supplemented with the expression relating heat flux $\boldsymbol{q}$, species mass fluxes $\boldsymbol{j}_j$ and stress $\boldsymbol{\sigma}$ to temperature, concentration or deformation. These relations, known as constitutive relations, are usually adopted in the following form

$$\boldsymbol{q}_k = -\lambda_k \nabla T \tag{3.19}$$

$$\boldsymbol{j}_{jk} = -\rho_k D_{jk} \nabla C_{jk} \tag{3.20}$$

where λ_k is thermal conductivity of kth phase, D_{jk} - mass diffusivity of jth species in kth phase. As the solid phase is usually assumed to be rigid the constitutive relation for the liquid phase is written as

$$\boldsymbol{\sigma}_l = -p_l \boldsymbol{1} + 2\mu_l \boldsymbol{e}(\boldsymbol{w}_l) \tag{3.21}$$

where p_l denotes pressure and $\boldsymbol{e}(\boldsymbol{w}_l) = (\nabla \boldsymbol{w}_l + \nabla^T \boldsymbol{w}_l)$ is deformation rate tensor.

The expression for the mass flux, eq. (3.20), can be modified by introducing $\widetilde{C}_{jk}$ and then it assumes the form

$$\boldsymbol{j}_{jk} = -\widetilde{D}_{jk} \nabla \widetilde{C}_{jk} + \widetilde{M}_{jk} \nabla T_k \tag{3.22}$$

where:

$$\widetilde{D}_{jk} = \rho_k D_{jk}, \qquad \widetilde{M}_{jk} = \widetilde{D}_{jk} \widetilde{C}_{jk} \ln[\partial_T M_{jk}]$$

The second term on right-hand-side of eq. (3.22) follows from dependence of solubility on temperature. It disappears when liquidus and solidus lines at the phase diagram (see Fig. 1) can be approximated by the straight lines.

In equations (3.1)-(3.8), (3.14)-(3.17) density, surface tension and solubility (and associated with it the partition factor) appear. For the case of internal equilibrium prevailing in any location inside the phase, these quantities are dependent on temperature and concentration and can be expressed, using linear approximations, as:

$$\rho_k(T, [\widetilde{C}_{jk}]) = \rho^o(T_r, [\widetilde{C}_{jk}]_r) \left[1 + \rho_{Tk}(T - T_r) + \sum_j \rho_{C_{jk}} (\widetilde{C}_{jk} - \widetilde{C}_{jkr}) \right] \tag{3.23}$$

$$\sigma(T, [\widetilde{C}_{jk}]) = \sigma^o(T_r, [\widetilde{C}_{jk}]_r) \left[1 + \sigma_{Tk}(T - T_r) + \sum_j \sigma_{C_{jk}} (\widetilde{C}_{jk} - \widetilde{C}_{jkr}) \right] \tag{3.24}$$

$$M_{jk}(T, [\widetilde{C}_{jk}]) = M^o(T_r, [\widetilde{C}_{jk}]_r) \left[1 + M_{Tk}(T - T_r) + \sum_j M_{C_{jk}} (\widetilde{C}_{jk} - \widetilde{C}_{jkr}) \right] \tag{3.25}$$

where symbols in square brackets (standing before temperature and concentration differences) denote coefficients describing temperature or concentration dependence of density, surface tension and solubility. Choice of the reference temperature T_r or concentration $\widetilde{C}_{jkr}$ in the above relations will be discussed in the further part of the text. The square brackets with symbols $[\widetilde{C}_{jk}]$ denote dependence of the respective quantity on the total set of mass concentration of dissolved species in an abbreviated form.

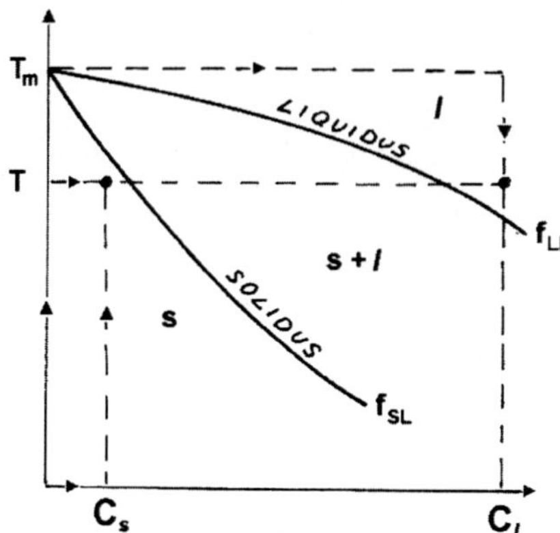

Figure 13. A scheme illustrating the way used in determination of the specific enthalpy of solid and liquid in a binary system.

Dependence of the volumetric enthalpy on temperature and concentration, $\rho_k h_k = \rho_k h_k (T,[\widetilde{C}_{jk}])$, needs more attention. Within the mushy zone a mixture of two phases exists, which in general may not be in thermodynamic equilibrium, and thus it is convenient to write respective expression for the specific enthalpy for each phase separately (see Fig. 13):

$$\rho_s h_s = \rho h_r + \int_{T_r}^{T} \rho c_s (T,0)\, dT + \sum_j \int_0^{\widetilde{C}_{js}} \rho h^*_{js} (T,\widetilde{C}_j)\, d\widetilde{C}_j \tag{3.26}$$

or

$$\rho_s h_s = \rho h_r + \sum_j \int_0^{\widetilde{C}_{js}} \rho h^*_{js} (T_r,\widetilde{C}_j)\, d\widetilde{C}_j + \int_{T_r}^{T} \rho c_s (T,[\widetilde{C}_{js}])\, dT \tag{3.27}$$

and

$$\rho_l h_l = \rho h_r + \int_{T_r}^{T_m} \rho c_s (T,0)\, dT + L_f (T_m,0) + \sum_j \int_0^{\widetilde{C}_{jl}} \rho h^*_{jl} (T_m,\widetilde{C}_j)\, d\widetilde{C}_j + \int_{T_m}^{T} \rho c_l (T,[\widetilde{C}_{jl}])\, dT \tag{3.28}$$

where ρh_r is the reference state of the volumetric enthalpy, ρc_k denotes the volumetric specific heats of kth phase and the symbol ρh^*_{jk} stands for the partial volumetric enthalpy of jth species in kth phase.

If the specific heats in both phases are constant and influence of the dissolved species on enthalpy is negligible (very dilute solution); these relations reduce to a simpler, well-known, form

$$\rho_s h_s = \rho h_r + \rho c_s (T - T_r) \tag{3.26'}$$

$$\rho_l h_l = \rho h_r + \rho c_s (T_m - T_r) + L_f (T_m) + \rho c_l (T - T_m) \tag{3.28'}$$

The volumetric concentration of jth species in kth phase that appears in eqs (3.14)-(3.16) is function of temperature and can be expressed as

$$\widetilde{\rho}_{jk} \widetilde{C}_{jk} = \rho_{jk} C_{jk} = \widetilde{\rho}_{jk} \widetilde{C}_{jk}{}^{o} (T_r) \left[1 + \rho_{Tk} (T - T_r) \right] \tag{3.29}$$

The set of the microscopic equations should be, finally, supplemented with relations describing conditions existing on the external boundaries of the solidifying material. The latter conditions may include specified values of pressure, velocity, temperature, and species concentration distributions on some boundaries that are open to liquid flow. For solid (mould) boundaries liquid velocity is equal to velocity of the wall (usually non-slip zero values is applied). If solid boundaries are not permeable to species then zero values of $\boldsymbol{j}_{jk}$ are assumed. On external free boundary conditions similar to eqs (3.5), (3.6) and (3.8) are used. On the rough solid walls (e.g. of the mould), bounding the solidifying material, care should be taken of thermal conditions. Continuity of heat flux in the material and the wall is here applied. In case of contact of the solidified material and the wall the second condition, corresponding to continuity of temperature, is here replaced by the following relation

$$\boldsymbol{q}_e \cdot \boldsymbol{n}_e = \alpha_c (T_s - T_e) \tag{3.30}$$

where $\boldsymbol{q}_e$ is the conductive flux on the mould side, T_e denotes the mould temperature at the interface and $\boldsymbol{n}_e$ the external unit vector perpendicular to the wall. Due to resistance caused by gaseous gap, temperature at the interface on the solidified material side, i.e. T_s is not equal to the mould temperature T_e. The heat transfer coefficient α_c describes local variation in this thermal resistance along the surface of the mould due to its roughness (see section 2.5).

The set of equations describing the dendritic solidification process can be solved by many methods. The finite element method with deforming mesh, a finite-difference enthalpy method, phase-field method (Warren and Boettinger, 1995, Tönhardt and Amberg, 1998) and front tracking methods based on mapping the governing equations into curvilinear co-ordinates (Shyy et al., 1996) or on the standard finite difference method using the immersed boundary technique and indicator function (Juric and Tryggvason, 1996) were used. Presentation of the phase-field method together with examples of its application will be given in the accompanying paper in this monograph.

4 Macro-modelling. Equations Describing Transport Phenomena in Solidifying Medium at the Macro-scale

4.1 Motivation for Macro-modelling

During unstable solidification in pure materials (e.g. pure metals) or solidification in the multi-component systems complicated microstructure is formed. This microstructure consists of a great number of columnar dendrites and floating equiaxed crystals. These microstructural elements are predominantly grown from an enormous number of nuclei located either in the liquid or on the mould wall. The dendrites growing at the wall and the growing crystal grains, which are floating in the melt, can interact in a complicated way. Moreover, the location of the nuclei of the solid phase bears statistical character. Also fluctuations of temperature and concentrations which lead to formation of dendrites and their secondary arms are random. As mentioned in the Section 2.5 the surfaces surrounding the solidifying material are usually rough and distribution of the asperities on their surface is also random in some sense. The properties of phases such as densities, thermal conductivities, viscosities, solubilities, diffusion coefficients, and many others are step-wise varying when crossing phase boundaries. As a result very complex, heterogeneous structure that develops in time is formed. The micro-modelling of the solidification in heterogeneous systems encounters problems when great number of interacting elements are present and is at present far from industrial applications where usually large scale systems occur (many orders of magnitude greater than dendrites dimensions, spacing or equiaxed grains size). There exist serious problems with handling both analytically or numerically such complicated structures. These are the reasons why the macroscopic approach is used.

4.2 A Concept of Macro-modelling

The idea of macroscopic description of phenomena stems from analysis of transfer phenomena in the classical heterogeneous systems like granular beds and porous media, fibrous insulation's, particle- and fibre-reinforced composite materials or suspensions of particles in the fluid. The theory of macroscopic modelling of heterogeneous media has been developing for decades and at the first glimpse it seems that it can be directly applied to study behaviour of solidifying heterogeneous materials. They are, however, significant differences. They should make the researcher to be very careful in direct use of concepts, developed within the classical macroscopic theory of heterogeneous media, to the solidifying systems. In the majority of the heterogeneous media listed above the microstructure is known *a priori*, the volume fractions and distribution of the constituents is also known and not varying in time. Moreover, not heat is usually generated at the interface of different constituents (phases) of the material. The transport processes in classical heterogeneous media are in general not conjugated and can be treated separately. Such particular aspects as: developing complicated microstructure, heat generation and numerous conjugated transfer processes make the problem of macroscopic modelling of the heterogeneous, solidifying materials especially challenging.

Macroscopic modelling of the transfer phenomena in heterogeneous systems is based on replacement of the phenomenon occurring in the medium with step-wise varying properties with the same phenomenon occurring in a fictitious medium which is characterised by constant or smoothly varying effective properties (see Fig. 14). On the macroscopic scale local variations in field variables, for example caused by presence of dendrites or equiaxed crystals, are smoothed out. The general conditions for this replacement should be, however, satisfied. They may be summarised in observation that the bulk conservation of mass, momentum, energy of the medium and conservation of mass of the dissolved species in the medium (if no chemical reactions are allowed) should not change upon this replacement.

Two versions of replacement are possible. In the first one the multi-phase (multi-component) heterogeneous medium is replaced by a single fictitious medium with continuously varying effective properties (*single continuum approach*). In the second one the multi-phase (multi-component) heterogeneous medium is replaced by a number fictitious media (phases) with continuously varying phase effective properties (*multi-continuum approach*). The number of these media usually corresponds to number of phases (or constituents) present in the original heterogeneous medium (see Fig. 14). In the case of binary alloy this replacement leads to two mutually intervening continua (*two-phase approach*). In this particular case two continuous phases liquid and solid are coexisting in every point of the mushy zone. When transfer processes occur in the heterogeneous medium there is an interaction between the respective fictitious continua associated with exchange of mass, momentum, energy and species between these continua.

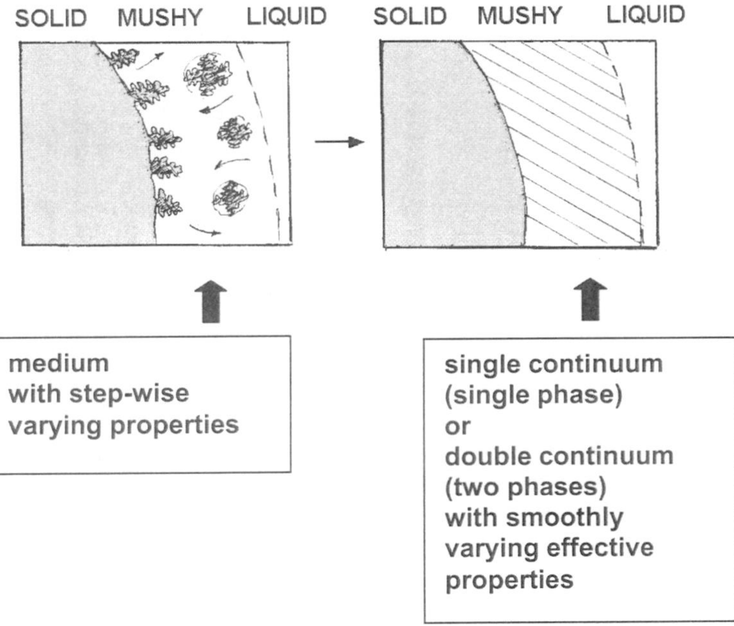

Figure 14. Schematic presentation of the idea of macro-modelling of phenomena occurring within the solidifying system.

Velocity, pressure, temperature or concentration distributions in the fictitious medium (or media) are called the macro-fields. The name micro-fields is here reserved to velocity, pressure, temperature and concentration distributions present in the original heterogeneous medium with step-wise varying properties.

Similar approach may be applied to the macroscopic modelling of the transfer phenomena occurring at the solidifying material – mould wall interface. The rough boundary (or smooth boundary with step-wise varying interface transfer coefficients) is replaced by a fictitious smooth boundary with continuously varying (along the interface) effective coefficients describing conditions of, for example, heat flow across the interface (see Fig. 15).

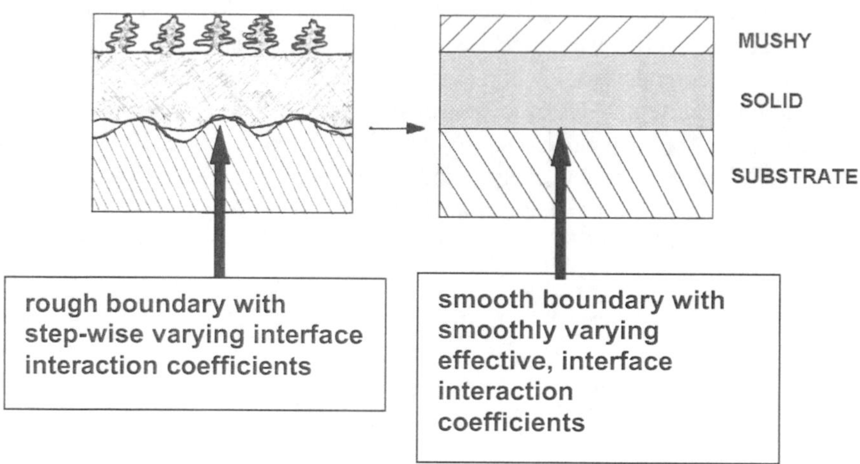

Figure 15. Schematic presentation of the idea of macro-modelling of phenomena occurring at the boundary of the solidifying system.

4.3 Micro-macroscopic Approach to Solution of Solidification in Heterogeneous Sytems

The effective, macroscopic properties or coefficients, either emerging from single phase (single continuum) or multi-phase (multi-continuum) approach, are dependent on details of the composition of the solidifying material, on the developing microstructure and transfer phenomena (and their coupling) occurring on the microscopic scale, i.e. on the scale of separate dendrites or equiaxed crystals. The microscopic phenomena are so directly affecting phenomena occurring on the macroscopic scale. It is thus clear that significant improvement in solidification modelling lies in the introduction of microscopic information to the equations describing the process on the macroscopic scale. In case of the classical heterogeneous media with fixed (composites, granular beds) or slowly varying (solid suspension) microstructure, it is sufficient to determine once the effective properties. They can be subsequently used in analysis of the whole process occurring on the macro-scale without need for their modification. In opposition to these media, in the solidifying multi-component systems or in pure materials, solidifying in an unstable way, proportion of solid and liquid phases is varying in time and space and the microstructure is changing. Effective properties are also varying not only from place to place but also in time following the microstructural changes. Thus their determination on each time step of the solidification process makes part of the problem of modelling. The ideal sequence of calculations should thus look in the following way (see Fig. 16). We solve the equations describing phenomena on the macroscopic scale knowing the effective properties. In this way, distributions of the macroscopic variables (i.e., macro-fields) like velocity, pressure, temperature and concentration are obtained. Then from knowledge of the macro-fields local solid fraction in the solidifying material is determined and the approximate microfields are calculated which allow finding evolution or other details of the material structure on the micro-scale. Information about variation in microstructure is then used to recalculate the effective properties in any place of the solidifying medium. Thus the loop of calculations is closed (see Fig. 16).

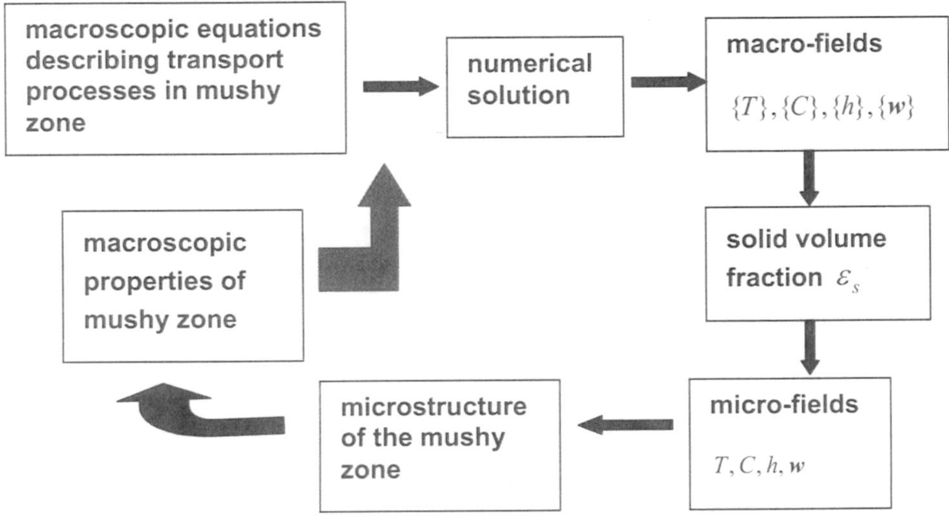

Figure 16. Schematic presentation of the idea of micro-macro modelling of phenomena occurring in the solidifying systems.

Possibility to find micro-fields, in an approximate way, from the macro-fields will make an alternative to purely microscopic approaches (see previous Section). Moreover, it will allow carrying out calculations for large-scale industrial processes retaining possibility to intervene into microstructure formed during the process. This ideal sequence of calculations is, however, far from contemporary stage of mathematical modelling of the solidification processes. At present only local solid volume fraction is determined in calculations. The effective, macroscopic properties are modified using this only information with no evolution in other microstructural parameters taken into account (like dendrite spacing, changes in radii of dendrite tips, formation of side arms during dendrite growth and their spacing, columnar–equiaxed transition, formation of solid skeleton from equiaxed crystals, etc.).

4.4 Methods of Macroscopic Modelling

Different methods (spectral, variational, volume averaging, ensemble averaging, etc.) were used in the past to obtain macroscopic description of phenomena occurring in heterogeneous media. The *volume averaging* method and the *ensemble averaging* are the most common and claim generality of application. It should be however stressed that they are different and, in general, lead to different results. The volume averaging method has been practically exclusively used up to this time in modelling of solidification processes. The ensemble averaging is presently gaining greater approval as it has certain clear advantages that enable many generalisations.

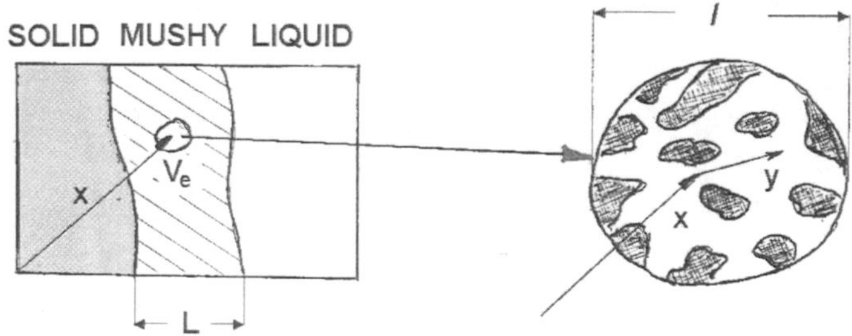

Figure 17. Schematic presentation of the idea of the volume averaging in the solidifying system.

Volume averaging. The volume averaging method is based on the concept of the *Representative Elementary Volume* (REV) – see Fig.17. Any macroscopic, time and space dependent, physical quantity appearing in the macroscopic equations obtained as a result of this method is understood as spatially integrated over volume V_e and related to the centroid of the REV

$$\langle f(\boldsymbol{x},\boldsymbol{t})\rangle = \boldsymbol{V_e}^{-1}\int_{V_e} f(\boldsymbol{x},\boldsymbol{y},t)\, d\boldsymbol{y} \tag{4.1}$$

The REV is strictly defined only for infinite media of periodic structure when it is identical with the repetitive element of the structure. The centroid point $\boldsymbol{x}$ is then the lattice node $\boldsymbol{x_n}$ and values of $\langle f(\boldsymbol{x}_n,\boldsymbol{t})\rangle$ make a discrete set of values. Transformation from the discrete to continuous form is carried out in a way similar to the way used in analysis of crystal atomic lattice and is based on the complex Fourier transform

$$\langle f(\boldsymbol{x},\boldsymbol{t})\rangle = (2\pi)^{-3}\int_B \sum_{n=1}^{\infty} \langle f(\boldsymbol{x_n},t)\rangle \exp[j\boldsymbol{k}\cdot(\boldsymbol{x}-\boldsymbol{x_n})\, d\boldsymbol{k}$$

where j is the imaginary unit, B denotes the Brillouin zone and the unit distance between lattice nodes was assumed.

For media which lack periodic regularity in their microstructure, the only information about the REV is that its dimension should be much greater than a characteristic size ℓ describing the interfacial structures, for example the interdendritic length, and at the same time much smaller than the macroscopic length scale L associated with the volume of the medium. The REV should also be chosen as the smallest differential volume that results in statistically meaningful local effective properties. Speaking about variation of the macroscopic quantities over distances smaller than dimensions of the REV makes no sense within this approach.

During derivation of the macroscopic equations, the volume averaging is applied to gradients of scalars or divergence of vectors. The procedure is based on the spatial averaging theorem and can be expressed as

$$\langle \nabla f \rangle = \nabla \langle f \rangle + V_e^{-1} \int_{A_i} f \, \boldsymbol{n} \, dA \tag{4.2}$$

$$\langle \nabla \cdot \boldsymbol{f} \rangle = \nabla \cdot \langle f \rangle + V_e^{-1} \int_{A_i} \boldsymbol{f} \cdot \boldsymbol{n} \, dA \tag{4.3}$$

$$\langle \partial_t f \rangle = \partial_t \langle f \rangle - V_e^{-1} \int_{A_i} f \, \boldsymbol{w}_i \cdot \boldsymbol{n} \, dA \tag{4.4}$$

where $\boldsymbol{n}$ is externally directed unit vector normal to the surface A_e of the REV. The theorem is based on an assumption of slow temporal and spatial variation in the field variables, thus placing another limitation on the procedure of deriving macroscopic equations.

One of the first complete macroscopic models using the volume averaging approach together with theory of mixtures was developed by Bennon and Incropera, 1987. The mixture theory assumes no differences in gradients of variables to exist between phases. Rigorous derivation of the macroscopic equations from the microscopic ones using the volume averaging technique and the theorems cited above was carried out by Beckermann and Viskanta, (1993); Ganesan et al., (1992). In the latter derivation, similar to the mixture theory approach, the phase interaction integral terms (corresponding to the second terms of the right hand side of eqs (4.2)-(4.4)) were also treated semi-empirical. Assumption of periodicity of the interfacial structures and quasi periodicity of the respective field variables allowed solving so-called closure problem associated with the phase interaction terms. It also allowed to give prescriptions for calculating effective transport properties under conditions that variation in the effective properties obey decreasing rate of geometry constraint (Bousquet-Melou et al., 2002).

Limitations of the volume averaging method in modelling of the mushy zone may be summarised in the following way. The elementary representative volume is here not well defined. In case of the columnar dendritic microstructure (with evolving heterogeneities) the average macroscopic properties are space-dependent. The REV should then vary in size and the problem with satisfying the condition, $\ell << L$, arises. Derivation of the respective macroscopic equations is complicated and often based on order of magnitude analysis. The technique is not appropriate for high temporal and spatial variation in the macro-fields. Presence of external boundaries and random nucleation causes troubles in derivation of the macroscopic equations from the microscopic equations. The thermodynamic relations and the constitutive relations between variables, needed for closure of the macroscopic models are, in majority of cases, postulated and so no possibility exists for finding microstructure from knowledge of macro-fields. Some progress in solution of this problem was attained in case when assumption of periodicity of the evolving interfacial structures was adopted.

Limitations of the volume averaging are realised by some investigators and extensions of this approach are sought which are based on introduction of some weighing functions and use of the theory of distributions (Quintard and Whitaker, 1993).

Ensemble averaging. A different approach for deriving macroscopic equation from the microscopic ones offers the ensemble averaging technique. The ensemble averaging comes from theory of random fields and stochastic processes where it has proved to be useful in finding solution of many problems that appear both in description and understanding of the transport phenomena at the macroscopic scale. The approach is based on the concept of an ensemble of realisations of the microstructure and does not need the notion of the REV. Any macroscopic, physical quantity appearing in the macroscopic equations obtained with this method is understood as statistically mean values (awaited values) defined in any point of the medium. In author's opinion it is more versatile and, what is also important, simpler to use than the volume averaging technique. The ensemble averaging has been applied to study heterogeneous media that are both homogeneous and nonhomogeneous at the macroscopic scale (Buyevich, 1992; Batchelor, 1994; Furmański, 1997). It will be used to present derivation of the macroscopic equations describing transport and phase change processes in the solidifying, heterogeneous systems in the further part of the text.

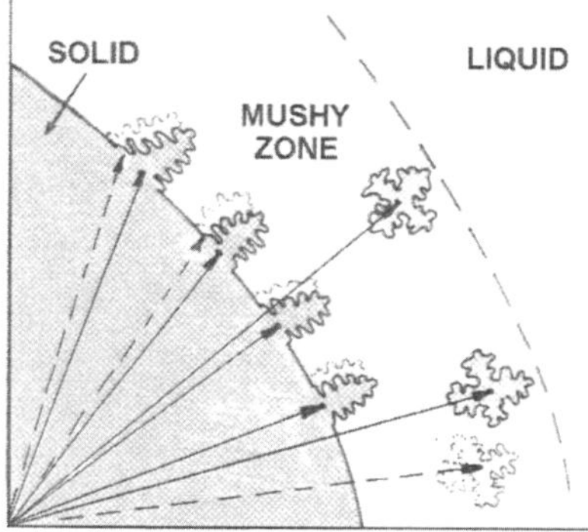

Figure 18. Schematic presentation of the idea of the ensemble averaging in the solidifying system.

In order to illustrate the ensemble averaging technique to more extent let us assume that solidification of the binary mixture occurs in a volume V containing two phases - solid (s) and liquid (l). A set of characteristic points, which may be understood as nucleation sites or places where the primary dendrite arms emerge, are spread all over the considered volume. A nucleation site, when activated, may give rise to a formation of the solid phase. Any spatial distribution of these points is considered to be a different configuration Ω. It can be specified for periodic or random distribution of interfacial structures. Any change in location of the characteristic points with respect to the external boundary is understood as a different configuration (see Fig. 18). Each configuration is usually one of many realisations of the microstructure of the heterogeneous medium (mushy zone) and can be treated as an element of the sample space (the ensemble of configurations) for which a probability measure, specified by the probability distribution function $p(\Omega)$, is given. It is assumed that $p(\Omega)$ is not seriously affected by movement of the characteristic points associated with centres of the equiaxed crystals in the melt. Many field variables like density, velocity, pressure, temperature, heat flux, solute concentration, etc. depend not only on time and location but also on the configuration thus becoming random functions of Ω. According to the theory of stochastic fields, many statistical averages can be introduced to describe their properties. Two of them are of special interest in this context and will be subsequently used in the further part of the presentation. The first is *the bulk (macroscopic)* average

$$\{f(\boldsymbol{x},t)\} = \int f(\boldsymbol{x},t|\Omega)\, p(\Omega)\, d\Omega \tag{4.5}$$

which should be understood as the awaited value in the statistical sense. The averaging is carried out over the set of all possible configurations.
The second one is the conditional statistical average that is defined as

$$\{f(\boldsymbol{x},t)\}^* = \int f(\boldsymbol{x},t|\Omega)\, p(\Omega|\boldsymbol{x}')\, d(\Omega|\boldsymbol{x}') \tag{4.6}$$

where $p(\Omega|\boldsymbol{x}')$ denotes the conditional probability density function corresponding to probability of occurrence of configuration Ω when the other random variable, e.g. associated with location of characteristic point of the constituent, is equal to $\boldsymbol{x}'$.

The use of the ensemble averaging has significant advantages over the volume averaging. It enables all derivations, leading to obtain equations for the macroscopic variables, to be carried out by the same method. There is thus no need to rely on any supplementary models in order to obtain the constitutive equations. There is also no need for additional hypotheses about the possible connections between volume averages and corresponding quantities obtained by averaging over other space objects like, e.g., over planes differently oriented in space. No apriori order of magnitude assumptions are necessary. One of the specific features of the ensemble averaging technique is that the operator of the ensemble averaging commutes with differentiation with respect to $\boldsymbol{x}$ and t. This occurs by virtue of independence of the probability density function on the space variables and time.

The ensemble averaging leads to results that have an evident meaning in terms of the probability theory. It conceptually defines mean quantities at a given point in space without having to define a volume over which these quantities should be integrated. Thus it eases the proper interpretation of the experimental results. It may be applied both to periodic and random heterogeneous media. It enables one to introduce information about the statistical description of the microstructure in an easy way. Its use to study bounded heterogeneous media follows from the lack of requirement of the presence of a great number of columnar dendrites or equiaxed grains in the physically small volumes of the system as in the case of the volume averaging. Thus it provides the means to study the inherent macroscopic heterogeneity of the heterogeneous medium and, associated with it, the variability of the effective, macroscopic properties as is the case when solidification occurs in the mushy zone (Furmański, 2000). It can also help in evaluation of the uncertainty of any method of estimation of these properties. At last, it provides a feasible way of generalising of all results to complex situation in which coupled phenomena occur in the medium and distributed heat sources are present. This is the case met during solidification in heterogeneous systems (Furmański, 1999).

The problem of relation between the ensemble and volume averaging has been discussed in a number of papers. In these cases, where the REV exists, it is possible to compare results coming from these two approaches by use of *the ergodic theorem*. When the ergodic theorem holds, i.e. when the macroscopic fields, obtained by statistical averaging, are stationary, then volume-averaged quantities are equal to ensemble averaged. For the fields that are slowly varying with $\boldsymbol{x}$, this theorem should be treated solely as a hypothesis. There are many cases when the condition of stationarity does not hold even if the REV can be distinguished in the medium. For instance this occurs in the unsteady states of heat transfer (when not whole of the REV is filled with a thermal disturbance), near the external boundaries or close to solidus or liquidus lines (where the microscopic properties of the material vary), in places where heat sources or sinks are present or the microstructure of the medium is varying (mushy zone). In all these cases, in order to apply the ergodic theorem properly, one should look for locations where the condition of stationarity holds. These may be, for example surfaces, parallel to solidus or liquidus lines over which spatial integration should be carried out.

5. Macroscopic description. Description of Microstructure and Types of Averages.

5.1. Description of the Microstructure of the Solidifying Heterogeneous Medium (Mushy Zone).

A complex microstructure of the mushy zone, which develops during solidification of multi-component media or during unstable solidification in pure materials, is the main feature that differs these systems from homogenous ones or from systems where stable solidification occurs in the pure material. A description of the microstructure is thus an important part of any analysis carried out in this field. The microstructure may be described by specifying to which phase each point in the medium belongs. For any moment of time t and any configuration Ω, distribution of the phases is described by the so-called *structure* (characteristic, phase or indicator*) function*

$$\theta_k = \theta_k(t, \boldsymbol{x}|\Omega) = \begin{cases} 1 & \text{for } \boldsymbol{x} \in \boldsymbol{V}_k \\ 0 & \text{for } \boldsymbol{x} \notin \boldsymbol{V}_k \end{cases} \tag{5.1}$$

where V_k denotes the volume filled with kth phase. The structure function satisfies the following relation

$$\sum_k \theta_k(t, \boldsymbol{x}) = 1 \tag{5.2}$$

where summation extends over all phases present in the system. Component of the heterogeneous system that does not undergo phase transformation can be here treated as separate phases.

In the solidifying system, where the interface moves, location and time in eq. (5.1) are related and the following expression holds

$$\partial_t \theta_k + \boldsymbol{w}_i \cdot \nabla \theta_k = 0 \tag{5.3}$$

which simply expresses convection of the structure function with movement of the interface.

The gradient of the structure function can be defined by the following expression

$$\nabla \theta_k = -\boldsymbol{n}_k \ \delta(\boldsymbol{x}, \boldsymbol{x}_i) \tag{5.4}$$

where symbol δ denotes Dirac function, $\boldsymbol{x}_i$ defines location of the interface between two phases while $\boldsymbol{n}_k$ is the unit vector perpendicular and externally directed with respect to the surface of kth phase.

If function f in eq. (4.5) is taken to be the structure function θ_k then its bulk average is equal to the volume fraction of the kth phase

$$\{\theta_k(t, \boldsymbol{x})\} = \varepsilon_k(t, \boldsymbol{x}) \tag{5.5}$$

In statistical sense the above average denotes probability of finding that $\boldsymbol{x}$ belongs to kth phase.

5.2. The Generalised Variables, Superficial and Intrinsic (Phase) Averages.

It is convenient to introduce a generalised variable defined as

$$f(t, \boldsymbol{x}|\Omega) = \sum_k \theta_k(t, \boldsymbol{x}|\Omega) f_k(t, \boldsymbol{x}|\Omega) \tag{5.6}$$

which allows to write the governing equations in one, unified way, taking into account all phases present in the system (Furmański, 2000).

It is also useful to introduce two additional ensemble averages, which are used in studies of transport processes in heterogeneous media. The first one is the superficial average defined by the relation

$$\{f(t, \boldsymbol{x})\}_k^s = \{\theta_k(t, \boldsymbol{x}) f(t, \boldsymbol{x})\}$$

The second one is *the phase (intrinsic) average* $\{f\}_k$ defined by the expression

$$\{f(t, \boldsymbol{x})\}_k = \{\theta_k(t, \boldsymbol{x}) \ f(t, \boldsymbol{x})\} / \varepsilon_k(t, \boldsymbol{x}) \tag{5.7}$$

which gives the mean value of the respective function in the *k*th phase.

The bulk, macroscopic averages of the basic thermodynamic variables like density ρ temperature T and the mass concentration $\widetilde{C}_j$ of the dissolved jth species in the case of two-phase solidifying system can thus be related, using eq. (5.6), to the respective intrinsic averages as

$$\{\rho(t, \boldsymbol{x})\} = \varepsilon_s(t, \boldsymbol{x})\rho_s + \varepsilon_l(t, \boldsymbol{x})\rho_l \tag{5.8}$$

$$\{T(t, \boldsymbol{x})\} = \varepsilon_s(t, \boldsymbol{x})\{T(t, \boldsymbol{x})\}_s + \varepsilon_l(t, \boldsymbol{x})\{T(t, \boldsymbol{x})\}_l \tag{5.9}$$

$$\{\widetilde{C}_j(t, \boldsymbol{x})\} = \varepsilon_s(t, \boldsymbol{x})\{\widetilde{C}_j(t, \boldsymbol{x})\}_s + \varepsilon_l(t, \boldsymbol{x})\{\widetilde{C}_j(t, \boldsymbol{x})\}_l \tag{5.10}$$

while the bulk velocity in the medium, when the solid phase is assumed to be stationary (i.e., $\boldsymbol{w}_s = 0$), is given by

$$\{\boldsymbol{w}(t,\boldsymbol{x})\} = \varepsilon_l(t,\boldsymbol{x})\{\boldsymbol{w}(t,\boldsymbol{x})\}_l \tag{5.11}$$

6. Macro-modelling of the Mass Conservation Equation

The bulk, macroscopic density of the solidifying heterogeneous system (with significantly different densities of separate phases) varies essentially due to change in proportion in volumes encompassed by each phase. Changes in density induced by variation of temperature and mass concentration of dissolved species are less important. In order to obtain equation that describes variation of the bulk density of the system on the macroscopic scale, eq. (3.1) was multiplied by the structure function θ_k. After some mathematical manipulations it took the following form

$$\partial_t(\theta_k \rho_k) + \nabla\cdot(\theta_k \boldsymbol{w}_k \rho_k) = \rho_k[\partial_t\theta_k + \boldsymbol{w}_k \cdot \nabla\theta_k] \tag{6.1}$$

The r.h.s. of this equation can be transformed, using eq. (5.3), to yield

$$\partial_t(\theta_k \rho_k) + \nabla\cdot(\theta_k \boldsymbol{w}_k \rho_k) = \rho_k(\boldsymbol{w}_k - \boldsymbol{w}_i)\cdot\nabla\theta_k \tag{6.2}$$

The term on the r.h.s. of the latter equation expresses local exchange of mass between phases due to solidification.

Equation (6.2), valid for each phase, were subsequently added and generalised variables, according to eq. (5.6), were introduced. This lead to the formula

$$\partial_t(\rho) + \nabla\cdot(\boldsymbol{w}\rho) = 0 \tag{6.3}$$

Note that the r.h.s. terms in eq. (6.3) disappeared due to condition of mass conservation at the interface (see eqs (3.5) and (5.4)).

A single continuum approach. Equation (6.3) was ensemble averaged and the macroscopic form of the mass conservation was then obtained

$$\partial_t\{\rho\} + \nabla\cdot\{\boldsymbol{w}\rho\} = 0 \tag{6.4}$$

If incompressibility of the phases is assumed (which is often the case) and condition of stationary ($\boldsymbol{w}_s = 0$) of the solid phase is used, then eq. (6.4) for two phase medium, can be rewritten in another form

$$(\rho_s - \rho_l)\partial_t\varepsilon_s + \rho_l\nabla\cdot[\varepsilon_l\{\boldsymbol{w}\}_l] = 0 \tag{6.5}$$

Equation (6.4) corresponds to replacement of the solidifying, heterogeneous medium with a single continuum on the macroscopic scale (see Section 4.2) where $\{\rho\}$ stands for the macroscopic, bulk density of the medium.

A multi-continuum approach. Another approach is also possible according to Section 4.2 Equation (6.2), valid in each phase, can be directly ensemble averaged to obtain

$$\partial_t(\varepsilon_k \rho_k) + \nabla \cdot [\varepsilon_k \{w\rho\}_k] = \Gamma_{mk} \tag{6.6}$$

where

$$\Gamma_{mk} = \{\rho_k (\boldsymbol{w_k} - \boldsymbol{w_i}) \cdot \nabla \theta_k\} \tag{6.7}$$

Equation (6.6) corresponds to replacement of the solidifying, heterogeneous medium with k number of fictitious continua on the macroscopic scale.

If we adopt as the reference temperature and the reference species concentration in eq. (3.23) as the intrinsic temperature $\{T\}_k$ and concentration $\{\widetilde{C}_j\}_k$ of the kth phase than the above equation can be cast in the form

$$\partial_t(\varepsilon_k \rho_k) + \nabla \cdot (\varepsilon_k \rho_k \{\boldsymbol{w}\}_k) = \Gamma_{mk} \tag{6.8}$$

where density of each phase is calculated for temperature and species concentrations corresponding to their intrinsic values.

If incompressibility of each phase and stationarity of the solid phase are assumed then eq. (6.6), for the two-phase medium, reduces to two coupled equations

$$\rho_s \, \partial_t \varepsilon_s = \Gamma_{ms} \tag{6.9}$$

$$\rho_l \, \partial_t \varepsilon_l + \rho_l \, \nabla \cdot (\varepsilon_l \{\boldsymbol{w}\}_l) = \Gamma_{ml} = -\Gamma_{ms} \tag{6.10}$$

Symbol Γ_{mk}, appearing in the cited equations, denotes the macroscopic mass flux exchanged between phases during the solidification process. If eq. (5.4) is used, than the rate of mass exchange between phases can be expressed as

$$\Gamma_{mk} = \{-\rho_k (\boldsymbol{w_k} - \boldsymbol{w_i}) \cdot \boldsymbol{n_k} \, \delta(\boldsymbol{x}, \boldsymbol{x_i})\} = a_i \{\rho_k (\boldsymbol{w}_i - \boldsymbol{w}_k) \cdot \boldsymbol{n_k})\}_A^* \tag{6.11}$$

where a_i is area of the interface per unit volume (specific interface area) of the heterogeneous medium while the symbol $\{\}_k^*$ denotes the conditional ensemble averaging over all configurations Ω assuming that the point $\boldsymbol{x}' = \boldsymbol{x}_i$, i.e. lies on the solid/liquid interface (see definition after eq. (5.6)).

7. Macro-modelling of the Species Transport

Variation in the species distribution in the phases is described by eq. (3.14). It is caused by species diffusion and advection. In order to obtain equation describing these changes on the macroscopic scale the microscopic equation, eq. (3.14) was multiplied by the structure function θ_k and rearranged. The resulting equation can be shown in the form

$$\partial_t(\theta_k \widetilde{\rho}_k \widetilde{C}_{jk}) + \nabla \cdot [\theta_k (\boldsymbol{w_k} \widetilde{\rho}_k \widetilde{C}_{jk} + \boldsymbol{j}_{jk}) = [\partial_t \theta_k + \boldsymbol{w_k} \cdot \nabla \theta_k] \widetilde{\rho}_k \widetilde{C}_{jk} + \nabla \theta_k \cdot \boldsymbol{j_{jk}} \tag{7.1}$$

The r.h.s. of this equation can be transformed, using relations from eq. (5.3), to yield

$$\partial_t(\theta_k\widetilde{\rho}_k\widetilde{C}_{jk})+\nabla\cdot[\theta_k(\boldsymbol{w}_k\widetilde{\rho}_k\widetilde{C}_{jk}+\boldsymbol{j}_{jk})=[\widetilde{\rho}_k\widetilde{C}_{jk}(\boldsymbol{w}_k-\boldsymbol{w}_i)+\boldsymbol{j}_{jk}]\cdot\nabla\theta_k \tag{7.2}$$

The term on the r.h.s. of the latter equation expresses local exchange of jth species between phases due to solidification, advection and diffusion.

Equations (7.2), valid for each phase, when added, lead to the following generalised form of jth species conservation equation

$$\partial_t(\widetilde{\rho}\widetilde{C}_j)+\nabla\cdot[\boldsymbol{w}\widetilde{\rho}\widetilde{C}_j+\boldsymbol{j}_j]=0 \tag{7.3}$$

The r.h.s. terms, in eq. (7.3), were again eliminated due to condition of jth species conservation at the interface (see eqs (3.15) and (5.4)).

7.1 Macroscopic Conservation Equation for jth Species

A single continuum approach. The equation (7.3) was ensemble averaged and the macroscopic form of jth species conservation was easy obtained

$$\partial_t\{\widetilde{\rho}\widetilde{C}_j\}+\nabla\cdot[\{\boldsymbol{w}\widetilde{\rho}\widetilde{C}_j\}+\{\boldsymbol{j}_j\}]=0 \tag{7.4}$$

Equation (7.4) again corresponds to replacement of the solidifying, heterogeneous medium with a single continuum on the macroscopic scale (see Section 4.2) If incompressibility of the phases is invoked and the solid phase is not moving then eq. (7.4) can be rewritten in another form

$$\partial_t\{\widetilde{\rho}\widetilde{C}_j\}+\nabla\cdot[\varepsilon_l\{\boldsymbol{w}\widetilde{\rho}\widetilde{C}_j\}_l+\{\boldsymbol{j}_j\}]=0 \tag{7.5}$$

A multi-continuum approach. Averaging directly conservation equation for jth species, i.e., eq. (7.2), corresponds to replacement of the solidifying, heterogeneous medium with a number of fictitious continua (see Section 4). This approach leads to the expression

$$\partial_t(\varepsilon_k\{\widetilde{\rho}\widetilde{C}_j\}_k)+\nabla\cdot[\varepsilon_k(\{\widetilde{\rho}\widetilde{C}_j\}_k+\{\boldsymbol{j}_j\}_k)]=\Gamma^c_{cjk}+\Gamma^d_{cjk} \tag{7.6}$$

where

$$\Gamma^c_{cjk}=\{\widetilde{\rho}_k\widetilde{C}_{jk}(\boldsymbol{w}_k-\boldsymbol{w}_i)\cdot\nabla\theta_k\} \tag{7.7}$$

corresponds to advective exchange of jth species between phases
and

$$\Gamma^d_{cjk}=\{\boldsymbol{j}_{jk}\cdot\nabla\theta_k\} \tag{7.8}$$

to the diffusive one.
If eq. (5.4) is used than the rate of exchange of jth species between phases can be expressed as

$$\Gamma^c_{cjk}=\{-\widetilde{\rho}_k\widetilde{C}_{jk}(\boldsymbol{w}_k-\boldsymbol{w}_i)\cdot\boldsymbol{n}_k\ \delta(\boldsymbol{x},\boldsymbol{x}_i)\}=a_i\{\widetilde{\rho}_k\widetilde{C}_{jk}(\boldsymbol{w}_i-\boldsymbol{w}_k)\cdot\boldsymbol{n}_k)\}^*_A \tag{7.9}$$

and

$$\Gamma^{d}_{cjk} = \{-\boldsymbol{j}_{jk} \cdot \boldsymbol{n}_k \, \delta(\boldsymbol{x}, \boldsymbol{x}_i)\} = -a_i \{\boldsymbol{j}_{jk} \cdot \boldsymbol{n}_k\}^{*}_{A} \tag{7.10}$$

7.2. Closure Problem. Constitutive Relations for jth Species

The macroscopic equations for conservation of jth species, presented in the Section 7.1, were easy obtained using the ensemble averaging technique. Their derivation is the place where most of the macro-modelling of transport phenomena in the solidifying systems ends (Bennon and Incropera, 1989, Beckermann and Viskanta, 1993). It is however obvious that in order to numerically solve conservation equations constitutive relations between variables are needed, i.e., *closure problem* should be solved. In the case of the single continuum approach we need to express the following terms $\{\widetilde{\rho}\widetilde{C}_j\}, \{\boldsymbol{w}\widetilde{\rho}\widetilde{C}_j\}$ and $\{\boldsymbol{j}_j\}$ as functions of the macroscopic temperature $\{T\}$ and macroscopic jth species concentration $\{\widetilde{C}_j\}$. In case of the multi-continuum approach relations between $\{\widetilde{\rho}\widetilde{C}_j\}_k, \{\boldsymbol{w}\widetilde{\rho}\widetilde{C}_j\}_k$, $\{\boldsymbol{j}_j\}_k$, Γ^{c}_{cjk}, Γ^{d}_{cjk} and the intrinsic temperatures $\{T\}_k$, intrinsic jth species concentration $\{\widetilde{C}_j\}_k$ and the interface rate of mass flux Γ_{mk} are needed. These relations are usually called *the macroscopic constitutive relations.* In most literature in the field of solidifying media their are postulated. The postulates use the respective relations valid in porous media with no phase transition present and characterised by fixed microstructure. Their validity in case of the solidifying, heterogeneous systems may be questioned.

The constitutive relations could be derived if relation between microscopic concentration $\widetilde{C}_j$ in any phase and its bulk value $\{\widetilde{C}_j\}$ was known. In order to derive these relations we start from the microscopic solute conservation equation, eq. (7.3), with included generalised form of the microscopic constitutive equation - see eq. (3.22)

$$\boldsymbol{j}_j = -\rho\widetilde{D}_j \nabla\widetilde{C}_j + \widetilde{M}\,\nabla T \tag{7.11}$$

Then the following equation for $\widetilde{C}_j$ results

$$\rho_r \partial_t \widetilde{C}_j + \rho_r \boldsymbol{w}_r \cdot \nabla\widetilde{C}_j - D_r \nabla^2 \widetilde{C}_j - F_c = 0 \tag{7.12}$$

where the source function F_c is defined as

$$F_c = \nabla \cdot (\widetilde{D}'_j \nabla\widetilde{C}_j - \widetilde{M}_j \nabla T) - \nabla \cdot (\boldsymbol{w}\widetilde{\rho}\widetilde{C}_j - \boldsymbol{w}_r \rho_r \widetilde{C}_j) - \partial_t (\widetilde{\rho}\widetilde{C}_j - \rho_r \widetilde{C}_j) \tag{7.13}$$

and

$$\widetilde{D}'_j = \widetilde{D}_j - D_r \tag{7.14}$$

$$\widetilde{D}_j = \sum_k \theta_k \widetilde{D}_{jk}\,,\ \widetilde{M}_j = \sum_k \theta_k \widetilde{M}_{jk} \tag{7.15}$$

Conditions satisfied by the mass concentration of jth species on the external boundaries of the system were discussed in Section 3. They can be expressed in the generalised form as

$$(\boldsymbol{j}_j + \boldsymbol{w}\widetilde{\rho}\widetilde{C}_j)\cdot \boldsymbol{n}_e = f_j \tag{7.16}$$

where function f_j depends on space and time but is independent of the configuration Ω. Particular form of the boundary conditions is chosen in relation to type of the problem in hand. For example, in case of solid boundaries the velocity, in eq. (7.16), is assumed to be zero. Finally the problem to be solved is completed with the initial condition

$$\widetilde{C}_j(t=0,\boldsymbol{x}|\Omega) = \widetilde{C}_{jo}(\boldsymbol{x}) \tag{7.17}$$

The Green function G was subsequently introduced. It was defined by the equations:

$$D_r\nabla^2 G + \rho_r \boldsymbol{w}_r \cdot \nabla G + \delta_x(\boldsymbol{x},\boldsymbol{y})\,\delta_t(t,\tau) = -\rho_r \partial_t G \qquad \text{for } \boldsymbol{x}\in V \tag{7.18}$$

$$(D_r\nabla G - \boldsymbol{w}\rho_r G)\cdot \boldsymbol{n}_e = 0 \qquad \text{for } \boldsymbol{x}\in A \tag{7.19}$$

$$G(t,\boldsymbol{x};\tau,\boldsymbol{y}) = 0 \qquad \text{for } t\le\tau \tag{7.20}$$

The quantities supplemented with index „r" denote to the reference values. In opposition to other variables the reference values are independent of the configuration Ω. However, the reference velocity $\boldsymbol{w}_r$ obeys condition of incompressible flow.

With use of the Green function theory the following formal expression for mass concentration of jth species was obtained

$$\widetilde{C}_j = \int_V \rho_r G_o \widetilde{C}_{jo} dV - \int_0^t\int_V G F_c\, dV'\, d\tau + \int_0^t\int_A [G(D_r\nabla\widetilde{C}_j - \widetilde{C}_j D_r \nabla G]\cdot \boldsymbol{n}_e\, dA'\, d\tau \tag{7.21}$$

where $G_o = G(t,\boldsymbol{x};\tau=0,\boldsymbol{y})$.

Equation (7.13) was subsequently introduced into eq. (7.21) and the conditions on the external boundary for the mass concentration and the Green function, i.e., eqs (7.16) and (7.19), used. Ensemble averaging of the resulting equation and subtracting the averaged equation from the original one gave the following formal equation for the local, microscopic concentration

$$\begin{aligned}\widetilde{C}_j = \{\widetilde{C}_j\} - \int_0^t\int_V \nabla G\cdot[(\widetilde{D}_j'\nabla\widetilde{C}_j - \{\widetilde{D}_j'\nabla\widetilde{C}_j\}) - (\widetilde{M}_j\nabla T - \{\widetilde{M}_j\nabla T\}) + \\ -(\boldsymbol{w}\widetilde{\rho}\widetilde{C}_j - \boldsymbol{w}_r\rho_r\widetilde{C}_j) + \{\boldsymbol{w}\widetilde{\rho}\widetilde{C}_j - \boldsymbol{w}_r\rho_r\widetilde{C}_j\})]\,dV'\,d\tau + \\ -\int_0^t G\partial_t[(\rho\widetilde{C}_j - \rho_r\widetilde{C}_j) - \{\rho\widetilde{C}_j - \rho_r\widetilde{C}_j\}]\,dV\,d\tau\end{aligned} \tag{7.22}$$

The term $\widetilde{\rho}\widetilde{C}_j$ is here dependent on temperature. Using expansion for this term in the form given in eq. (3.29) and adopting the reference temperature equal to the bulk temperature of the medium, it can be expressed as a function of local difference between the microscopic and the macroscopic temperature.

The form of the resulting integro-differential equation, i.e. eq. (7.22), together with the corresponding equation for the temperature (see Section 8) suggest that the microscopic mass concentration of jth species and the microscopic temperature should be sought in the following form:

$$\begin{aligned}\widetilde{C}_j &= \{\widetilde{C}_j\} + S_{Cj} + \int_0^t\int_V \varphi_{Cj}\cdot\nabla'\{\widetilde{C}_j\}\,dV'\,d\tau + \int_0^t\int_V \psi_{Cj}\ \partial_\tau\{\widetilde{C}_j\}\,dV'\,d\tau + \\ &+ \int_0^t\int_V \varphi_{TCj}\cdot\nabla'\{T\}\,dV'\,d\tau + \int_0^t\int_V \psi_{TCj}\ \partial_\tau\{T\}\,dV'\,d\tau + \\ &+ \int_0^t\int_{A_c} \chi_{TCj}(\{T\}-T_e)\,dA\,d\tau\end{aligned} \tag{7.23}$$

$$\begin{aligned}T &= \{T\} + S_T + \int_0^t\int_V \varphi_T\cdot\nabla'\{T\}\,dV'\,d\tau + \int_0^t\int_V \psi_T\ \partial_\tau\{T\}\,dV'\,d\tau + \\ &+ \sum_j\int_0^t\int_V \varphi_{CTj}\cdot\nabla'\{\widetilde{C}_j\}\,dV'\,d\tau + \sum_j\int_0^t\int_V \psi_{CTj}\ \partial_\tau\{\widetilde{C}_j\}\,dV'\,d\tau + \\ &+ \int_0^t\int_{A_c} \chi_T(\{T\}\text{-}T_e)\,dA'\,d\tau\end{aligned} \tag{7.24}$$

where A_c denotes this part of the external boundary of the solidifying medium where the solid comes in contact with the mould surface and symbol T_e stands for (configuration independent) temperature of the mould surface.

The unknown functions $\varphi_{Cj}, \psi_{Cj}, \varphi_{TCj}, \psi_{TCj}, S_{Cj}, \chi_{TCj}$ satisfy the following integro-differential equations, which were obtained through substitution of eqs (7.23) and (7.24) into eq. (7.22):

$$\begin{aligned}\varphi_{Cj} &= -\int_0^t\int_V \nabla G\cdot[(\widetilde{D}'_j(\mathbf{1}\delta_t\delta_{\mathbf{x}}+\nabla\varphi_{Cj}) - \{\widetilde{D}'_j(\mathbf{1}\delta_t\delta_{\mathbf{x}}+\nabla\varphi_{Cj})\}) - (\widetilde{M}_j\nabla\varphi_{CTj} - \{\widetilde{M}_j\nabla\varphi_{CTj}\}) + \\ &\qquad - (\mathbf{w}\rho_T^o\widetilde{\rho}\widetilde{C}_j^o\varphi_{CTj} - \{\mathbf{w}\rho_T^o\widetilde{\rho}\widetilde{C}_j^o\varphi_{CTj}\}) + \mathbf{w}_r\rho_r\varphi_{Cj}]\,dV'\,d\tau + \\ &- \int_0^t\int_V G[(\rho_T^o\widetilde{\rho}\widetilde{C}_j^o\partial_\tau\psi_{CTj} - \{\rho_T^o\widetilde{\rho}\widetilde{C}_j^o\partial_\tau\psi_{CTj}\}) + \\ &\qquad + ((\mathbf{w}_i\cdot\nabla\theta_s)\varphi_{CTj} - \{(\mathbf{w}_i\cdot\nabla\theta_s)\varphi_{CTj}\})\Delta(\rho_T^o\widetilde{\rho}\widetilde{C}_j^o) - \rho_r\partial_\tau\varphi_{Cj}]\,dV'\,d\tau\end{aligned} \tag{7.25}$$

$$\begin{aligned}\psi_{Cj} &= -\int_0^t\int_V \nabla G\cdot[(\widetilde{D}'_j\nabla\psi_{Cj} - \{\widetilde{D}'_j\nabla\psi_{Cj}\}) - (\widetilde{M}_j\nabla\psi_{CTj} - \{\widetilde{M}_j\nabla\psi_{CTj}\}) + \\ &\qquad - (\mathbf{w}\rho_T^o\widetilde{\rho}\widetilde{C}_j^o\psi_{CTj} - \{\mathbf{w}\rho_T^o\widetilde{\rho}\widetilde{C}_j^o\psi_{CTj}\}) + \mathbf{w}_r\rho_r\psi_{Cj}]\,dV'\,d\tau + \\ &- \int_0^t\int_V G[(\rho_T^o\widetilde{\rho}\widetilde{C}_j^o(\delta_t\delta_{\mathbf{x}}+\partial_\tau\psi_{CTj}) - \{\rho_T^o\widetilde{\rho}\widetilde{C}_j^o(\delta_t\delta_{\mathbf{x}}+\partial_\tau\psi_{CTj})\}) + \\ &\qquad + ((\mathbf{w}_i\cdot\nabla\theta_s)\psi_{CTj} - \{(\mathbf{w}_i\cdot\nabla\theta_s)\psi_{CTj}\})\Delta(\rho_T^o\widetilde{\rho}\widetilde{C}_j^o) - \rho_r\partial_\tau\psi_{Cj}]\,dV'\,d\tau\end{aligned} \tag{7.26}$$

$$\varphi_{TCj} = -\int_0^t\int_V \nabla G\cdot[(\widetilde{D}_j'\nabla\varphi_{TCj} - \{\widetilde{D}_j'\nabla\varphi_{TCj}\}) - (\widetilde{M}_j(\mathbf{1}\delta_t\delta_{\mathbf{x}} + \nabla\varphi_T) - \{\widetilde{M}_j(\mathbf{1}\delta_t\delta_{\mathbf{x}} + \nabla\varphi_T)\}) +$$
$$-(\mathbf{w}\rho_T^o\widetilde{\rho}\widetilde{C}_j^o\varphi_T - \{\mathbf{w}\rho_T^o\widetilde{\rho}\widetilde{C}_j^o\varphi_T\}) + \mathbf{w_r}\rho_r\varphi_{TCj}]\,dV'\,d\tau + \tag{7.27}$$
$$-\int_0^t\int_V G[(\rho_T^o\widetilde{\rho}\widetilde{C}_j^o\partial_\tau\varphi_T - \{\rho_T^o\widetilde{\rho}\widetilde{C}_j^o\partial_\tau\varphi_T\}) +$$
$$+((\mathbf{w}_i\cdot\nabla\theta_s)\varphi_T - \{(\mathbf{w}_i\cdot\nabla\theta_s)\varphi_T\})\Delta(\rho_T^o\widetilde{\rho}\widetilde{C}_j^o) - \rho_r\partial_\tau\varphi_{TCj}]\,dV'\,d\tau$$

$$\psi_{TCj} = -\int_0^t\int_V \nabla G\cdot[(\widetilde{D}_j'\nabla\psi_{TCj} - \{\widetilde{D}_j'\nabla\psi_{TCj}\}) - (\widetilde{M}_j\nabla\psi_T - \{\widetilde{M}_j\nabla\psi_T\}) +$$
$$-(\mathbf{w}\rho_T^o\widetilde{\rho}\widetilde{C}_j^o\psi_T - \{\mathbf{w}\rho_T^o\widetilde{\rho}\widetilde{C}_j^o\psi_T\}) + \mathbf{w_r}\rho_r\psi_{TCj}]\,dV'\,d\tau + \tag{7.28}$$
$$-\int_0^t\int_V G[(\rho_T^o\widetilde{\rho}\widetilde{C}_j^o(\delta_t\delta_{\mathbf{x}} + \partial_\tau\psi_T) - \{\rho_T^o\widetilde{\rho}\widetilde{C}_j^o(\delta_t\delta_{\mathbf{x}} + \partial_\tau\psi_T)\}) +$$
$$+((\mathbf{w}_i\cdot\nabla\theta_s)\psi_T - \{(\mathbf{w}_i\cdot\nabla\theta_s)\psi_T\})\Delta(\rho_T^o\widetilde{\rho}\widetilde{C}_j^o) - \rho_r\partial_\tau\psi_{TCj}]\,dV'\,d\tau$$

$$S_{Cj} = -\int_0^t\int_V \nabla G\cdot[(\widetilde{D}_j'\nabla S_{Cj} - \{\widetilde{D}_j'\nabla S_{Cj}\}) - (\widetilde{M}_j\nabla S_T - \{\widetilde{M}_j\nabla S_T\}) +$$
$$-(\mathbf{w}\widetilde{\rho}\widetilde{C}_j^o - \{\mathbf{w}\widetilde{\rho}\widetilde{C}_j^o\}) - (\mathbf{w}\rho_T^o\widetilde{\rho}\widetilde{C}_j^o S_T - \{\mathbf{w}\rho_T^o\widetilde{\rho}\widetilde{C}_j^o S_T\}) +$$
$$+\mathbf{w_r}\rho_r S_{Cj}]\,dV'\,d\tau + \tag{7.29}$$
$$-\int_0^t\int_V G[(\rho_T^o\widetilde{\rho}\widetilde{C}_j^o\partial_\tau S_T - \{\rho_T^o\widetilde{\rho}\widetilde{C}_j^o\partial_\tau S_T\}) +$$
$$+((\mathbf{w}_i\cdot\nabla\theta_s) - \{(\mathbf{w}_i\cdot\nabla\theta_s)\})\Delta\widetilde{\rho}\widetilde{C}_j^o +$$
$$+((\mathbf{w}_i\cdot\nabla\theta_s)S_T - \{(\mathbf{w}_i\cdot\nabla\theta_s)S_T\})\Delta(\rho_T^o\widetilde{\rho}\widetilde{C}_j^o) - \rho_r\partial_\tau S_{Cj}]\,dV'\,d\tau$$

$$\chi_{TCj} = -\int_0^t\int_V \nabla G\cdot[(\widetilde{D}_j'\nabla\chi_{TCj} - \{\widetilde{D}_j'\nabla\chi_{TCj}\}) - (\widetilde{M}_j\nabla\chi_{TCj} - \{\widetilde{M}_j\nabla\chi_{TCj}\}) +$$
$$-(\mathbf{w}\rho_T^o\widetilde{\rho}\widetilde{C}_j^o\chi_T - \{\mathbf{w}\rho_T^o\widetilde{\rho}\widetilde{C}_j^o\chi_T\}) + \mathbf{w_r}\rho_r\chi_{TCj}]\,dV'\,d\tau + \tag{7.30}$$
$$-\int_0^t\int_V G[(\rho_T^o\widetilde{\rho}\widetilde{C}_j^o\partial_\tau\chi_T - \{\rho_T^o\widetilde{\rho}\widetilde{C}_j^o\partial_\tau\chi_T\}) +$$
$$+((\mathbf{w}_i\cdot\nabla\theta_s)\chi_T - \{(\mathbf{w}_i\cdot\nabla\theta_s)\chi_T\})\Delta(\rho_T^o\widetilde{\rho}\widetilde{C}_j^o) - \rho_r\partial_\tau\chi_{TCj}]\,dV'\,d\tau$$

where ρ_T^o is temperature coefficient of density, the upper index „o" denotes evaluation of the respective value at the bulk temperature and the symbol $\Delta(\rho_T^o\widetilde{\rho}\widetilde{C}_j^o)$ denotes differences between respective quantities in solid and liquid phases.

Substituting eq. (7.23) into the ensemble averaged terms, that appear in eq. (7.4), allows writing these terms in the following form

$$\{\boldsymbol{w}\widetilde{\rho}\widetilde{C}_j + \boldsymbol{j}_j\} = \{\boldsymbol{w}\widetilde{\rho}\widetilde{C}_j\} - \boldsymbol{j}_{SCj} + \\ - \int_0^t \int_V D_{Cefj} \cdot \nabla'\{\widetilde{C}_j\}\, dV'\, d\tau - \int_0^t \int_V \beta_{Cefj}\, \partial_\tau \{\widetilde{C}_j\}\, dV'\, d\tau + \\ - \int_0^t \int_V D_{Tefj} \cdot \nabla'\{T\}\, dV'\, d\tau - \int_0^t \int_V \beta_{Tefj}\, \partial_\tau \{T\}\, dV'\, d\tau + \\ - \int_0^t \int_{A_c} \alpha_{TCefj} (\{T\} - T_e)\, dA\, d\tau \tag{7.31}$$

$$\{\widetilde{\rho}\widetilde{C}_j\} = \{\rho^o \widetilde{C}_j^o\} + h_{SCj} + \int_0^t \int_V \varsigma_{Cefj} \cdot \nabla'\{\widetilde{C}_j\}\, dV'\, d\tau + \int_0^t \int_V \mu_{Cefj} \partial_\tau \{\widetilde{C}_j\}\, dV'\, d\tau + \\ + \int_0^t \int_V \varsigma_{CTefj} \cdot \nabla'\{T\}\, dV'\, d\tau + \int_0^t \int_V \mu_{CTefj} \partial_\tau \{T\}\, dV'\, d\tau + \\ + \int_0^t \int_{A_c} \xi_{CTefj} (\{T\} - T_e)\, dA\, d\tau \tag{7.32}$$

where

$$D_{Cefj} = \{\widetilde{D}_j (\boldsymbol{1}\delta_x \delta_t + \nabla \varphi_{Cj}) - \widetilde{M}_j \nabla \varphi_{TCj} - \boldsymbol{w}\rho_T^o \widetilde{\rho}\widetilde{C}_j^o \varphi_{TCj}\} \tag{7.33}$$

$$D_{Tefj} = \{\widetilde{D}_j \nabla \varphi_{TCj} - \widetilde{M}_j (\boldsymbol{1}\delta_x \delta_t + \nabla \varphi_T) - \boldsymbol{w}\rho_T^o \widetilde{\rho}\widetilde{C}_j^o \varphi_T\} \tag{7.34}$$

$$\beta_{Cefj} = \{\widetilde{D}_j \nabla \psi_{Cj} - \widetilde{M}_j \nabla \psi_{CTj} - \boldsymbol{w}\rho_T^o \widetilde{\rho}\widetilde{C}_j^o \psi_{CTj}\} \tag{7.35}$$

$$\beta_{Tefj} = \{\widetilde{D}_j \nabla \psi_{TCj} - \widetilde{M}_j \nabla \psi_T - \boldsymbol{w}\rho_T^o \widetilde{\rho}\widetilde{C}_j^o \psi_T\} \tag{7.36}$$

$$\boldsymbol{j}_{SCj} = \{\widetilde{D}_j \nabla S_{Cj} - \widetilde{M}_j \nabla S_T - \boldsymbol{w}\rho_T^o \widetilde{\rho}\widetilde{C}_j^o S_T\} \tag{7.37}$$

$$\alpha_{TCefj} = \{\widetilde{D}_j \nabla \chi_{TCj} - \widetilde{M}_j \nabla \chi_T - \boldsymbol{w}\rho_T^o \widetilde{\rho}\widetilde{C}_j^o \chi_T\} \tag{7.38}$$

$$h_{SCj} = \{\rho_T^o \widetilde{\rho}\widetilde{C}_j^o S_T\} \tag{7.39}$$

$$\varsigma_{Cefj} = \{\rho_T^o \widetilde{\rho}\widetilde{C}_j^o \varphi_{CTj}\} \tag{7.40}$$

$$\mu_{Cefj} = \{\rho_T^o \widetilde{\rho}\widetilde{C}_j^o \psi_{CTj}\} \tag{7.41}$$

$$\varsigma_{CTefj} = \{\rho_T^o \widetilde{\rho}\widetilde{C}_j^o \varphi_T\} \tag{7.42}$$

$$\mu_{CTefj} = \{\rho_T^o \widetilde{\rho}\widetilde{C}_j^o \psi_T\} \tag{7.43}$$

$$\xi_{CTefj} = \{\rho_T^o \widetilde{\rho}\widetilde{C}_j^o \chi_T\} \tag{7.44}$$

The relation between the considered averaged terms and the bulk concentration and temperature, given by eqs (7.31) and (7.32), are thus nonlocal. The nonlocality denotes that the macroscopic solute flux in the considered point in the medium depends not only on the macroscopic solute concentration in the same point and moment of time but on the macroscopic concentration and temperature distribution in the whole medium both in the considered moment of time as well as those existing in the past. The terms in eqs (7.31) and (7.32), that contain time derivatives of $\partial_t\{\widetilde{C}_j\}$, $\partial_t\{T\}$, correspond to relaxation phenomena associated with the bulk solute flux $\{\boldsymbol{j}_j\}$ and the bulk volumetric concentration $\{\widetilde{\rho C}_j\}$. It should also be noted that some part of the advective solute flux is dispersed on the macroscopic scale and on the level of the macroscopic description the fluid velocity enters diffusive terms (see eqs (7.33)-(7.38)). Moreover, due to dependence of density and solubility on temperature the cross terms, known as Soret effect, are present in the eqs (7.31) and (7.32). The last term on the r.h.s. of these relations represents influence of the solidifying material/mould interface on the bulk volumetric concentration of jth species and its transport. This effect is usually limited to small distances from the common boundary. The term $\boldsymbol{j}_{SCj}, h_{SCj}$, present in eqs (7.31) and (7.32), account for influence of the liquid/solid interface movement on the transport and species accumulation processes in the heterogeneous medium on the macroscopic scale.

Symbols $D_{Ceff}, D_{Teff}, \beta_{Ceff}, \beta_{Teff}, \varsigma_{Ceff}, \varsigma_{CTeff}, \mu_{Ceff}, \mu_{CTeff}, \alpha_{Ceff}, \xi_{CTeff}$ stand for the effective parameters (properties) describing intensity of species transport and accumulation in the mushy zone or at the solidifying material/mould interface. According to eqs (7.33) and (7.44), they can be determined from knowledge of microscopic properties of solid and fluid and microstructure of the mushy zone hidden in the microstructure functions as for example $\varphi_{Cj}, \psi_{Cj}, \varphi_{TCj}, \psi_{TCj}, S_{Cj}, \chi_{TCj}$.

Many length-scales are usually observed in the mushy region. There are connected with details of microstructure of the considered medium and with variation of the species macroscopic concentration. The specific feature of the non-local form of the constitutive relations given above and in the subsequent Section 8 is that they take into account all of the mentioned scales and no scale order analysis is necessary.

Nonlocal phenomena describe locally non-equilibrium processes. A non-equilibrium process in case of solute transport denotes that, on the macroscopic level, the solute is exchanged between the phases. One of the ways to approximately account for chemical non-equilibrium is the so-called, diffusion model that relates the intrinsic solute concentration in solid and liquid (or the bulk solute concentration and the intrinsic solute concentration in the liquid). This kind of a model that is often additionally adopted in literature bears postulative character. Another way of accounting for chemical non-equilibrium is, the so-called, two-equation (two-fluid) model in which averaging, in the mushy zone, is carried out separately for each phase and exchange of the solute between phases taken into account on the macroscopic level (see Beckermann and Viskanta, 1993).

8. Macro-modelling of Energy Transport

Heat conduction and advection, heat accumulation and generation of heat due to phase transformation affect variation in temperature distribution. The macroscopic energy conservation equation can be obtained by multiplying the microscopic equation, eq. (3.3), with the structure function. After some mathematical manipulations the following expression follows

$$\partial_t(\theta_k \rho_k h_k) + \nabla \cdot [\theta_k (\boldsymbol{w}_k \rho_k h_k + \boldsymbol{q}_k) = [\partial_t \theta_k + \boldsymbol{w}_k \cdot \nabla \theta_k]\rho_k h_k + \nabla \theta_k \cdot \boldsymbol{q}_k \tag{8.1}$$

The r.h.s. of this equation can be transformed, using relations from eq. (5.3), to yield

$$\partial_t(\theta_k \rho_k h_k) + \nabla \cdot [\theta_k (\boldsymbol{w}_k \rho_k h_k + \boldsymbol{q}_k) = [\rho_k h_k (\boldsymbol{w}_k - \boldsymbol{w}_i) + \boldsymbol{q}_k] \cdot \nabla \theta_k \tag{8.2}$$

The term on the r.h.s. of the above equation expresses local exchange of energy between phases due to solidification, advection and conduction.

Equation (8.2) can be summed over the phases present in the system to yield the generalised microscopic energy conservation equation for the whole medium

$$\partial_t(\rho h) + \nabla \cdot [\boldsymbol{w} \rho h + \boldsymbol{q}] = 0 \tag{8.3}$$

Again the r.h.s. terms were eliminated due to condition of energy conservation at liquid/solid interface (see eqs (3.7) and (5.4)).

Equation (8.3) represents enthalpy formulation of the energy conservation equation. If the volumetric specific heats of phases are assumed constant and differences between volumetric specific enthalpies of liquid and solid (at the phase transformation temperature) are treated as weakly dependent on temperature and species concentration then another formulation of eq. (8.3) is possible

$$\partial_t(\rho c T) + \nabla \cdot [(\boldsymbol{w} \rho c T + \boldsymbol{q})] = \boldsymbol{w}_i L_f \nabla \theta_l \tag{8.4}$$

8.1 Macroscopic Energy Conservation Equation

A single continuum approach. The equation (8.3) can be ensemble averaged and the macroscopic form of the energy conservation takes then the form

$$\partial_t\{\rho h\} + \nabla \cdot [\{\boldsymbol{w} \rho h\} + \{\boldsymbol{q}\}] = 0 \tag{8.5}$$

Equation (8.5) corresponds to a single continuum description of the solidifying, heterogeneous medium on the macroscopic scale (see Section 4.2). For incompressible phases with no solid phase movement it can be rewritten in another form

$$\partial_t\{\rho h\} + \nabla \cdot [\varepsilon_l \{\boldsymbol{w} \rho h\}_l + \{\boldsymbol{q}\}] = 0 \tag{8.6}$$

In this case averaging of eq. (8.4) leads to the following expression

$$\partial_t\{\rho c T\} + \nabla \cdot [\{\boldsymbol{w} \rho c T\} + \{\boldsymbol{q}\}] = -L_f \, \partial_t \varepsilon_l \tag{8.7}$$

When the enthalpy formulation of the energy conservation equation is used, the location of the liquid/solid interface during stable, equilibrium solidification of pure materials is determined from knowledge of the melting temperature T_m. In case of the unstable solidification of pure materials or multi-component mixtures (alloys) the mushy zone is usually formed. Location of solid/mushy zone or mushy zone/liquid interfaces is more complicated in this case. If processes occur close to equilibrium then the solidus and liquidus temperature, corresponding to the macroscopic solute concentration (in case of binary mixtures) can be used (see Fig. 1). For solidification processes occurring far from thermodynamic equilibrium eq. (2.5) gives temperature at the liquid/solid interface. The equation should be ensemble averaged in order to obtain the respective condition for location of the interface in the macroscopic approach

$$\{T_i\} = T^e - \{\frac{\sigma(\vartheta_o, s)\kappa}{\Delta S_{ls}}\} - \{\mu_k(\vartheta_o, s)\, \boldsymbol{w}_i \cdot \boldsymbol{n}_s\} \tag{8.8}$$

A multi-continuum approach. Direct averaging of eq. (8.2) for energy conservation in each kth phase allows obtaining macroscopic formulation of this equation within multi-continuum approach (see Section 4.2)

$$\partial_t(\varepsilon_k\{\rho h\}_k) + \nabla\cdot[\varepsilon_k(\{\rho h\}_k + \{\boldsymbol{q}\}_k)] = \Gamma^c_{Tk} + \Gamma^d_{Tk} \tag{8.9}$$

where

$$\Gamma^c_{Tk} = \{\rho_k h_k(\boldsymbol{w_k} - \boldsymbol{w_i})\cdot\nabla\theta_k\} \tag{8.10}$$

corresponds to advective of energy exchange between phases
and

$$\Gamma^d_{Tk} = \{\boldsymbol{q}_k \cdot \nabla\theta_k\} \tag{8.11}$$

to the conductive one.

If eq. (5.4) is used then the rate energy exchanges between phases can be expressed as

$$\Gamma^c_{Tk} = \{-\rho_k h_k(\boldsymbol{w_k} - \boldsymbol{w_i})\cdot\boldsymbol{n_k}\ \delta(\boldsymbol{x},\boldsymbol{x_i})\} = a_i\{\rho_k h_k(\boldsymbol{w}_i - \boldsymbol{w}_k)\cdot\boldsymbol{n_k})\}^*_A \tag{8.12}$$

and

$$\Gamma^d_{Tk} = \{-\boldsymbol{q}_k\cdot\boldsymbol{n_k}\ \delta(\boldsymbol{x},\boldsymbol{x_i})\} = -a_i\{\boldsymbol{q}_k\cdot\boldsymbol{n_k}\}^*_A \tag{8.13}$$

where the symbol a_i was defined in respective part of Section 6.

8.2. Closure Problem. Constitutive Relations for Energy Transport

In order to numerically solve conservation equations constitutive relations between variables are needed, i.e., closure problem should be again solved. In the case of the single continuum approach we need to express such terms as $\{\rho h\}$, $\{\boldsymbol{w}\rho h\}$ and $\{\boldsymbol{q}\}$, see eq. (8.6), as functions of the macroscopic temperature $\{T\}$ and macroscopic concentration $\{\widetilde{C}_j\}$ of jth species. In case of the multi-continuum approach relations between $\{\rho h\}_k, \{\boldsymbol{w}\rho h\}_k, \{\boldsymbol{q}\}_k$, $\Gamma^c_{Tk}, \Gamma^d_{Tk}$, see eq. (8.9), and the intrinsic temperatures $\{T\}_k$, intrinsic jth species concentration $\{\widetilde{C}_j\}_k$ and the interface rate of mass flux Γ_{mk} are necessary. In most of the available literature these relations are postulated using respective relations for no solidifying heterogeneous media with fixed microstructure.

The macroscopic constitutive relations could be derived if relation between microscopic temperature T in any phase and its bulk value $\{T\}$ was known. In order to formally derive this relation the generalised form the microscopic constitutive equation (Fourier law)

$$q = -\lambda \nabla T \tag{8.14}$$

was introduced into the microscopic energy conservation equation, eq. (8.3). The generalised thermal conductivity λ was here defined as

$$\lambda = \sum_k \theta_k (t, \boldsymbol{x}|\Omega)\, \lambda_k \tag{8.15}$$

The microscopic energy equation together with the Fourier law were subsequently rewritten in the following form

$$\rho_r c_r \partial_t T + \rho_r c_r \boldsymbol{w}_r \cdot \nabla T - \lambda_r \nabla^2 T - F_T = 0 \tag{8.16}$$

where the source function F_T is defined as

$$F_T = \nabla \cdot (\lambda' \nabla T) - \nabla \cdot (\boldsymbol{w}\rho h - \boldsymbol{w}_r \rho_r c_r T) - \partial_t (\rho h - \rho_r c_r h) \tag{8.17}$$

and

$$\lambda' = \lambda - \lambda_r \tag{8.18}$$

The thermal conditions on the external boundaries of the system were discussed in Section 3. They can be rewritten in the following, generalised, form

$$(\boldsymbol{q} + \boldsymbol{w}\rho h) \cdot \boldsymbol{n}_e - \alpha_C (T - T_e) = f_T \tag{8.18}$$

where the mould temperature T_e and function f_T are space and time dependent but they do not depend on the configuration Ω. For solid boundaries the value of velocity $\boldsymbol{w}$ is assumed to be zero. The symbol α_C stands for the locally varying heat transfer coefficient at the solidifying material/mould interface. It is dependent on the chosen configuration Ω. The initial condition for the problem was adopted in the form

$$T(t = 0, \boldsymbol{x}|\Omega) = T_o(\boldsymbol{x}) \tag{8.19}$$

Subsequently the Green function G was defined by the set of equations:

$$\lambda_r \nabla^2 G + \rho_r c_r \boldsymbol{w}_r \cdot \nabla G + \delta_x (\boldsymbol{x}, \boldsymbol{y})\, \delta_t (t, \tau) = -\rho_r c_r \partial_t G \quad \text{for } \boldsymbol{x} \in V \tag{8.20}$$

$$(\lambda_r \nabla G - \boldsymbol{w}\rho_r G) \cdot \boldsymbol{n}_e - \alpha_r G = 0 \quad \text{for } \boldsymbol{x} \in A \tag{8.21}$$

$$G(t, \boldsymbol{x}; \tau, \boldsymbol{y}) = 0 \qquad \text{for } t \le \tau \tag{8.22}$$

Other symbols appearing in the above equations have been previously described in Section 7.

With use of the Green function theory the following formal expression for temperature was then obtained

$$\begin{aligned} T = &\int_V \rho_r c_r G_o \widetilde{C}_{jo} dV - \int_0^t \int_V \nabla G \cdot [\lambda' \nabla T - (\boldsymbol{w}\rho \boldsymbol{h} - \boldsymbol{w}_r \rho_r c_r T)]\, dV'\, d\tau + \\ &- \int_0^t \int_V G\, \partial_\tau (\rho h - \rho_r c_r T)\, dV'\, d\tau + \int_0^t \int_A G [\boldsymbol{w}_r \rho_r c_r T \cdot \boldsymbol{n}_e - \alpha' T - f_T]\, dA'\, d\tau \end{aligned} \tag{8.23}$$

where $\alpha' = \alpha - \alpha_r$ and α_r denotes the reference value of the heat transfer coefficient at the mould surface that is configuration independent.

Ensemble averaging of the above equation and subtracting the averaged equations from the original one gives the following relation between local, microscopic and the macroscopic temperatures

$$
\begin{aligned}
T = \{T\} - \int_0^t \int_V \nabla G \cdot [(\lambda' \nabla T - \{\lambda' \nabla T\}) + \\
- (\boldsymbol{w}\rho h - \boldsymbol{w}_r \rho_r c_r T - \{\boldsymbol{w}\rho h - \boldsymbol{w}_r \rho_r c_r T\})]\, dV'\, d\tau + \\
- \int_0^t \int_V G\, \partial_t [(\rho h - \rho_r c_r T) - \{\rho h - \rho_r c_r T\}]\, dV'\, d\tau + \\
- \int_0^t \int_{A_c} G [\alpha (T - T_e) - \{\alpha (T - T_e)\} - \alpha_r (T - \{T\})]\, dA\, d\tau
\end{aligned}
\tag{8.24}
$$

The volumetric enthalpy ρh is dependent on temperature and species concentrations. Using expansions similar to eq. (3.29) and adopting the reference temperature and concentration equal to the bulk temperature and the bulk concentration in the medium, respectively, the terms mentioned can be expressed as functions of local differences between the microscopic and the macroscopic temperature and macroscopic concentration of jth species.

Expressions (7.23) and (7.24) give the formal solution of eq. (8.24). After substituting these expressions into eq. (8.24) equations for the unknown functions $\varphi_T, \psi_T, \varphi_{CT}, \psi_{CT}, S_T, \chi_T$ were obtained in the following form:

$$
\begin{aligned}
\varphi_T = -\int_0^t \int_V \nabla G \cdot [(\lambda' (\boldsymbol{1}\delta_t \delta_{\boldsymbol{x}} + \nabla \varphi_T) - \{\lambda' (\boldsymbol{1}\delta_t \delta_{\boldsymbol{x}} + \nabla \varphi_T)\}) + \\
- (\boldsymbol{w}\rho^o c^o \varphi_T - \{\boldsymbol{w}\rho^o c^o \varphi_T\}) + \boldsymbol{w}_r \rho_r c_r \varphi_T + \sum_j (\boldsymbol{w}\rho h_j^* \varphi_{TCj} - \{\boldsymbol{w}\rho h_j^* \varphi_{TCj}\})]\, dV'\, d\tau + \\
- \int_0^t \int_V G [(\rho^o c^o \partial_\tau \varphi_T - \{\rho^o c^o \partial_\tau \varphi_T\}) + \sum_j (\rho h_j^* \partial_\tau \varphi_{TCj} - \{\rho h_j^* \partial_\tau \varphi_{TCj}\}) + \\
+ ((\boldsymbol{w}_i \cdot \nabla \theta_s) \varphi_T - \{(\boldsymbol{w}_i \cdot \nabla \theta_s) \varphi_T\}) \Delta(\rho^o c^o) - \rho_r c_r \partial_\tau \varphi_T + \\
+ \sum_j ((\boldsymbol{w}_i \cdot \nabla \theta_s) \varphi_{TCj} - \{(\boldsymbol{w}_i \cdot \nabla \theta_s) \varphi_{TCj}\}) \Delta(\rho h_j^*)]\, dV'\, d\tau + \\
- \int_0^t \int_{A_c} G [\alpha' \varphi_T - \{\alpha' \varphi_T\}]\, dA\, d\tau
\end{aligned}
\tag{8.25}
$$

$$\begin{aligned}\psi_T = &-\int_0^t\int_V \nabla G\cdot[(\lambda'\nabla\psi_T - \{\lambda'\nabla\psi_T\}) + \\ &-(\boldsymbol{w}\rho^o c^o\psi_T - \{\boldsymbol{w}\rho^o c^o\psi_T\}) + \boldsymbol{w}_r\rho_r c_r\psi_T + \sum_j(\boldsymbol{w}\rho h_j^*\psi_{TCj} - \{\boldsymbol{w}\rho h_j^*\psi_{TCj}\})]\,dV'\,d\tau + \\ &-\int_0^t\int_V G[(\rho^o c^o(\delta_x\delta_t + \partial_\tau\psi_T) - \{\rho^o c^o(\delta_x\delta_t + \partial_\tau\psi_T)\}) + \sum_j(\rho h_j^*\partial_\tau\psi_{TCj} - \{\rho h_j^*\partial_\tau\psi_{TCj}\}) + \\ &\qquad + ((\boldsymbol{w}_i\cdot\nabla\theta_s)\psi_T - \{(\boldsymbol{w}_i\cdot\nabla\theta_s)\psi_T\})\Delta(\rho^o c^o) - \rho_r c_r\partial_\tau\psi_T + \\ &+\sum_j((\boldsymbol{w}_i\cdot\nabla\theta_s)\psi_{TCj} - \{(\boldsymbol{w}_i\cdot\nabla\theta_s)\psi_{TCj}\})\Delta(\rho h_j^*)]\,dV'\,d\tau + \\ &-\int_0^t\int_{A_c} G[\alpha'\psi_T - \{\alpha'\psi_T\}]\,dA\,d\tau\end{aligned} \tag{8.26}$$

$$\begin{aligned}\varphi_{CTj} = &-\int_0^t\int_V \nabla G\cdot[(\lambda'\nabla\varphi_{CTj}) - \{\lambda'\nabla\varphi_{CTj})\}) + \\ &-(\boldsymbol{w}\rho^o c^o\varphi_{CTj} - \{\boldsymbol{w}\rho^o c^o\varphi_{CTj}\}) + \boldsymbol{w}_r\rho_r c_r\varphi_{CTj} + (\boldsymbol{w}\rho h_j^*\varphi_{Cj} - \{\boldsymbol{w}\rho h_j^*\varphi_{Cj}\})\,dV'\,d\tau + \\ &-\int_0^t\int_V G[(\rho^o c^o\partial_\tau\varphi_{CTj} - \{\rho^o c^o\partial_\tau\varphi_{CTj}\}) + (\rho h_j^*\partial_\tau\varphi_{Cj} - \{\rho h_j^*\partial_\tau\varphi_{Cj}\}) + \\ &\qquad + ((\boldsymbol{w}_i\cdot\nabla\theta_s)\varphi_{CTj} - \{(\boldsymbol{w}_i\cdot\nabla\theta_s)\varphi_{CTj}\})\Delta(\rho^o c^o) - \rho_r c_r\partial_\tau\varphi_{CTj} + \\ &+((\boldsymbol{w}_i\cdot\nabla\theta_s)\varphi_{Cj} - \{(\boldsymbol{w}_i\cdot\nabla\theta_s)\varphi_{Cj}\})\Delta(\rho h_j^*)]\,dV'\,d\tau + \\ &-\int_0^t\int_{A_c} G[\alpha'\varphi_{CTj} - \{\alpha'\varphi_{CTj}\}]\,dA\,d\tau\end{aligned} \tag{8.27}$$

$$\begin{aligned}\psi_{CTj} = &-\int_0^t\int_V \nabla G\cdot[(\lambda'\nabla\psi_{CTj}) - \{\lambda'\nabla\psi_{CTj})\}) + \\ &-(\boldsymbol{w}\rho^o c^o\psi_{CTj} - \{\boldsymbol{w}\rho^o c^o\psi_{CTj}\}) + \boldsymbol{w}_r\rho_r c_r\psi_{CTj} + (\boldsymbol{w}\rho h_j^*\psi_{Cj} - \{\boldsymbol{w}\rho h_j^*\psi_{Cj}\})\,dV'\,d\tau + \\ &-\int_0^t\int_V G[(\rho^o c^o\partial_\tau\psi_{CTj} - \{\rho^o c^o\partial_\tau\psi_{CTj}\}) + (\rho h_j^*(\delta_x\delta_t + \partial_\tau\psi_{Cj}) - \{\rho h_j^*(\delta_x\delta_t + \partial_\tau\psi_{Cj})\}) + \\ &\qquad + ((\boldsymbol{w}_i\cdot\nabla\theta_s)\psi_{CTj} - \{(\boldsymbol{w}_i\cdot\nabla\theta_s)\psi_{CTj}\})\Delta(\rho^o c^o) - \rho_r c_r\partial_\tau\psi_{CTj} + \\ &+((\boldsymbol{w}_i\cdot\nabla\theta_s)\psi_{Cj} - \{(\boldsymbol{w}_i\cdot\nabla\theta_s)\psi_{Cj}\})\Delta(\rho h_j^*)]\,dV'\,d\tau + \\ &-\int_0^t\int_{A_c} G[\alpha'\psi_{CTj} - \{\alpha'\psi_{CTj}\}]\,dA\,d\tau\end{aligned} \tag{8.28}$$

$$S_T = -\int_0^t\int_V \nabla G\cdot[(\lambda'\nabla S_T)-\{\lambda'\nabla S_T)\})-(\boldsymbol{w}\rho^o h^o-\{\boldsymbol{w}\rho^o h^o\})+$$
$$-(\boldsymbol{w}\rho^o c^o S_T-\{\boldsymbol{w}\rho^o c^o S_T\})+\boldsymbol{w}_r\rho_r c_r S_T+\sum_j(\boldsymbol{w}\rho h_j^* S_{Cj}-\{\boldsymbol{w}\rho h_j^* S_{Cj}\})]\,dV'\,d\tau+$$
$$-\int_0^t\int_V G[(\rho^o c^o\partial_\tau S_T-\{\rho^o c^o\partial_\tau S_T\})+\sum_j(\rho h_j^*\partial_\tau S_{Cj}-\{\rho h_j^*\partial_\tau S_{Cj}\})+$$
$$+((\boldsymbol{w}_i\cdot\nabla\theta_s)-\{(\boldsymbol{w}_i\cdot\nabla\theta_s)\})\Delta(\rho^o h^o)+((\boldsymbol{w}_i\cdot\nabla\theta_s)S_T-\{(\boldsymbol{w}_i\cdot\nabla\theta_s)S_T\})\Delta\rho^o c^o+$$
$$-\rho_r c_r\partial_\tau S_T+\sum_j((\boldsymbol{w}_i\cdot\nabla\theta_s)S_{Cj}-\{(\boldsymbol{w}_i\cdot\nabla\theta_s)S_{Cj}\})\Delta(\rho h_j^*)]\,dV'\,d\tau+$$
$$-\int_0^t\int_{A_c} G[\alpha' S_T-\{\alpha' S_T\}]\,dA\,d\tau \tag{8.29}$$

$$\chi_T = -\int_0^t\int_V \nabla G\cdot[(\lambda'\nabla\chi_T)-\{\lambda'\nabla\chi_T)\})+$$
$$-(\boldsymbol{w}\rho^o c^o\chi_T-\{\boldsymbol{w}\rho^o c^o\chi_T\})+\boldsymbol{w}_r\rho_r c_r\chi_T]\,dV'\,d\tau+$$
$$-\int_0^t\int_V G[(\rho^o c^o\partial_\tau\chi_T-\{\rho^o c^o\partial_\tau\chi_T\})+$$
$$+((\boldsymbol{w}_i\cdot\nabla\theta_s)\chi_T-\{(\boldsymbol{w}_i\cdot\nabla\theta_s)\chi_T\})\Delta(\rho^o c^o)-\rho_r c_r\partial_\tau\chi_T]\,dV'\,d\tau+$$
$$-\int_0^t\int_{A_c} G[\alpha'\chi_T-\{\alpha'\chi_T\}-(\alpha-\{\alpha\})\,\delta_t\delta_{xA}]\,dA\,d\tau \tag{8.30}$$

In eq. (8.25) the upper index „o” denotes evaluation of the respective value at the bulk temperature and the symbols $\Delta(\rho^o c^o)$, $\Delta(\rho h_j^*)$ stand for differences between volumetric specific heats and the volumetric partial enthalpies in solid and liquid phases, respectively. When eq. (7.24) is substituted into the ensemble averaged terms, appearing in eq. (8.5), the following relations can be written

$$\{\boldsymbol{w}\rho h+\boldsymbol{q}\} = \{\boldsymbol{w}\rho^o h^o\}-\boldsymbol{q}_{ST}+$$
$$-\int_0^t\int_V \lambda_{Tef}\cdot\nabla'\{T\}\,dV'\,d\tau-\int_0^t\int_V \gamma_{Tef}\,\partial_\tau\{T\}\,dV'\,d\tau+$$
$$-\sum_j\int_0^t\int_V \lambda_{Cefj}\cdot\nabla'\{\widetilde{C}_j\}\,dV'\,d\tau-\sum_j\int_0^t\int_V \gamma_{Cefj}\,\partial_\tau\{\widetilde{C}_j\}\,dV'\,d\tau+$$
$$-\int_0^t\int_{A_c}\alpha_{Tef}(\{T\}-T_e)\,dA\,d\tau \tag{8.31}$$

$$\{\rho h\} = \{\rho^o h^o\} + h_{ST} + \int_0^t \int_V \varsigma_{Tef} \cdot \nabla'\{T\}\, dV'\, d\tau + \int_0^t \int_V \mu_{T\,eff} \partial_\tau \{T\}\, dV'\, d\tau +$$
$$+ \sum_j \int_0^t \int_V \varsigma_{TCeff} \cdot \nabla'\{\widetilde{C}_j\}\, dV'\, d\tau + \sum_j \int_0^t \int_V \mu_{TC\,eff} \partial_\tau \{\widetilde{C}_j\}\, dV'\, d\tau + \tag{8.32}$$
$$+ \int_0^t \int_{A_c} \xi_{Tef} (\{T\} - T_e)\, dA\, d\tau$$

where

$$\lambda_{Tef} = \{\lambda(\mathbf{1}\delta_x \delta_t + \nabla \varphi_T) - w\rho^o c^o \varphi_T - \sum_j w\rho h_j^* \varphi_{TCj}\} \tag{8.33}$$

$$\lambda_{Ceff} = \{\lambda \nabla \varphi_{CTj} - w\rho^o c^o \varphi_{CTj} - w\rho h_j^* \varphi_{Cj}\} \tag{8.34}$$

$$\gamma_{Tef} = \{\lambda \nabla \psi_T) - w\rho^o c^o \psi_T - \sum_j w\rho h_j^* \psi_{TCj}\} \tag{8.35}$$

$$\gamma_{Ceff} = \{\lambda \nabla \psi_{CTj} - w\rho^o c^o \psi_{CTj} - w\rho h_j^* \psi_{Cj}\} \tag{8.36}$$

$$\mathbf{q}_{ST} = \{\lambda \nabla S_T) - w\rho^o c^o S_T - \sum_j w\rho h_j^* S_{Cj}\} \tag{8.37}$$

$$\alpha_{Tef} = \{\lambda \nabla \chi_T - w\rho^o c^o \chi_T\} \tag{8.38}$$

$$h_{ST} = \{\rho^o c^o S_T\} \tag{8.39}$$

$$\varsigma_{Tef} = \{\rho^o c^o \varphi_T + \sum_j \rho h_j^* \varphi_{TCj}\} \tag{8.40}$$

$$\mu_{Tef} = \{\rho^o c^o \psi_T + \sum_j \rho h_j^* \psi_{TCj}\} \tag{8.41}$$

$$\varsigma_{CTeff} = \{\rho^o c^o \varphi_{CTj} + \sum_j \rho h_j^* \varphi_{Cj}\} \tag{8.42}$$

$$\mu_{CTeff} = \{\rho^o c^o \psi_{CTj} + \sum_j \rho h_j^* \psi_{Cj}\} \tag{8.43}$$

$$\xi_{Tef} = \{\rho^o c^o \chi_T + \sum_j \rho h_j^* \chi_{TCj}\} \tag{8.44}$$

It should be noted that both relations, eqs (8.31) and (8.32), are nonlocal. They also show that the bulk heat flux and the bulk enthalpy depend not only on the macroscopic temperature but also on the macroscopic concentration of the dissolved species. This conclusion follows from dependence of enthalpy on the solute concentration. The nonlocality denotes that the macroscopic heat flux and enthalpy in the considered point in the medium depends not only on the macroscopic temperature and solute concentration in the same point and moment of time but on the macroscopic temperature and concentration distributions in the whole medium both in the considered moment of time as well as those existing in the past. Moreover, some of the energy, due to tortuous path followed by liquid phase in the mushy region, is dispersed and enters diffusive terms. The latter conclusion is justified by presence of fluid velocity in the expressions for the effective parameters $\lambda_{Tef}, \lambda_{Ceff}, \gamma_{Tef}, \gamma_{CTeff}$. The term $\boldsymbol{q}_{TS}$, appearing in eq. (8.31), arises due to presence of temperature-dependent local heat sources at the solid-liquid interface, while the term containing α_{Tef} is only important at the solidified material/mould interface. The terms containing effective parameters h_{ST}, ς_{Tef}, μ_{Tef}, ς_{CTeff} μ_{CTeff} and ξ_{Tef} account for additional accumulation of energy in the medium. Heat sources, occurring at the liquid/solid interface which are associated with phase change, locally deform temperature fields and cause additional transport of heat and heat accumulation.

9. Macro-modelling of the Momentum Transport

In order to obtain equation that describes variation in momentum of the solidifying medium on the macroscopic scale eq. (3.2) was multiplied by the structure function θ_k and transformed to the form

$$\partial_t(\theta_k \rho_k \boldsymbol{w}_k) + \nabla \cdot [\theta_k(\boldsymbol{w}_k \rho_k \boldsymbol{w}_k - \boldsymbol{\sigma}_k)] - \theta_k \rho_k \boldsymbol{f}_m = \rho_k \boldsymbol{w}_k [\partial_t \theta_k + \boldsymbol{w}_k \cdot \nabla \theta_k] - \nabla \theta_k \cdot \boldsymbol{\sigma}_k \tag{9.1}$$

If eq. (5.3) is substituted into it, eq. (9.1) it gives

$$\partial_t(\theta_k \rho_k \boldsymbol{w}_k) + \nabla \cdot [\theta_k(\boldsymbol{w}_k \rho_k \boldsymbol{w}_k - \boldsymbol{\sigma}_k)] - \theta_k \rho_k \boldsymbol{f}_m = [\rho_k \boldsymbol{w}_k(\boldsymbol{w}_k - \boldsymbol{w}_i) - \boldsymbol{\sigma}_k] \cdot \nabla \theta_k \cdot \tag{9.2}$$

The term on the r.h.s. of the latter equation presents force-interaction at the interface leading to variation of momentum across it.

When eqs (9.2) are summed over phases, conditions satisfied at the interface (see eq. (3.6)) are included, and the generalised variables are introduced, the following form the momentum equation is obtained

$$\partial_t(\rho \boldsymbol{w}) + \nabla \cdot (\boldsymbol{w} \rho \boldsymbol{w} - \boldsymbol{\sigma}) - \rho \boldsymbol{f}_m = -(2\sigma\kappa \boldsymbol{n}_s + \nabla_t \sigma)(\boldsymbol{n}_s \cdot \nabla \theta_s) \tag{9.3}$$

Please note that in this case the r.h.s. terms do not disappeared and are associated with presence of the surface tension.

9.1 Macroscopic Conservation Equation of Momentum

A single continuum approach. Equation (9.3) when ensemble averaged gives the macroscopic form of the momentum conservation equation

$$\partial_t\{\rho \boldsymbol{w}\} + \nabla \cdot [\{\boldsymbol{w} \rho \boldsymbol{w}\} - \{\boldsymbol{\sigma}\}] - \{\rho\} \boldsymbol{f}_m = -\{(2\sigma\kappa \boldsymbol{n}_s + \nabla_t \sigma)(\boldsymbol{n}_s \cdot \nabla \theta_s)\} \tag{9.4}$$

Equation (9.4) corresponds to replacement of the solidifying, heterogeneous medium with a single continuum on the macroscopic scale (see Section 4). If incompressibility of the phases is invoked and the solid phase is not moving then eq. (9.4) can be rewritten in another form

$$\partial_t(\varepsilon_l \rho_l \{\boldsymbol{w}\}_l) + \nabla\cdot[\varepsilon_l \rho_l \{\boldsymbol{ww}\}_l - \{\boldsymbol{\sigma}\}] - \{\rho\}\boldsymbol{f}_m = -\{(2\sigma\kappa \boldsymbol{n}_s + \nabla_t \sigma)(\boldsymbol{n}_s \cdot \nabla\theta_s)\} \tag{9.5}$$

A multi-continuum approach. If direct averaging of eq. (9.2) is carried out then the following equation is obtained

$$\partial_t(\varepsilon_k \rho_k \{\boldsymbol{w}\}_k) + \nabla\cdot[\varepsilon_k(\rho_k \{\boldsymbol{ww}\}_k - \{\boldsymbol{\sigma}\}_k)] - \varepsilon_k \rho_k \boldsymbol{f}_m = \boldsymbol{\Gamma}^c_{wk} + \boldsymbol{\Gamma}^d_{wk} \cdot \tag{9.6}$$

This equation corresponds to replacement of the solidifying, heterogeneous medium with k number of fictitious continua (see Section 4). Terms appearing on the r.h.s. of eq. (9.6) are defined as

$$\boldsymbol{\Gamma}^c_{wk} = \{\rho_k \boldsymbol{w}_k(\boldsymbol{w}_k - \boldsymbol{w}_i)\cdot\nabla\theta_k\} \tag{9.7}$$

where $\boldsymbol{\Gamma}^c_{wk}$ corresponds to *advective momentum exchange* between phases

and

$$\boldsymbol{\Gamma}^d_{wk} = -\{\boldsymbol{\sigma}_k \cdot \nabla\theta_k\} \tag{9.8}$$

where $\boldsymbol{\Gamma}^d_{wk}$ is *the diffusive momentum exchange* between phases.

If eq. (5.4) is used than the momentum exchange between phases, for the solidification process occurring in the mushy zone, can be expressed as

$$\boldsymbol{\Gamma}^c_{wk} = \{-\rho_k \boldsymbol{w}_k(\boldsymbol{w}_k - \boldsymbol{w}_i)\cdot\boldsymbol{n}_k\ \delta(\boldsymbol{x},\boldsymbol{x}_i)\} = a_i\{\rho_k \boldsymbol{w}_k(\boldsymbol{w}_i - \boldsymbol{w}_k)\cdot\boldsymbol{n}_k)\}^*_A \tag{9.9}$$

and

$$\boldsymbol{\Gamma}^d_{wk} = \{\boldsymbol{\sigma}_k \cdot \boldsymbol{n}_k\ \delta(\boldsymbol{x},\boldsymbol{x}_i)\} = a_i\{\boldsymbol{\sigma}_k \cdot \boldsymbol{n}_k\}^*_A \tag{9.10}$$

9.2. Closure Problem. Constitutive Relations for Momentum Transport

In order to numerically solve conservation equations constitutive relations between variables are needed, i.e., closure problem should be solved. In the case of the single continuum approach we need to express such terms as $\{\rho\boldsymbol{w}\}, \{\boldsymbol{w}\rho\boldsymbol{w}\}$ and $\{\boldsymbol{\sigma}\}$ as functions of the macroscopic velocity $\{\boldsymbol{w}\}$ and macroscopic pressure $\{p\}$. In case of the multi-continuum approach relations between $\{\boldsymbol{ww}\}_k, \{\boldsymbol{\sigma}\}_k$, $\boldsymbol{\Gamma}^c_{wk}, \boldsymbol{\Gamma}^d_{wk}$ and the intrinsic velocities $\{\boldsymbol{w}\}_k$, intrinsic pressures $\{p\}_k$ and the interface rate of mass flux Γ_{mk} are necessary. In most of the available literature these relations are postulated using respective relations for no solidifying heterogeneous media with the fixed microstructure.

The constitutive relations could be derived if relations between microscopic temperature $\boldsymbol{w}$ in any phase and its bulk value $\{\boldsymbol{w}\}$ or the intrinsic velocities $\{\boldsymbol{w}\}_k$ were known. The problem is more complicated then in case of species or energy transport as the solid phase is usually treated as rigid. In further part of this Section attention will be devoted to the case when solid phase is also stationary. This assumption corresponds primarily to the columnar solidification. At first eq. (9.6) is rewritten in the form

$$\partial_t(\varepsilon_l\rho_l\{\boldsymbol{w}\}_l)+\nabla\cdot(\varepsilon_l\rho_l\{\boldsymbol{ww}\}_l)=\nabla\cdot(\varepsilon_l\{\boldsymbol{\sigma}\}_l)+\varepsilon_l\rho_l\boldsymbol{f}_m+ \\ +\{(\boldsymbol{w}_l-\boldsymbol{w}_i)\rho_l\boldsymbol{w}_l\cdot\nabla\theta_l\}-\{\sigma_l\cdot\nabla\theta_l\} \tag{9.11}$$

Then the microscopic constitutive relation for the liquid phase, eq. (3.21), was multiplied by the structure function θ_l and subsequently ensemble averaged giving

$$\varepsilon_l\{\boldsymbol{\sigma}\}_l=-\varepsilon_l\{p\}_l\boldsymbol{1}+2\mu_l\boldsymbol{e}(\varepsilon_l\{\boldsymbol{w}\}_l\}-\mu_l\{\nabla\theta_l\boldsymbol{w}_l+\boldsymbol{w}_l\nabla\theta_l\} \tag{9.12}$$

where $\boldsymbol{e}(\boldsymbol{w})=(\nabla\boldsymbol{w}+\nabla^{\boldsymbol{T}}\boldsymbol{w})/2$ is the deformation rate tensor.

After introducing eqs (9.12) and (3.21) into eq. (9.11) the following relation was obtained

$$\partial_t(\varepsilon_l\rho_l\{\boldsymbol{w}\}_l)+\nabla\cdot(\varepsilon_l\rho_l\{\boldsymbol{ww}\}_l)=-\nabla(\varepsilon_l\{p\}_l)+\mu_l\nabla^2(\varepsilon_l\{\boldsymbol{w}\}_l)-\mu_l\nabla\cdot\{\nabla\theta_l\boldsymbol{w}_l+\boldsymbol{w}_l\nabla\theta_l\}+ \\ +\varepsilon_l\rho_l\boldsymbol{f}_m+\{(\boldsymbol{w}_l-\boldsymbol{w}_i)\rho_l\boldsymbol{w}_l\cdot\nabla\theta_l\}+\{[p_l\boldsymbol{1}-2\mu_l\boldsymbol{e}(\boldsymbol{w}_l)]\cdot\nabla\theta_l\} \tag{9.13}$$

If the microscopic constitutive relation, eq. (3.21), is substituted directly into the momentum conservation equation, eq. (3.2), valid for the liquid phase then the latter equation takes the form

$$\partial_t(\rho_l\boldsymbol{w}_l)+\nabla\cdot(\rho_l\boldsymbol{w}_l\boldsymbol{w}_l)=-\nabla p_l+\mu_l\nabla^2\boldsymbol{w}_l+\rho_l\boldsymbol{f}_m \tag{9.14}$$

The latter equation was multiplied by the volume fraction of the liquid phase and then eq. (9.13) was subtracted from it. When these manipulations had been carried out, and the eq. (6.6) used, the following expression could be written

$$\varepsilon_l\partial_t(\rho_l\boldsymbol{w}_l')+\varepsilon_l\nabla\cdot(\rho_l\{\boldsymbol{w}\}_l\boldsymbol{w}_l')+\varepsilon_l\nabla\cdot(\rho_l\boldsymbol{w}_l'\{\boldsymbol{w}\}_l)+\varepsilon_l\nabla\cdot(\rho_l\boldsymbol{w}_l'\boldsymbol{w}_l')-\nabla\cdot(\varepsilon_l\rho_l\{\boldsymbol{w}'\boldsymbol{w}'\}_l)= \\ =-\varepsilon_l\nabla p_l'+\varepsilon_l\mu_l\nabla^2(\boldsymbol{w}_l')+\{[-p_l'\boldsymbol{1}+2\mu_l\boldsymbol{e}(\boldsymbol{w}_l')]\cdot\nabla\theta_l\}+\mu_l\nabla\cdot\{\nabla\theta_l\boldsymbol{w}_l'+\boldsymbol{w}_l'\nabla\theta_l\}+ \\ -\{(\boldsymbol{w}_l-\boldsymbol{w}_i)\rho_l\boldsymbol{w}_l'\cdot\nabla\theta_l\} \tag{9.15}$$

where

$$p_l'=p_l-\{p\}_l, \tag{9.16}$$

$$\boldsymbol{w}_l'=\boldsymbol{w}_l-\{\boldsymbol{w}\}_l \tag{9.17}$$

Equation (9.15) was subsequently rewritten in the following form

$$\partial_t(\rho_l\boldsymbol{w}_l')+\rho_l\{\boldsymbol{w}\}_l\cdot\nabla\boldsymbol{w}_l'-\mu_l\nabla^2(\boldsymbol{w}_l')-\boldsymbol{F}_w=0 \tag{9.18}$$

where

$$\boldsymbol{F}_w=-\varepsilon_l^{-1}\{(\{\boldsymbol{w}\}_l-\boldsymbol{w}_i)\rho_l\boldsymbol{w}_l'\cdot\nabla\theta_l\}-\nabla p_l'-\rho_l\boldsymbol{w}_l'\,\nabla\{\boldsymbol{w}\}_l+ \\ +\varepsilon_l^{-1}\{[-p_l'\boldsymbol{1}+2\mu_l\boldsymbol{e}(\boldsymbol{w}_l')]\cdot\nabla\theta_l\}+\varepsilon_l^{-1}\mu_l\nabla\cdot\{\nabla\theta_l\boldsymbol{w}_l'+\boldsymbol{w}_l'\nabla\theta_l\} \tag{9.19}$$

In order to solve eq. (9.18) knowledge of velocity on the external boundaries of the system where liquid is in contact with the surroundings and on the liquid/solid interface is necessary.

In the present case it was assumed that velocity values are known on the external boundaries and that they are independent of the configuration Ω, i.e. equal to the intrinsic mean values. They may be; for example, equal to zero when the solidifying system is in contact with the stationary solid wall. On the liquid/solid interface the normal component of velocity follows from eq. (3.5) and is equal to

$$\boldsymbol{w}_l \cdot \boldsymbol{n}_l = (1 - \rho_s / \rho_l)\boldsymbol{w}_i \cdot \boldsymbol{n}_l + \rho_s / \rho_l \, \boldsymbol{w}_s \cdot \boldsymbol{n}_l \tag{9.20}$$

while the tangential component is equal to

$$\boldsymbol{w}_l \cdot \boldsymbol{t} = \boldsymbol{w}_s \cdot \boldsymbol{t} \tag{9.21}$$

For stationary solid eqs (9.20) and (9.21) reduce to

$$\boldsymbol{w}_l \cdot \boldsymbol{n}_l = (1 - \rho_s / \rho_l)\, \boldsymbol{w}_i \cdot \boldsymbol{n}_l, \qquad \boldsymbol{w}_l \cdot \boldsymbol{t} = 0 \tag{9.22}$$

and the first of the conditions is solely dependent on difference in densities of the phases and on the normal component of the interface velocity. For the particular case of equal densities the normal component of liquid velocity is equal to zero.

The initial condition for the problem was assumed in the following form

$$\boldsymbol{w}_l(t=0, \boldsymbol{x}|\Omega) = \{\boldsymbol{w}(t=0, \boldsymbol{x}\}_l = \boldsymbol{w}_{lo}(\boldsymbol{x}) \tag{9.23}$$

Let the Green function to be solution of the problem:

$$\mu_l \nabla^2 G + \rho_l \{\boldsymbol{w}\}_l \cdot \nabla G + \delta_x(\boldsymbol{x}, \boldsymbol{y})\, \delta_t(t, \tau) = -\rho_l \partial_t G \qquad \text{for } \boldsymbol{x} \in V_l \tag{9.24}$$

$$G(t, \boldsymbol{x}; \tau, \boldsymbol{y}) = 0 \qquad \text{or } \boldsymbol{x} \in A \tag{9.25}$$

$$G(t, \boldsymbol{x}; \tau, \boldsymbol{y}) = 0 \qquad \text{for } t \leq \tau \tag{9.26}$$

With use of the Green function theory the following formal expression for liquid velocity was obtained from eq. (9.18)

$$\boldsymbol{w}'_l = \int_0^t \int_{V_l} G\, \boldsymbol{F}_w \, dV'\, d\tau - \int_0^t \int_{A_i} \mu_l [(\boldsymbol{w}'_l \cdot \boldsymbol{n}_l)\boldsymbol{n}_l + (\boldsymbol{w}'_l \cdot \boldsymbol{t})\boldsymbol{t}] \nabla G \cdot \boldsymbol{n}_l \, dA'\, d\tau \tag{9.27}$$

where the symbol A_i stands for the interface area and the unit vector $\boldsymbol{t}$ lies in plane tangential to the interface.

Substitution of $\boldsymbol{F}_w$ and use of the boundary conditions in eq. (9.27) leads to the equation

$$\begin{aligned} \boldsymbol{w}'_l = \int_o^t \int_{V_l} & G[\nabla p'_l + \rho_l \boldsymbol{w}'_l \cdot \nabla \{\boldsymbol{w}\}_l + \varepsilon_l^{-1} \{((\{\boldsymbol{w}\}_l - \boldsymbol{w}_i)\rho_l \boldsymbol{w}'_l + \\ & + p'\boldsymbol{1}_l - 2\mu_l \boldsymbol{e}(\boldsymbol{w}'_l)) \cdot \nabla \theta_l\} - \varepsilon_l^{-1} \mu_l \nabla \cdot \{\nabla \theta_l \boldsymbol{w}'_l + \boldsymbol{w}'_l \nabla \theta_l\}] dV'\, d\tau + \\ - \int_0^t \int_{A_i} & \mu_l [((1 - \rho_s / \rho_l)\boldsymbol{w}'_i \cdot \boldsymbol{n}_l + \rho_s / \rho_l)\boldsymbol{w}'_s \cdot \boldsymbol{n}_l)\boldsymbol{n}_l + (\boldsymbol{w}'_s \cdot \boldsymbol{t})\boldsymbol{t}] \nabla G \cdot \boldsymbol{n}_l \, dA'\, d\tau \end{aligned} \tag{9.28}$$

The formal solution of eq. (9.28) is sought in the form

$$p_l' = \int_0^t \int_{V_l} \varphi_p \cdot \{\boldsymbol{w}\}_l \, dV' \, d\tau \tag{9.29}$$

$$\boldsymbol{w}_l' = \boldsymbol{S}_w + \int_0^t \int_{V_l} \psi_w \cdot \{\boldsymbol{w}\}_l \, dV' \, d\tau \tag{9.30}$$

where functions $\boldsymbol{S}_w, \varphi_p, \varphi_w$ satisfy the following integro-differential equation

$$\begin{aligned} \boldsymbol{S}_w = \int_o^t \int_{V_l} & G[\rho_l \boldsymbol{S}_w \cdot \nabla\{\boldsymbol{w}\}_l + \varepsilon_l^{-1}\{((\{\boldsymbol{w}\}_l - \boldsymbol{w}_i)\rho_l \boldsymbol{S}_w + \\ & -2\mu_l \boldsymbol{e}(\boldsymbol{S}_w)) \cdot \nabla\theta_l\} - \varepsilon_l^{-1}\mu_l \nabla \cdot \{\nabla\theta_l \boldsymbol{S}_w + \boldsymbol{S}_w \nabla\theta_l\}] \, dV' \, d\tau + \\ - \int_0^t \int_{A_i} & \mu_l [((1-\rho_s/\rho_l)\boldsymbol{w}_i' \cdot \boldsymbol{n}_l + \rho_s/\rho_l)\boldsymbol{w}_s' \cdot \boldsymbol{n}_l)\boldsymbol{n}_l + (\boldsymbol{w}_s' \cdot \boldsymbol{t})\boldsymbol{t}] \nabla G \cdot \boldsymbol{n}_l \, dA' \, d\tau \end{aligned} \tag{9.31}$$

$$\begin{aligned} \psi_w = \int_o^t \int_{V_l} & G[\nabla\varphi_p + \rho_l \psi_w \cdot \nabla\{\boldsymbol{w}\}_l + \varepsilon_l^{-1}\{((\{\boldsymbol{w}\}_l - \boldsymbol{w}_i)\rho_l \psi_w + \\ & + \varphi_p \boldsymbol{1} - 2\mu_l \boldsymbol{e}(\psi_w)) \cdot \nabla\theta_l\} - \varepsilon_l^{-1}\mu_l \nabla \cdot \{\nabla\theta_l \psi_w + \psi_w \nabla\theta_l\}] \, dV' \, d\tau \end{aligned} \tag{9.32}$$

The macroscopic momentum equation, i.e. eq. (9.13), can then be rewritten, using (6.6), and presented in the following form

$$\rho_l \partial_t \{\boldsymbol{w}\}_l + \rho_l \{\boldsymbol{w}\}_l \cdot \nabla\{\boldsymbol{w}\}_l = -\nabla\{p\}_l + \mu_l \nabla^2 \{\boldsymbol{w}\}_l + \rho_l \boldsymbol{f}_m + \boldsymbol{F}_{mi} \tag{9.33}$$

where

$$\boldsymbol{F}_{mi} = \varepsilon_l^{-1}\left[-\mu_l \nabla \cdot \{\nabla\theta_l \boldsymbol{w}_l' + \boldsymbol{w}_l' \nabla\theta_l\} + \{[(\{\boldsymbol{w}\}_l - \boldsymbol{w}_i)\rho_l \boldsymbol{w}_l' + p_l' \boldsymbol{1} - 2\mu_l \boldsymbol{e}(\boldsymbol{w}_l')] \cdot \nabla\theta_l\}\right] \tag{9.34}$$

and the non-linear terms with averages of products of the velocity deviations were dropped. The term $\boldsymbol{F}_{mi}$ corresponds to the additional resistance, which the developing solid microstructure exerts on the liquid flow. Using eqs (9.29) and (9.30) this term can be expressed as function of gradient of the intrinsic velocity $\{\boldsymbol{w}\}_l$

$$\boldsymbol{F}_{mi} = \boldsymbol{v}_{Sw} + \int_0^t \int_{V_l} \boldsymbol{v}_{wef} \cdot \{\boldsymbol{w}\}_l \, dV' \, d\tau \tag{9.35}$$

where

$$\boldsymbol{v}_{Sw} = \varepsilon_l^{-1}\left[-\mu_l \nabla \cdot \{\nabla\theta_l \boldsymbol{S}_w + \boldsymbol{S}_w \nabla\theta_l\} + \{[(\{\boldsymbol{w}\}_l - \boldsymbol{w}_i)\rho_l \boldsymbol{S}_w - 2\mu_l \boldsymbol{e}(\boldsymbol{S}_w)] \cdot \nabla\theta_l\}\right] \tag{9.36}$$

$$\boldsymbol{v}_{wef} = \varepsilon_l^{-1}\left[-\mu_l \nabla \cdot \{\nabla\theta_l \psi_w + \psi_w \nabla\theta_l\} + \{[(\{\boldsymbol{w}\}_l - \boldsymbol{w}_i)\rho_l \psi_w + \varphi_p \boldsymbol{1} - 2\mu_l \boldsymbol{e}(\psi_w)] \cdot \nabla\theta_l\}\right] \tag{9.37}$$

10. Modelling of Solidification Process in Mushy Zone for Slowly Varying Macroscopic Fields

Microstructure of the mushy zone, and associated with it variation in local properties of the mushy zone treated as a heterogeneous medium, can be described by many characteristic length-scales. To these length-scales belong transverse dimensions of dendrites, diameters of the equiaxed grains, radii of dendrite tips, spacing of the primary and secondary dendrite arms, characteristic lengths associated with variation of the correlation functions (see Furmański, 1997), etc. If it happens that the greatest micro-length ℓ (e.g. the primary dendrite spacing) is smaller than the smallest of the characteristic macro-lengths describing variations in the macroscopic velocity, temperature or species concentration then significant simplifications in the constitutive relations can be attained. These simplifications can be performed by expansion of the pertinent microstructure functions, like $\varphi_T, \varphi_C, \varphi_w, \varphi_p, \ldots; \psi_T, \psi_C, \ldots; \chi_T, \ldots$ in an infinite series in growing powers of the micro-dimension ℓ, i.e.,

- for all φ functions

$$\varphi(t,\boldsymbol{x};\tau,\boldsymbol{y}|\Omega)/\ell = \varphi_0(t,\boldsymbol{x}|\Omega)\,\delta(t,\tau)\,\delta(\boldsymbol{x},\boldsymbol{y}) + \ell\,\varphi_1(t,\boldsymbol{x}|\Omega)\,\delta(t,\tau)\,\nabla\delta(\boldsymbol{x},\boldsymbol{y}) + O(l^2) \tag{10.1}$$

- for all χ functions

$$\chi(t,\boldsymbol{x};\tau,\boldsymbol{y}|\Omega)/\ell = \chi_0(t,\boldsymbol{x}|\Omega)\,\delta(t,\tau)\,\delta(\boldsymbol{x},\boldsymbol{y}) + \ell\,\chi_1(t,\boldsymbol{x}|\Omega)\,\delta(t,\tau)\,\nabla\delta(\boldsymbol{x},\boldsymbol{y}) + O(l^2) \tag{10.2}$$

- for all ψ functions

$$\psi(t,\boldsymbol{x};\tau,\boldsymbol{y}|\Omega)/\ell^2 = \psi_0(t,\boldsymbol{x}|\Omega)\,\delta(t,\tau)\,\delta(\boldsymbol{x},\boldsymbol{y}) + \ell\,\psi_1(t,\boldsymbol{x}|\Omega)\,\delta(t,\tau)\,\nabla\delta(\boldsymbol{x},\boldsymbol{y}) + O(l^2) \tag{10.3}$$

This kind of expansions, when introduced into definition of the effective parameters and in the integro-differential equations presented in Sections 7, 8 and 9, allow to carry out order of magnitude analysis between scales characterising microstructure of the mushy zone and scales characterising energy, momentum and species transport processes in the medium on the macroscopic scale. Since the microstructure functions describe how the microstructure helps the fictitious macroscopic continuum at the point $\boldsymbol{x}$ to react for a disturbance applied at the point $\boldsymbol{y}$, these expansions can be understood as an approximation of the microstructure functions at the neighbourhood of the point $\boldsymbol{x}$ by their spatial and time derivatives taken at the considered point. The expansions discussed may be also interpreted as expansions in series of a small parameter $\in = \ell / L$ where L is the smallest of characteristic macro-lengths associated with space or time variation in the macroscopic fields of variables. Retaining the first r.h.s. term in the above expansions corresponds to an assumption that gradients (and time derivatives) of the bulk concentration are slowly spatially (or temporally) varying, i.e., scales associated with these variations are much greater than the micro-dimension ℓ. In fact this assumption is equivalent to assumption of the well separation of a spectrum of length-scales describing the microstructure of the mushy zone and a spectrum of length-scales describing variation in the macroscopic velocity, concentration and temperature fields.

When expansions eqs (10.1)-(10.3), are introduced into eqs (7.23), (7.24), (9.29) and (9.30) the following relations between microscopic and the macroscopic fields are obtained (see Furmański, 1997, 2000)

$$T = \{T\} + \ell S_T + \ell\left[\varphi_{T0}\cdot\nabla\{T\} + \sum_j \varphi_{CTj0}\cdot\nabla\{\widetilde{C}_j\} + \chi_{T0}[\{T\} - T_e]\delta(\boldsymbol{x},\boldsymbol{x}_A)\right] + O(\ell^2) \tag{10.4}$$

$$\widetilde{C}_j = \{\widetilde{C}_j\} + \ell S_{cj} + \ell\left[\varphi_{Cj0}\cdot\nabla\{\widetilde{C}_j\} + \varphi_{TCj0}\cdot\nabla\{T\} + \chi_{TCj0}[\{T\} - T_e]\delta(\boldsymbol{x},\boldsymbol{x}_A)\right] + O(\ell^2) \tag{10.5}$$

$$p_l = \{p\}_l + \ell\varphi_{p0}\cdot\{\boldsymbol{w}\}_l + O(\ell^2) \tag{10.6}$$

$$\boldsymbol{w}_l = \{\boldsymbol{w}\}_l + \ell^2 \boldsymbol{S}_w + \ell^2\varphi_{w0}\cdot\{\boldsymbol{w}\}_l + O(\ell^2) \tag{10.7}$$

Note that not all microstructure functions appear in the above relations.

The expansions, given in eqs (10.1) and (10.2), were also introduced into the respective integro-differential equations for the microstructure functions (shown in Section 7, 8 and 9). After order of magnitude analysis (see for example Furmański, 1997) the equations were simplified to the form

$$\varphi_{T0} = -\int_V \nabla G\cdot[\lambda'(\boldsymbol{1}+\nabla\varphi_{T0}) - \{\lambda'(\boldsymbol{1}+\nabla\varphi_{T0})\}]\,dV' \tag{10.8}$$

$$\varphi_{CTj0} = 0 \tag{10.9}$$

$$\chi_{T0} = -\int_V \nabla G\cdot[\lambda'\nabla\chi_{T0}) - \{\lambda'\nabla\chi_{T0})\}]\,dV' + \int_{A_c}[\alpha - \{a\}]\,\delta\, dA \tag{10.10}$$

$$\begin{aligned} S_{T0} = &-\int_0^t\int_V \nabla G\cdot[(\lambda'\nabla S_{T0}) - \{\lambda'\nabla S_{T0})\}) - (\boldsymbol{w}\rho^o h^o - \{\boldsymbol{w}\rho^o h^o\})]\,dV'\,d\tau + \\ &- \int_0^t\int_V G\ [(\boldsymbol{w}_i\cdot\nabla\theta_s) - \{(\boldsymbol{w}_i\cdot\nabla\theta_s)\}]\Delta(\rho^o h^o)\,dV' d\tau \end{aligned} \tag{10.11}$$

$$\varphi_{Cj0} = -\int_V \nabla G\cdot[D_j(\boldsymbol{1}+\nabla\varphi_{Cj0}) - \{\lambda'(\boldsymbol{1}+\nabla\varphi_{Cj0})\}]\,dV' \tag{10.12}$$

$$\begin{aligned} \varphi_{TCj0} = &-\int_V \nabla G\cdot[D_j\nabla\varphi_{TCj0} - \{D_j\nabla\varphi_{TCj0}\}]\,dV' + \\ &+ \int_V \nabla G\cdot[M_j(\boldsymbol{1}+\nabla\varphi_{T0}) - \{M_j(\boldsymbol{1}+\nabla\varphi_{T0})\}]\,dV' \end{aligned} \tag{10.13}$$

$$\begin{aligned} \chi_{TCj0} = &-\int_V \nabla G\cdot[D_j\nabla\chi_{TCj0} - \{D_j\nabla\chi_{TCj0}\}]\,dV' + \\ &+ \int_V \nabla G\cdot[M_j\nabla\chi_{T0} - \{M_j\nabla\chi_{T0}\}]\,dV' \end{aligned} \tag{10.14}$$

$$S_{Cj0} = -\int_0^t \int_V \nabla G \cdot [(\lambda' \nabla S_{Cj0}) - \{\lambda' \nabla S_{Cj0})\}) + (M_j \nabla S_{T0} - \{M_j \nabla S_{T0}\}) + \\ + (\boldsymbol{w}\tilde{\rho}^o \tilde{C}_j^o - \{\boldsymbol{w}\tilde{\rho}^o \tilde{C}_j^o\})] \, dV' \, d\tau + \tag{10.15}$$

$$-\int_0^t \int_V G \; [(\boldsymbol{w}_i \cdot \nabla \theta_s) - \{(\boldsymbol{w}_i \cdot \nabla \theta_s)\}] \Delta(\tilde{\rho}^o \tilde{C}_j^o) \, dV' d\tau$$

$$S_{w0} = 0 \tag{10.16}$$

$$\psi_{w0} = \int_{V_l} G[\nabla \varphi_{p0} + \varepsilon_l^{-1} \{(\varphi_{p0} \boldsymbol{1} - 2\mu_l \boldsymbol{e}(\psi_{w0})) \cdot \nabla \theta_l\} + \\ - \varepsilon_l^{-1} \mu_l \nabla \cdot \{\nabla \theta_l \psi_{w0} + \psi_{w0} \nabla \theta_l\}] \, dV' \tag{10.17}$$

Similarly introducing the expansions of eqs (10.1)-(10.3) into the nonlocal constitutive relations of eqs (7.31), (7.32), (8.31), (8.32) and (9.35) the following local forms of the macroscopic constitution relations to order $O(\ell)$ are found

$$\{\boldsymbol{w}\rho \boldsymbol{h} + \boldsymbol{q}\} = \{\boldsymbol{w}\rho^o \boldsymbol{h}^o\} - \boldsymbol{\lambda}_{Tef0} \cdot \nabla \{T\} + \\ - \sum_j \boldsymbol{\lambda}_{Cefj0} \cdot \nabla \{\tilde{C}_j\}) + \alpha_{Tef0}[\{T\} - T_e] \delta(\boldsymbol{x}, \boldsymbol{x}_A) \tag{10.18}$$

$$\{\boldsymbol{w}\tilde{\rho}_j \tilde{C}_j + \boldsymbol{j}_j\} = \{\boldsymbol{w}\tilde{\rho}_j^o \tilde{C}_j^o\} - \boldsymbol{D}_{Cefj0} \cdot \nabla \{\tilde{C}_j\} + \\ - \boldsymbol{D}_{Tefj0} \cdot \nabla \{T\} + \alpha_{TCefj0}[\{T\} - T_e] \delta(\boldsymbol{x}, \boldsymbol{x}_A) \tag{10.19}$$

$$\boldsymbol{F}_{mi} = \mu_l \, \boldsymbol{K}^{-1} \{\boldsymbol{w}\}_l \tag{10.20}$$

where: $$\boldsymbol{K}^{-1} = \boldsymbol{v}_{wef0} \tag{10.20'}$$

$$\{\rho h\} = \{\rho^o h^o\} \tag{10.21}$$

$$\{\tilde{\rho}_j \tilde{C}_j\} = \{\tilde{\rho}_j^o \tilde{C}_j^o\} \tag{10.22}$$

In these relations the superscript „o" denotes evaluation of the respective value at temperature $\{T\}$ and concentrations $\{\tilde{C}_j\}$. Note that the last terms on the r.h.s. of eqs (10.18) and (10.19) appear only in locations where the solidified material is in contact with the mould. The nonlocal relations are thus reduced to the local form, which is valid for the macroscopic field of variables weakly varying over distances comparable with the micro-dimension ℓ. These relations are much simpler than the nonlocal ones. They resemble respective relations for homogeneous media but contain additional terms: Dufour coefficient $\boldsymbol{\lambda}_{Cefj0}$ in eq. (10.18) and Soret coefficient $\boldsymbol{D}_{Tefj0}$ in eq. (10.19). These terms are a consequence of coupling between temperature and concentration fields on the microscopic scale. Equation (10.20) is identical to Darcy term in modelling of fluid flow in porous media. All effective properties appearing in the local form of the constitutive relations for the solidifying system are defined by the following formulae:

$$\boldsymbol{\lambda}_{Tef0} = \{\lambda(\boldsymbol{1} + \nabla\varphi_{T0})\} - \ell\,\{\boldsymbol{w}[\rho^o c^o \varphi_{T0}\} + \sum_j \rho h_j^* \varphi_{TCj0}]\} \tag{10.23}$$

$$\boldsymbol{\lambda}_{Cefj0} = -\ell\,\{\boldsymbol{w}\rho h_j^* \varphi_{Cj0}\} \tag{10.24}$$

$$\boldsymbol{D}_{Cefj0} = \{\widetilde{D}_j(\boldsymbol{1} + \nabla\varphi_{Cj0})\} - \{\boldsymbol{w}\widetilde{\rho}_j \varphi_{Cj0}\} \tag{10.25}$$

$$\boldsymbol{D}_{Tefj0} = \{\widetilde{D}_j \nabla\varphi_{TCj0}\} - \{\widetilde{M}_j(\boldsymbol{1} + \nabla\varphi_{T0})\} - \ell\,\{\boldsymbol{w}\partial_T \rho^o \widetilde{\rho}_j^o \widetilde{C}_j^o \varphi_{T0}\} \tag{10.26}$$

$$\alpha_{Tef0} = \{\lambda\nabla\chi_{T0}\} \tag{10.27}$$

$$\alpha_{TCefj0} = \{\widetilde{D}_j \nabla\chi_{TCj0})\} - \{\widetilde{M}_j(\boldsymbol{1} + \nabla\varphi_{T0})\} \tag{10.28}$$

In some terms describing the macroscopic transport property terms containing liquid velocity $\boldsymbol{w}$ appear. These terms are known as dispersion terms. They indicate that some part of the advective transport that occurs on the microscopic scale is transformed into diffusive transport on the macroscopic scale. This transformation is due to tortuous path that liquid follows among evolving dendrites.

Coefficient α_{Tef0} is known as the thermal contact conductance (or the interface heat transfer coefficient) and is one of the major controlling factors during cooling of the solidifying system from the wall. This macroscopic coefficient is responsible for drop in the macroscopic temperature at the solidifying system/mould interface. The coefficient α_{TCefj0} results from coupling between temperature and concentration fields and shows that jth species flow between the solidifying system and the bounding wall can be caused by drop of the macroscopic temperature at this interface.

All macroscopic properties of the solidifying medium can be found from knowledge of the respective microstructure function. It should be noted that differential form of these equations, allowing to find the microstructure functions and corresponding to eqs (10.8) – (10.19), can easily be written (Furmański, 1997). For example the differential form of eq. (10.17) can be presented as

$$\rho_l \{\boldsymbol{w}_l\} \cdot \nabla\boldsymbol{\psi}_{w0} = -\nabla\varphi_{p0} + \mu_l \nabla^2 \boldsymbol{\psi}_{w0} \text{ in } V_l \tag{10.29}$$

$$\nabla \cdot \boldsymbol{\psi}_{w0} = 0 \qquad \text{in } V_l \tag{10.30}$$

$$\boldsymbol{\psi}_{w0} = 0 \qquad \text{on } A_i \tag{10.31}$$

and it is similar to the classical Stokes problem in the fluid flow.

It is worth to say at the end of this section that it may be proved, see Furmański (1994), that equations and the effective properties for the two-phase (double continuum) macroscopic model (see Section 4.2) can be obtained by retaining the second term (of order ℓ) in expansion given by eqs (10.1)-(10.3). According to the classification introduced by Kunin (1984) the two-phase macroscopic model describes weakly nonlocal phenomena.

Local thermodynamic equilibrium in the mushy zone of binary solidifying systems

One of the problems raised in analysis of heterogeneous, solidifying systems is existence of thermodynamic equilibrium within the mushy zone. In order to discuss the problem in more detail let us multiply eqs (10.4) and (10.5) by the respective structure function θ_k and calculate intrinsic averages (see Section 5.2). This leads, after noting eq. (10.9), to the following formulae

$$\{T\}_k = \{T\} + \ell\{S_T\}_k + \ell[\{\varphi_{T0}\}_k \cdot \nabla\{T\}] \tag{10.32}$$

$$\{\widetilde{C}_j\}_k = \{\widetilde{C}_j\} + \ell\{S_{cj}\}_k + \ell[\{\varphi_{Cj0}\}_k \cdot \nabla\{\widetilde{C}_j\} + \{\varphi_{TCj0}\}_k \cdot \nabla\{T\} \tag{10.33}$$

The local thermodynamic equilibrium, at the macroscopic scale, exists if there is no energy and species net exchange between the phases. Thus, in the case of thermal equilibrium the intrinsic liquid and solid temperatures are equal to the bulk temperature $\{T\}_s = \{T\}_l = \{T\}$ while in the case of chemical equilibrium the modified intrinsic liquid and solid concentrations are equal to the bulk concentration, i.e., $\{\widetilde{C}_j\}_s = \{\widetilde{C}_j\}_l = \{\widetilde{C}_j\}$. The local thermodynamic equilibrium on the macroscopic scale should be properly understood. No net exchange of energy and species between phases does not mean that no flow of heat or species is present on the microscopic scale. It only denotes that inflow and outflow balance each other. For the thermodynamic equilibrium to be held on the macroscopic scale, see eqs (10.32) and (10.33), the characteristic micro-dimension ℓ should be very small or all intrinsic averages: $\{\varphi_{T0}\}_k, \{\varphi_{Cj0}\}_k, \{\varphi_{TCj0}\}_k$ should be equal to zero. The latter happens if symmetrical (from the statistical point of view) distribution of phases in the mushy zone is present and at a distance from boundaries of the mushy zone. Moreover, the averages $\{S_T\}_k, \{S_{Cj}\}_k$ should also be zero. This happens, see eqs (10.11) and (10.15), if velocity of liquid flow $\boldsymbol{w}_l$ in the mushy zone and the progress of solidification, i.e. interface velocity $\boldsymbol{w}_i$, is very low. If all above conditions are met then the approximate microscopic temperature and species concentrations distributions, given by eqs (10.4) and (10.5), correspond to local thermodynamic equilibrium on the macroscopic scale.

It is sometimes convenient to represent relations of eqs (10.19) and (10.22) directly as function of species concentrations C_j instead of $\widetilde{C}_j$. For a binary mixture it can be done by introducing the so-called equilibrium function P_{eq} (see Furmański, 2000). Using condition for chemical equilibrium this function can be proven to have the form

$$P_{eq}(t, \boldsymbol{x}|\Omega) = \theta_s(t, \boldsymbol{x}|\Omega)\,\kappa_p/(\varepsilon_l + \kappa_p\,\varepsilon_s) + \theta_l(t, \boldsymbol{x}|\Omega)\,/(\varepsilon_l + \kappa_p\,\varepsilon_s) \tag{10.34}$$

The constitutive relations, eqs (10.19) and (10.22), take then the following form in terms of the solute concentration

$$\{\boldsymbol{w}\widetilde{\rho}\widetilde{C}+\boldsymbol{j}\}=\{\boldsymbol{w}\}(\rho P_{eq})_L\{C\}-\boldsymbol{D}_{ef}\cdot\nabla\{C\} \tag{10.35}$$

$$\{\widetilde{\rho}\widetilde{C}\}=\{\rho P_{eq}\}\{C\} \tag{10.36}$$

where

$$\boldsymbol{D}_{Cef}=\{DP_{eq}(\boldsymbol{1}+\nabla\varphi_{C0})\}-\{\boldsymbol{w}(\rho/M)\varphi_{C0}\} \tag{10.37}$$

From the chemical equilibrium conditions, valid on the macroscopic scale, (see Furmański, 2000) it follows that that the intrinsic concentrations for the solid and liquid phase can expressed in relation to the bulk concentration by the formulae

$$\{C\}_s=\frac{\kappa_p}{(\varepsilon_l+\kappa_p\,\varepsilon_s)}\{C\},\qquad \{C\}_l=\frac{1}{(\varepsilon_l+\kappa_p\,\varepsilon_l)}\{C\} \tag{10.38}$$

Porosity of the mushy zone in the case of thermodynamic equilibrium and the binary systems

In order to close the set of equations for the case of the thermodynamic equilibrium in the binary mixture, it is necessary to derive expression for the volume fraction of the liquid phase ε_l. This can be done in the following way. Relations of eqs (5.10) and (10.38) relate the intrinsic phase concentrations to the bulk concentration. Elimination of $\{C\}_l$, $\{C\}_s$ between these two equations leads to the cited below expression for volume fractions of the phases in the mushy region

$$\varepsilon_l(t,\boldsymbol{x})=1-\varepsilon_s(t,\boldsymbol{x})=[\{C\}-f_{SL}^{-1}(\{T\})]/[f_{LL}^{-1}(\{T\})-f_{SL}^{-1}(\{T\})=\varepsilon_l(\{T\},\{C\}) \tag{10.39}$$

Here f_{SL}^{-1} and f_{LL}^{-1} stand for the inverse functions describing solidus and liquidus line (Fig.13), respectively, and $\kappa_p=f_{SL}/f_{LL}$.

The bulk enthalpy of the mushy zone in the case of thermodynamic equilibrium and the binary system

The bulk volumetric enthalpy, eq. (10.21), is related to the volume fractions and the phase enthalpies by the formula (see Section 5.2)

$$\{\rho^o h^o\}=\varepsilon_s(t,\boldsymbol{x})\,\rho_s h_s(\{T\},\{C\})+\varepsilon_l(t,\boldsymbol{x})\,\rho_l h_l(\{T\},\{C\}) \tag{10.40}$$

The phase enthalpies, as potentials, can be evaluated, in relation to a certain reference state $(T_r,0)$ of temperature and concentration, in a different way. In the case of the liquid enthalpy $(\rho^o h^o)_l=\rho_l h_l(\{T\},\{C\})$ it is advisable not to cross the two-phase (mushy) region when carrying out calculations as enthalpy in this region depends on the composition and the liquid fraction is not *a priori* known. Instead it is convenient to follow the way presented in Section 3 (see Fig.13).

According to eq. (10.39), the liquid and solid volume fractions are dependent solely on the bulk temperature and concentration and when introduced, together with eqs (3.26) and (3.28), into eq. (10.40) allow to express the bulk volumetric enthalpy only as a function of $\{T\}$ and $\{C\}$. An (Swaminathan and Voller, alternative to this analytical procedure is to adopt an empirical relation between temperature and concentration for the binary mixture in question 1992).

11. Effective properties and other macroscopic coefficients. Their influence the solidification in binary alloys

11.1. Selected effective properties of the mushy zone

At present the effective properties are not obtained from definitions of eqs (10.20), (10.23) or (10.26). Instead certain, approximate formulae for estimating of their values were proposed in literature. For the effective thermal conductivity and species diffusivity the mixture formulae given by the expressions

$$\lambda_{Tef} = \lambda_s \varepsilon_s + \lambda_l \varepsilon_l \tag{11.1}$$

$$\widetilde{D}_{cef} = \rho_s D_s \varepsilon_s + \rho_l D_l \varepsilon_l \tag{11.2}$$

are often used. Some other formulae used for the effective thermal conductivity, species diffusivity of the mushy zone were gathered in Table 1.

Table 1. Effective thermal conductivity and species diffusivity of the mushy zone.

$\lambda''_{Tef} = \lambda_s \varepsilon_s + \lambda_l \varepsilon_l$ $\lambda^{\perp}_{Tef} = [\varepsilon_s / \lambda_s + \varepsilon_l / \lambda_l]^{-1}$	Parallel to columnar dendrites growth Perpendicular to columnar dendrites growth	Banaszek and Furmański, 2000
$\lambda_{Tef} = \sqrt{\lambda''_{Tef} \lambda^{\perp}_{Tef\,Tef}}$	Isotropic mushy zone	Banaszek and Furmański, 2000
$\lambda_{Tef} = \lambda_l \dfrac{1 + \varepsilon_s^{2/3} (\lambda_s / \lambda_l - 1)}{1 + (\varepsilon_s^{2/3} - \varepsilon_s)(\lambda_s / \lambda_l - 1)}$	Isotropic mushy zone	Simpson et al., 1998
$\widetilde{D}_{cef} = \rho_s D_s \varepsilon_s + \rho_l D_l \varepsilon_l / \kappa_p$	Isotropic mushy zone	Voller, Brent and Prakash, 2000
$D_{cef} = D_s + (D_l - D_s) \dfrac{\varepsilon_l}{\varepsilon_l + \kappa_p \varepsilon_s}$	Isotropic mushy zone	Beckerman et al., 1999

More refined models were proposed for the permeability of the mushy zone. The generalised expression for the permeability was given by Ganesan et al. (1992)

$$K'' = \frac{f(\varepsilon_l / \varepsilon_s)}{a_i^2} \tag{11.3}$$

which is claimed to be valid for the liquid volume fraction in the range: $0.16 < \varepsilon_l < 0.49$. The symbol a_i denotes here liquid/solid interface area per unit volume of the mushy zone (specific interface area).

Table 2. Effective viscosity and permeability of the mushy zone.

$\mu_{ef} = m\lvert(\boldsymbol{e}:\boldsymbol{e})/2\rvert^{n-1}$ where: $m = \exp(9.783\varepsilon_s + 1.435)$ $n = 0.105 + 0.41\varepsilon_s$ valid for $\varepsilon_s < 0.3$	Isotropic mushy zone	Mat and Ilegbushi, 2002
$\mu_{ef} = \mu_l(1-\varepsilon_s)^{-m}$ m depends on the mean shape and orientation of suspended particles	Isotropic mushy zone	Sältzer and Schultz, 1983
$\mu_{ef} = \mu_l(1 - \frac{\varepsilon_s}{\varepsilon_{s,\max}})^{-2.5\varepsilon_{s,\max}}$	Isotropic mushy zone	Ganesan et al., 1992
$K'' = \frac{m\varepsilon_l^3}{(1-\varepsilon_l)^2}$ $K^{\perp} = \frac{m\varepsilon_l^3}{(1-\varepsilon_l)^{3/4}}$	Parallel to columnar dendrites growth Perpendicular to columnar dendrites growth	Poirier et al., 1991
$K = K_o\left[\frac{\varepsilon_l}{\varepsilon_{lo}}\right]^n$ subscript o -values before solidification, $n = 1 - 14$	Three phase medium. One of them not undergoing phase transformation	Matsumoto et al., 1993
$K^{\perp} = K_o \frac{\varepsilon_l}{(1-\varepsilon_l)^{1/2}}$	Perpendicular to columnar mushy zone	Naterer and Schneider, 1995
$K = 1.6 \cdot 10^{-10}\varepsilon_l^{3.3}$ $0.27 < \varepsilon_l < 0.48$	Isotropic mushy zone	Murakami and Okamoto, 1984

The specific interface area can be estimated either from expression

$$a_i = \varepsilon_s \varepsilon_l \tag{11.4}$$

or from relation

$$a_i = \frac{4\varepsilon_s}{\Lambda} \tag{11.5}$$

where Λ denotes the mean dendrite arms spacing.

The most popular expression used in modelling of the permeability of the mushy zone is the Carman-Kozeny formula originally used for a bed of uniformly sized spheres

$$K = \frac{\varepsilon_l^{\ 3}}{180(1-\varepsilon_l)^2}\Lambda^2 \tag{11.6}$$

Some other formulae used for the effective viscosity and permeability of the mushy zone were given in Table 2.

11.2. Thermal contact conductance at solidified material/mould interface

At the solid/mould interface, due to roughness of the mould and due evolution of gases (see Section 2.5) heat flux is varying along the interface. When the averaging procedure is applied it leads to the macroscopic boundary condition at this interface in the form given by eq. (10.18)

$$\{\boldsymbol{q}\} = \alpha_{Tef0}[\{T\} - T_e] \tag{11.7}$$

The reciprocal of the thermal contact conductance known as the thermal contact resistance leads to observable temperature drop at the solid/mould interface. The thermal contact conductance for solidification processes was investigated by Loulou et al. (1999), O'Mahoney and Browne (2000), Browne and O'Mahoney (2001), Wang and Qiu (2002). It was concluded that the higher roughness of the mould, latent heat of solidification, surface tension and the contact angle the smaller the decrease in the thermal contact conductance. High liquid superheat, improving wettability, and the thermal conductivity of the solidifying material increase the thermal contact conductance. It was also found that the thermal contact resistance at the solid/mould interface varies in time (see Fig. 19a). This variation is connected with microscopic processes occurring at the interface, which were described in detail in Section 2.5. Note that time variation of the thermal contact resistance does not last longer than several seconds and then assumes quasi-stationary values. Note also that it may be assisted by large drops of temperature at the solid/mould interface (see Fig. 19b).

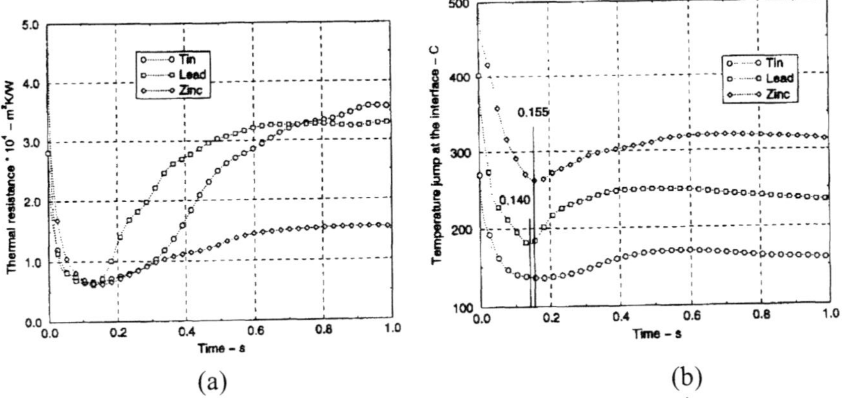

Figure 19. Thermal contact resistance (a) and temperature jump at the solidifying medium/mould interface (b) – (after Lolou et al., 1999).

11.3. Examples of application of the macroscopic approach

Equilibrium solidification. The effective properties in general are anisotropic. Influence of directionality on the effective properties of the mushy zone was studied by Banaszek and Furmański, 2001, for the case of solidification of water-ammonium chloride solution in a square cavity with upper and bottom walls thermally insulated. The other walls were kept at temperature exceeding liquidus value for the assumed ammonium chloride concentration.

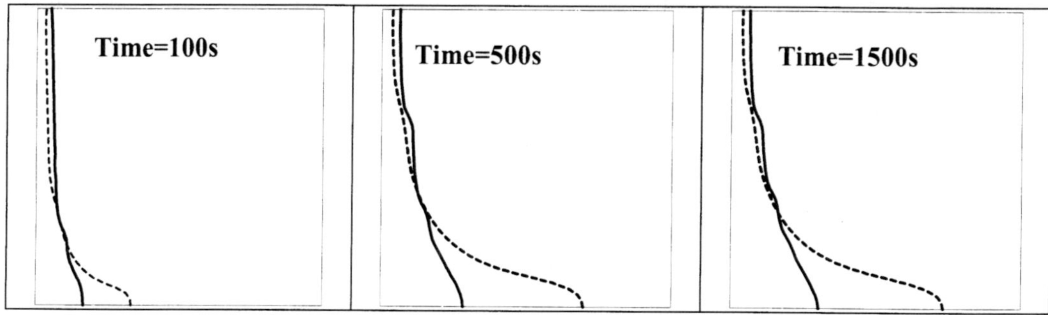

Figure 20. Temporal shape of the mushy zone for isotropic (dashed line) and anisotropic permeability for the case when only mushy and liquid phase can be formed (Banaszek and Furmański, 2000).

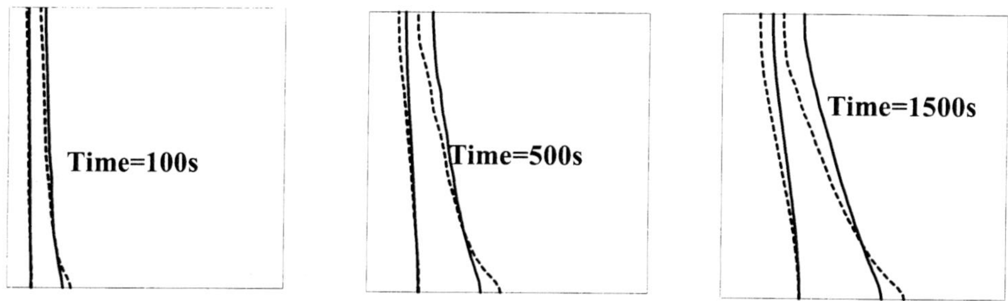

Figure 21. Temporal shape of the mushy zone for isotropic (dashed line) and anisotropic permeability for the case when solid, mushy and liquid phases can be formed (Banaszek and Furmański, 2000).

The process of solidification was initialised by drop of temperature of the left wall below the liquidus value but greater than the eutectic temperature. Some of the results are presented in Figures 20-23. It seems that variation in directional permeability exerts greater influence on the shape of the mushy zone than the respective variation in effective thermal conductivity for the investigated case. In all presented results it was assumed that the solidification process is sufficiently slow for thermodynamic equilibrium at the liquid/solid interface to be valid. Equal densities of the phases and stationarity of the solid phase were assumed. The solidification process has been described by the following set of equations following from eqs (3.23), (3.26), (3.28'), (6.5), (8.5), (9.33), (10.18), (10.20), and (10.21):

$$\nabla \cdot (\varepsilon_l \{\boldsymbol{w}\}_l) = 0 \tag{11.8}$$

$$\rho_l \, \partial_t \{\boldsymbol{w}\}_l + \rho_l \{\boldsymbol{w}\}_l \cdot \nabla \{\boldsymbol{w}\}_l = -\{p\}_l + \mu_l \nabla^2 \{\boldsymbol{w}\}_l + \rho_l \boldsymbol{f_m} + \boldsymbol{F_{mi}} \tag{11.9}$$

$$\partial_t \{\rho h\} + \nabla \cdot \{\boldsymbol{w} \rho h + \boldsymbol{q}\} = 0 \tag{11.10}$$

where

$$\boldsymbol{F_{mi}} = \mu_l \, \boldsymbol{K}^{-1} \{\boldsymbol{w}\}_l \tag{11.11}$$

$$\{\boldsymbol{w} \, \rho h + \boldsymbol{q}\} = \varepsilon_l \{\boldsymbol{w}\}_l \, \rho h_l^o - \boldsymbol{\lambda}_{ef} (\{\boldsymbol{w}\}_l) \cdot \nabla \{T\} \tag{11.12}$$

$$\boldsymbol{f_m} = g \, \rho_{Tl} \, (T - T_r) \tag{11.13}$$

The averaged volumetric enthalpy was defined as

$$\{\rho h\} = \varepsilon_s \, \rho h_s^o + \varepsilon_l \, \rho h_l^o \tag{11.14}$$

where

$$\rho h_s^o = \rho c_s \{T\}, \qquad \rho h_l^o = \rho c_s \, T_s + L_f (T_s) + \rho c_l \, (\{T\} - T_s) \tag{11.15}$$

and the symbol T_s denotes solidus temperature.

Ideal thermal contact between solidified material and the mould, i.e. $\alpha_{Tef0} \to \infty$ - see eq. (11.7), was additionally assumed.

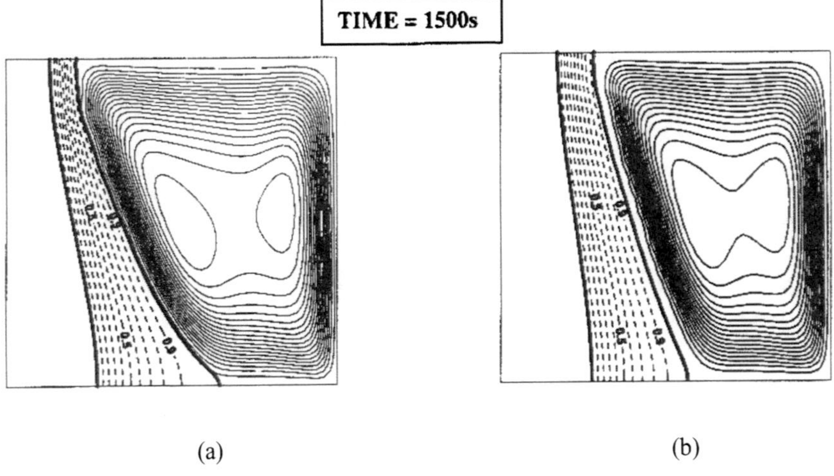

Figure 22. Flow pattern in the liquid phase and liquid volume fraction distribution in the mushy zone for isotropic (a) and anisotropic (b) permeability (Banaszek and Furmański, 2000).

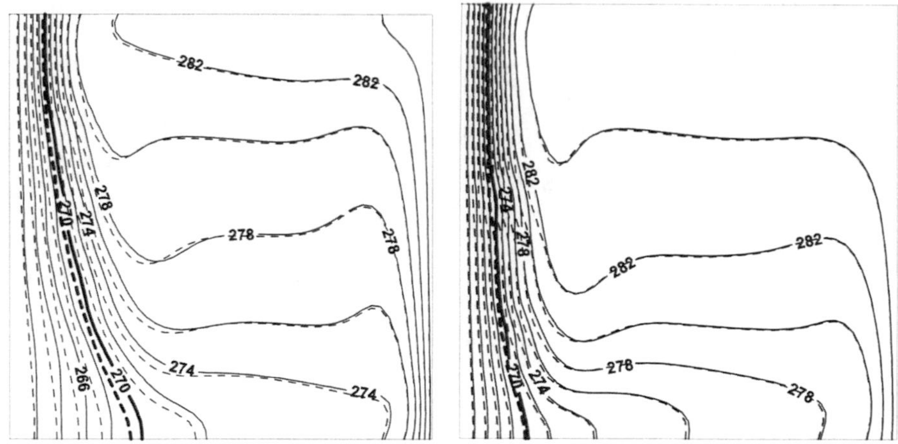

Figure 23. Temperature field for anisotropic (solid line) and isotropic (dashed line) models of thermal conductivity in the mushy zone for the case when only mushy and liquid phase can be formed (Banaszek and Furmański, 2000).

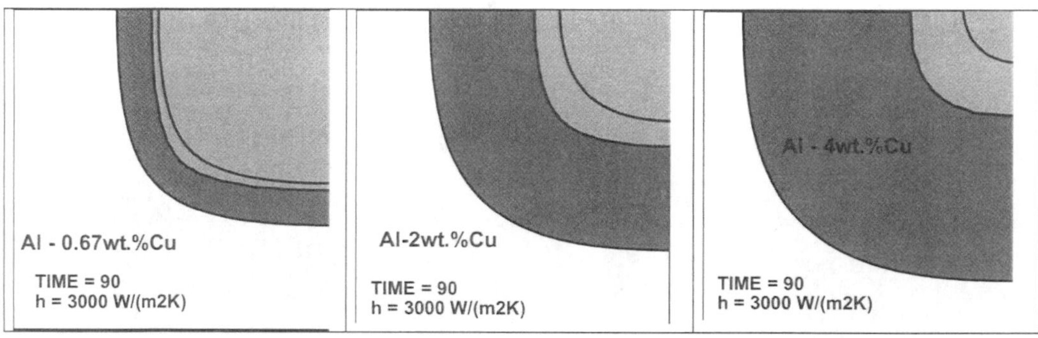

Figure 24. Influence of composition on the solid, columnar mushy zone and undercooled region (equiaxed mushy zone) for Al.-Cu alloy: (a) Al – 0.67 wt.% Cu (C1=0.3), (b) Al – 2 wt.% Cu (C1=0.1), (c) Al – 4 wt.% Cu (C1=0.05) (Banaszek, Browne and Furmański, 2002).

Non-equilibrium solidification. The case of non-equilibrium solidification within the macroscopic approach can be studied if use of eq. (8.8) is made. The curvature κ of the solid/liquid interface can assume negative and positive values depending on location on the dendrite surface. Similarly, the vector $\boldsymbol{n}_s$, normal to the interface, can assume different directions within the mushy zone. Basing on these observations it is reasonable to assume that the second and the third terms on the r.h.s. of eq. (8.8) are

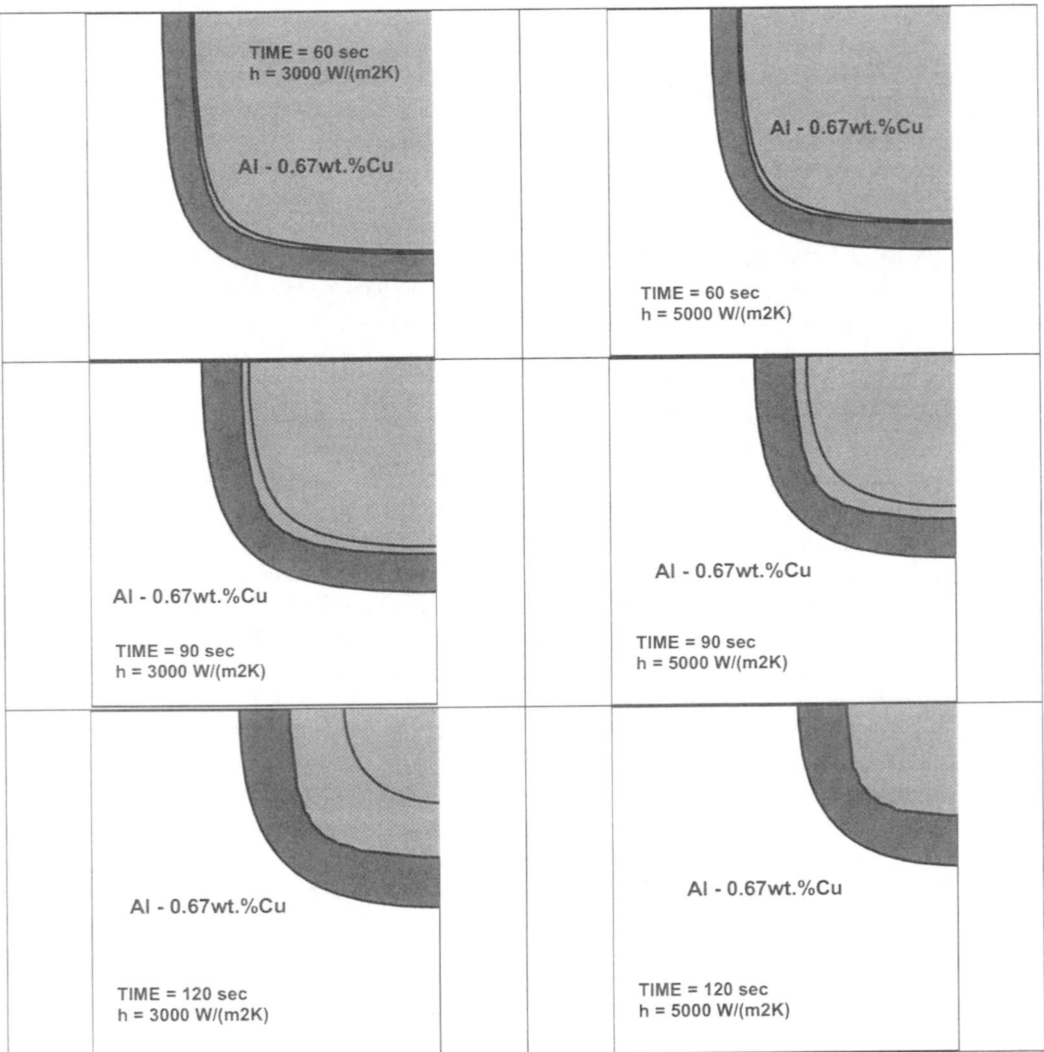

Figure 25. Influence of time and rate of cooling on the extent of different regions in Al – 0.67wt. %Cu (C=0.3) alloy (white – solid phase, dark grey – columnar dendrites region, medium grey – undercooled liquid region, grey – superheated liquid region) (Browne, Banaszek and Hunt, 2002).

equal to zero everywhere in the mushy except at *the solid/ columnar mushy zone*, *columnar mushy zone/ equiaxed mushy zone* or *columnar mushy zone/undercooled liquid* interfaces. The most important is the mushy zone/liquid interface where the solidification is being initiated. So in eq. (8.8) curvature κ_t and velocity w_t of the dendrite tip should be substituted for κ and w_i

$$\{T_i\} = T^e - \frac{T^e}{L} \gamma \kappa_t - \mu_k \, \boldsymbol{w}_t \cdot \boldsymbol{n}_s \tag{11.16}$$

Subsequently, relation between tip radius r_t ($\kappa_t = 1/r_t$) and dendrite tip velocity w_t can be used to eliminate curvature from eq. (11.16). In general the latter relation can be approximated by the expression (see Section 2.2) of the form

$$r_t^2 w_t = f(Pe) \tag{11.17}$$

where the symbol Pe stands for the Peclet number depending on the tip radius, liquid thermal diffusivity and liquid velocity (close to the dendrite tip). For slow solidification the last term on the r.h.s. of eq. (11.16) may be abandoned. Then the expression (11.17) was used to eliminate curvature from eq. (11.16) and the following relation, between velocity of the dendrite tip and undercooling, followed

$$w_t = C1[T^e - \{T_i\}]^2 \tag{11.18}$$

where T^e denotes equilibrium liquidus temperature and $C1$ is generally dependent on the Peclet number and composition of the alloy. Equation (11.18) describes a non-equilibrium condition at the considered interface on the macroscopic scale. Thus in the whole columnar mushy zone thermal equilibrium is assumed except for the surface running through tips of the columnar dendrites. In this way we can use the front tracking method to locate interface of the columnar mushy zone/equiaxed mushy zone or interface of the columnar mushy zone/undercooled liquid zone. The procedure is carried out by calculating progress of the columnar mushy zone from eq. (11.18) for the known undercooling of the liquid. This derivation makes a justification for application of the technique proposed by Browne and Hunt (2000). Burden and Hunt (1974) earlier derived relation similar in form to eq. (11.18) from solution of the growth of non-isothermal, isolated dendrite in a binary alloy.

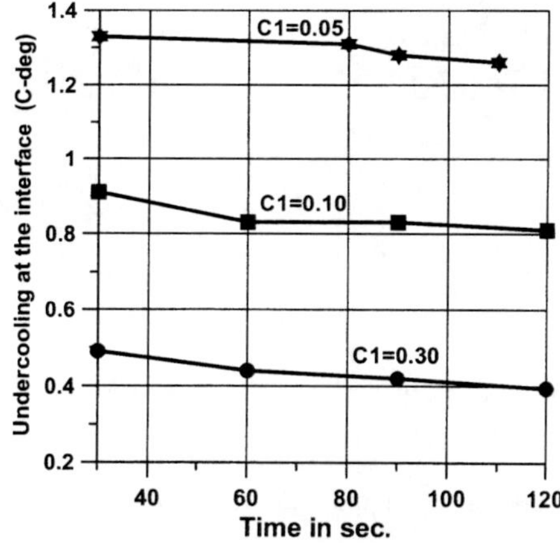

Figure 26. Variation in undercooling at the interface for three different Al-Cu alloys (Banaszek, Browne and Furmański, 2002).

In the case when no convection is present in the solidifying system the parameter $C1$ is independent on velocity of flow close to the dendrite tip. This allows to study columnar/equiaxed transition as ahead of the columnar mushy zone the undercooled liquid layer is formed which is the origin of the equiaxed crystal nucleation (see Section 2). Variation of this undercooled layer with time, cooling rate and composition of the binary alloy in case of solidification in a square cavity that is symmetrically cooled from all sides is shown in Figures 24 and 25. The calculations correspond to pure conduction in the melt. It can be observed that higher solute concentration gives a larger solidification range and therefore a wider mushy zone. This behaviour is expected.

Additionally, in Fig.26, respective undercooling of the columnar mushy zone/liquid interface was shown as a function of time and composition of the binary alloy. Higher level of solute results in greater growth undercooling and larger regions of the undercooled liquid, thus favouring equiaxed growth.

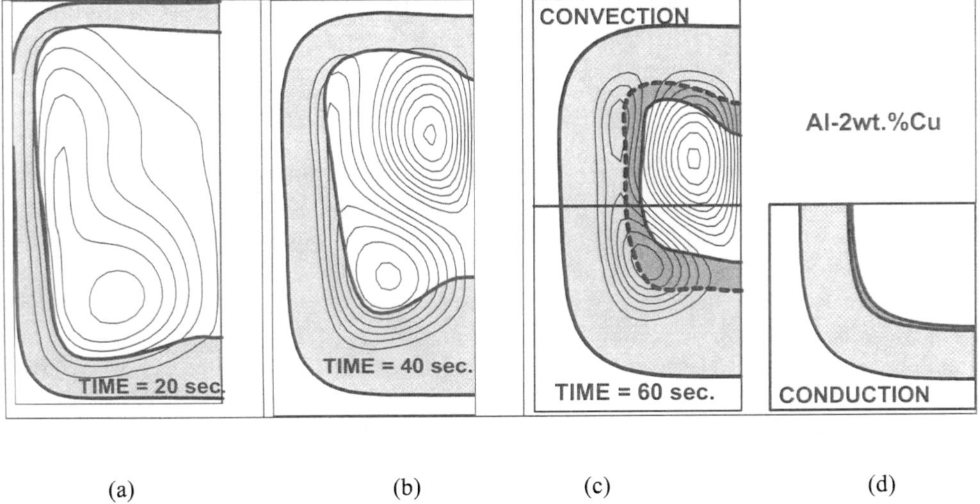

(a) (b) (c) (d)

Figure 27. Flow pattern and the solidus and liquidus isotherms (which bound the grey mushy zone) for solidification with natural thermal convection. Note influence of convection on the extent of the undercooled liquid region (two last pictures on the r.h.s. of the figure) (Browne, Banaszek and Hunt, 2002).

An attempt was made to study influence of convection in the liquid on the evolution of the columnar mushy zone and the subcooled layer. A simplified assumption that convection does not significantly influence dendrite tip temperature is made here. Results of calculations are shown in Fig.27. Local temperature gradient is much lower here than in pure conduction and the undercooled zone is much larger (dark grey region). The latter is shown in Fig. 27c where the dashed line of the dendrite tips is given. The undercooled zone can be compared with that of a pure conduction in Fig. 27d. Lower temperature gradient is also observed near bottom of the mould leading to wider mushy zone and the undercooled liquid region (see Fig. 27c).

Concluding Remarks

The macro modelling of the solidification processes in heterogeneous system like mushy zone or liquid saturated granular media is an alternative to the micro-modelling where details of the distribution of phases and their topological details are taken into account. It is especially useful in case when the microstructure is very complex and contains many structural elements like columnar dendrites or equiaxed crystals or when knowledge of the microstructure bears statistical character. These are the reasons why the macroscopic modelling is practically exclusively used in numerical modelling of large-scale systems.

Essentially two techniques are followed in deriving equations describing solidification phenomena on the macroscopic scale. One is based on the volume averaging of the respective variables over the representative elementary volume (REV) while the second the statistical averaging over an ensemble of configurations of nuclei present in the melt. The latter technique, also less popular in modelling of solidification in heterogeneous systems, is more flexible as it can easily treat statistically nonhomogeneous systems (with varying liquid volume fraction) and systems with internal heat generation due to phase transformations. Both techniques give the same results if gradients of the macroscopic variables are constant over the REV, which is seldom the case.

The equations for field variables like velocity; concentration or temperature are obtained from substituting the constitutive relations into equations describing conservation of momentum, energy and species within the solidifying heterogeneous system. The form of the conservation equations, as universal, can not be altered when the microscopic description of the phenomena is replaced by the macroscopic one. However, the constitutive relations are changed during this transformation. The constitutive relations between macroscopic variables need not to be postulated but can be formally derived from respective microscopic relations irrespective of the assumptions as to the length-scales influencing the process. The constitutive relations contain effective properties of the mushy zone. These effective properties can be essentially calculated from the microstructure functions if microstructure of the mushy zone and the microscopic properties of solid and liquid phase are known.

The macroscopic constitutive relations are generally nonlocal. The nonlocality is generally associated with the non-equilibrium processes occurring in the mushy zone. If the spectrum of scales, associated with microstructure of the mushy zone, is well separated from the spectrum of scales, associated with spatial distributions and time variation of macroscopic temperature (species concentration), then it is possible to obtain simplified, local form of the constitutive relations which resemble the known Fourier and Fick laws valid for homogeneous media or Darcy law as in porous media. The latter relations describe, under certain conditions, thermodynamic equilibrium processes within the mushy zone.

Macroscopic approach when used in modelling of alloy solidification is still in preliminary stage of development. Neither all microscopic phenomena are included nor relations between microstructure of the mushy zone and the effective properties are well known. Attempts should be undertaken to find micro-fields from the macro-fields in the mushy zone and thus to predict the microstructure formation and its transformations in this way. Such procedure would allow closing the loop, shown in Figure 16, to the end. For fast solidification processes macroscopic thermodynamic non-equilibrium in the mushy zone should be investigated more deeply. Further development of the multi-continuum (multi-phase) macroscopic approach to macro modelling of the mushy zone will here be helpful. Moreover, thermal phenomena at the mould/ solidified alloy interface should studied in more detail as they significantly affect heat removal from the alloy.

References

Amberg, G. (2002). Solidification, microstructure and convection. *Lecture Notes from CISM Course "Phase Change with Convection: Modelling and Validation*, Coordinated by T. Kowalewski and D. Gobin, September 2-6, CISM, Udine.

Arnberg, L., Chai, G., and Backerud, L. (1993). Determination of dendritic coherency in solidifying melts by rheological measurements. *Materials Science and Engineering*, A173: 101-103.

Aziz, M. J. (1996). Interface attachment kinetics in alloy solidification. *Metallurgical and Materials Trans.*, 27A: 671-686.

Banaszek, J., Jaluria, Y., Kowalewski, T.A., and Rebow, M. (1999). Semi-implicit FEM analysis of natural convection in freezing water. *Num. Heat Transfer,* A36: 449-472.

Banaszek, J., and Furmański, P. (2000). FEM analysis of binary dilute system solidification using the anisotropic porous medium model of a mushy zone. *Computer Assisted Mechanics and Engineering Sciences*, 7: 343-362.

Banaszek, J., Browne, D.J., and Furmański, P. (2002). Some aspects of modelling of binary system solidification on a fixed grid. *Proceedings of International Conference on Phase Change Processes*, Kielce, June, Poland.

Batchelor, G.K. (1974). Transport properties of two-phase materials with random structure. *Ann. Rev. Fluid Mech*., 6: 227-255.

Beckermann, C., and Viskanta, R. (1993). Mathematical modelling of transport phenomena during alloy solidification. *Appl. Mech. Rev*., 46:1-27.

Beckermann, C., Diepers, H.-J., Steinbach, I., Karma, A., and Tong, X. (1999). Modelling melt convection in phase-field simulations of solidification. *J. of Computational Physics*, 154: 468-496.

Bennon, W.D., and Incropera, F.P. (1987). A continuum model for momentum, heat and species transport in binary solid-liquid phase change systems - I and - II. *Int. J Heat & Mass Transfer*, 30: 2161-2187.

Bouissou, Ph., and Pelcé, P. (1989). Effect of a forced flow on dendritic growth. *Physical Review A*, 40: 6673-6680.

Bousquet-Melou, P., Goyeau, B., Quitard, M., Fichot, F., and Gobin, D. (2002). Average momentum equation for interdendritic flow in a solidifying columnar mushy zone. *Inter. J. of Heat and Mass Transfer*, 45: 3651-3665.

Browne, D. J., Hunt, J. D. (2000). A model of columnar growth using a front tracking technique. *Modelling of Casting, Welding and Advanced Solidification Processes IX*, Springer Verlag, Aachen, Germany, 437-444.

Browne, D. J., Banaszek, J., and Hunt, J. D. (2002). Front tracking method on a fixed grid versus enthalpy approach in modelling of binary alloy solidification. *Proceedings of IMECE 2002: International Mechanical Engineering Congress and Exhibition*, November 17-22, New Orleans, USA.

Browne, D.J., and O'Mahoney, D. (2001). Interface heat transfer in investment casting of aluminium alloys. *Metallurgical and Materials Trans.* , 32A: 3055-3063.

Burden, M. .H., Hunt, J. D. (1974). Cellular and dendritic growth. *Journal of Crystal Growth*, 22: 99-116.

Buyevich, Yu. A. (1992). Heat and mass transfer in disperse media. I. Averaged Field Equations. *Int. J. Heat & Mass Transfer*, 35: 2445-2452.

Chung, J.D., Lee, J.S., and Yoo, H. (1998). An extended similarity solution for the alloy solidification system. *Proceedings of 11th IHTC*, 7: 187-192.

Furmański P. (1994). A mixture theory for heat conduction in heterogeneous media. *Int. J. Heat & Mass Transfer*, 37: 2993-3002.

Furmański, P. (1995). Influence of laminar convection of fluid on effective thermal conductivity of some porous media. *Advances in Engineering Heat Transfer*. Computational Mechanics Publications, 513-524.

Furmański, P. (1997). Heat conduction in composites. Homogenization and macroscopic behavior. *Appl. Mech. Rev.*, 50: 327-356.

Furmański, P.(1999). Thermal properties and local heat sources in composite materials. *Thermal Conductivity*, 24: 581-594.

Furmański, P. (2000). Modelling of transport phenomena during solidification of binary systems. *Computer Assisted Mechanics and Engineering Sciences*, 7: 391-402.

Ganesan, S., Chan, C.L., and Poirier, D.R. (1992). Permeability for flow parallel to primary dendrite arms. *Mater. Sci.and Eng.,* A151: 97-105.

Garda, B., Hodaj, F., Durand, F. (1993). Semi-analytical calculation of equiaxed grain-size in recoalescence conditions. *Mater. . and Eng.*, A173: 105-108.

Geindreau, C., and Auriault, J-L. (2001). Transport phenomena in saturated porous media undergoing liquid-solid phase change. *Computer Assisted Mechanics and Engineering Sciences*, 8: 391-402.

Hills, R.N., Loper, D.E., and Roberts, P.H. (1992). On continuum models for momentum, heat and species transport in solid-liquid phase change systems. *Int. Comm. Heat & Mass Transfer*, 19: 585-594.

Hunt, J. .D., and Lu, S. Z. (1996). Numerical modelling of cellular/dendritic array growth: spacing and structure predictions. *Metallurgical and Materials Transactions*, 27A: 611-623.

Juric, D. and Tryggvason, G. (1996). A front-tracking method for dendritic solidification. *Journal of Computational Physics*, 123: 127-148.

Kaviany, M. (1995). *Principles of Heat Transfer in Porous Media*. Springer Verlag, 2nd edition, New York.

Khan, J.A., and Tong, X. (1998). Unidirectional infiltration and solidification/remelting of Al-Cu alloy. *J. of Thermophysics and Heat Transfer*. 12: 100-106.

Koch, D.L., and Brady, J.F. (1987). A non-local description of advection-diffusion with application to dispersion in porous media. *J. Fluid Mech.*, 387-403.

Kunin, I.A. (1984). On foundations of the theory of elastic media with microstructure. *Int. J.Engng Sci.*, 22: 969-978.

Krieger, I.M. (1972). Rheology of monodispersed lattices. *Advances in Colloid Interface Science*, 3: 111-136.

Langer, J.S. (1989). Dendrites, viscous fingers, and the theory of pattern formation. *Science*, 243:1150-1156.

Loulou, T., Artyukhin, E.A., and Bardon, J.P. (1999). Estimation of thermal contact resistance during the first stages of metal solidification process: I – experiment principle and modelisation. *International J. of Heat & Mass Transfer*, 42: 2119-2127.

Loulou, T., Artyukhin, E.A., and Bardon, J.P. (1999). Estimation of thermal contact resistance during the first stages of metal solidification process: II – experimental set-up and results. *International J. of Heat & Mass Transfer*, 42: 2129-2142.

Mat, M.D., and Ilegbusi, O.J. (2002). Application of a hybrid model of mushy zone to macrosegregation in alloy solidification. *Int. J. of Heat and Mass Transfer*, 45: 279-289.

Matsumoto, K., Okada, M., Murakami, M., and Yabushita, Y. (1193). Solidification of porous medium saturated with aqueous solution in a rectangular cell. Int. J. Heat and Mass Trans., 36: 2869-2880.

Mortensen, A., and Flemings, M.C. (1996). Solidification of binary hypoeutectic alloy matrix composite castings. *Metallurgical and Materials Trans.*, 27A: 595-609.

Murakami, K., and Okamoto, T. (1984). Fluid flow in the mushy zone composed of granular grains. *Acta Metall.*, 32: 1741-1744.

Naterer, G.F., and Schneider, G.E. (1995). Phases model for binary-constituent solid-liquid phase transition. Part 1*: Numerical Methods. Numerical Heat Transfer*, Part B, 28:111-126.

Naumann, R. (1995). Marangoni convection around voids in Bridgman growth. *J. of Crystal Growth*, 154: 156-162.

Okada, M., Ochi, M., and Tateno, M. (1998). Solidification of a supercooled aqueous solution in a porous medium. *Proceedings of 11th IHTC*, Kongjiu, Korea, 7: 169-174.

O'Mahoney, D., and Browne, D.J. (2000). Use of experiment and an inverse method to study interface heat transfer during solidification in the investment casting process. *Experimental Thermal and Fluid Science*, 22: 111-122.

Poirier, D.R. (1987). Permeability for flow of interdendritic liquid in columnar-dendritic alloys. *Metallurgical Trans.,* 18B: 245-255.

Poirier, D.R., Nandapurkar, P.J., and Ganesan, S. (1991). The energy and solute conservation equations for dendritic solidification. *Metallurgical. Trans*., 22B: 889-900.

Prakash , C., and Voller, V. (1989). On the numerical solution of continuum mixture model equations describing binary solid-liquid phase change. *Numerical Heat Transfer B*, 15: 171-189.

Prescott, P.J., Incropera, F.P., and Bennon, W.D. (1991). Modelling of dendritic solidification systems: reassessment of the continuum momentum equation. *Int. J. of Heat and Mass Transfer*, 34: 2351-2359.

Quintard, M., and Whitaker, S. (1993). One and two equations models for transient diffusion processes in two-phase systems. *Advances in Heat Transfer*, 23: 369-464.

Rappaz, M. (1989). Modelling of microstructure formation in solidification processes. *International Materials Reviews*, 34: 93-123.

Rappaz, M., and Voller, V.R. (1990). Modelling of micro-macrosegregation in solidification processes. *Metall. Trans*., 21A: 749-753.

Rappaz, M., Gandin, Ch. A., Desbiolles, J. L., and Thevoz, Ph.. (1996). Prediction of grain structures in various solidification processes. *Metallurgical and Materials Trans.*, 27A: 695-705.

Sältzer, W.D., and Schultz, B. (1983). Theory and measurement of the viscosity of suspensions. *High Temperatures – High Pressures*, 15: 289-298.

Santoli, L., Cumo, F., and Menno, I. (1998). Freezing of liquid-saturated porous media of building materials. *Proceedings of 11th IHTC*, Kyongiu, Korea, 4: 387-391.

Schrage, D.S. (1999). A simplified model of dendritic growth in the presence of natural convection. *J. of Crystal Growth*, 205: 410-426.

Shyy, W., Udaykumar, H. S., Rao, M. .M., and Smith, R. W. (1996), *Computational Fluid Dynamics with Moving Boundaries*, Taylor and Francis, Washington DC, USA.

Simpson, J.E., Garimella, S. V., and Guslick, M.M. (1998). Interface propagation in the presence of a fibrous phase in alloy solidification. *Proceedings of 11th IHTC*, Kyongiu, Korea, 7: 235-240.

Sinha, S.K., Sundararajan, T., and Garg, V.K. (1992). A variable property analysis of alloy solidification using the anisotropic porous medium approach. *Int. J. of Heat and Mass Transfer*: 2865-2877.

Swaminathan, C.R., and Voller, V.R. (1992). General enthalpy method for modelling solidification processes. *Metall. Trans., 23B*: 651-65.

Swaminathan, C.R., and Voller, V.R. (1997). Towards a general numerical scheme for solidification systems. *Int. J. of Heat and Mass Transfer*, 40:2959-2868.

Timchenko, V.,. Chen, P.Y.P., de Vahl Davies, G, and Leonardi, E. (1998). Directional solidification in microgravity. *Proceedings of 11th IHTC*, , Kyongju, Korea, 7: 241-246.

Tönhardt, R., and Amberg, G. (1998). Phase-field simulation of dendritic growth in a shear flow. *J. of Crystal Growth*, 194: 406-425.

Wang, W., and Qiu, H.H. (2002). Interfacial thermal conductance in rapid contact solidification process. *International J. of Heat & Mass Transfer*, 45: 2043-2053.

Warren, J. A., and Boettinger, W. J. (1995). Prediction of dendritic microsegregation patterns using a diffuse interface phase field model. *Modelling of Casting, Welding and Advanced Solidification Processes VII*, M. Cross and J. Campbell eds, TMS, Warrendale, PA, USA, 601-607.

Voller, V.R., Brent, A.D., Prakash, C. (1989). The modelling of heat, mass and solute transport in solidification systems. *Int. J. Heat & Mass Transfer*, 32: 1719-1731.

Voller, V.R., and Swaminathan, C.R. (1991). General source-based method for solidification phase change. *Numerical Heat Transfer*, 19: 175-189.

Viskanta, R. (1988). Heat transfer during melting and solidification of metals. *Trans. of ASME, J. of Heat Transfer*, 110:1205-1219.

ACKNOWLEDGMENTS

This work was financially supported by State Committee for Scientific Research of Poland (KBN) under the grant No 8T10B 07220.

Natural Convection at a Solid-Liquid Phase Change Interface

Dominique Gobin *

* FAST (UMR 7608) - Campus Universitaire, Bât. 502, 91405 Orsay (France)

Abstract In this course "Phase Change and Convection - Modelling and Validation", the influence of convection on solid-liquid phase change processes is considered at different levels. The local effects of convection at the local scale on dendritic growth is addressed in the chapter by G. Amberg and the scaling-up from the local equations to the averaged *macroscopic* equations in a *mushy* zone is presented in the chapter by P. Furmanski. The present chapter deals with the interaction between a "smooth" phase change interface and laminar natural convection in the melt. It addresses the main features of solid-liquid phase change in situations where the interface between the solid and the liquid phase is clearly defined. The presentation is focused on the consequence of buoyancy forces in the liquid phase on heat and mass transfer at a solid-liquid interface through convective flows: *thermal* natural convection for pure substances, or *thermosolutal* convection in multi-component fluids.

1 Fundamentals of laminar natural convection

The purpose of this section is to briefly recall the main features of natural convection flows in closed cavities, insisting on the necessary elements required to understand the further developments concerning the coupling of natural convection and solid-liquid phase change processes. It is assumed that the reader is familiar with the basic conservation equations and the dimensionless numbers currently used in this field. Beyond the formal presentation of the equations, our purpose is to give some insight in the main orders of magnitude for heat and mass transfer in natural convection. For a more detailed presentation, the reader may refer to the book by Adrian Bejan (1995). The case of laminar free convection driven by thermal gradients is first presented, and since melting/solidification processes of mixtures involve also composition gradients, the case of the so-called "double-diffusive" convection is then considered.

1.1 Thermal Natural Convection

Many different practical situations may appear when natural convection is generated by buoyancy forces in a fluid system. A large range of boundary conditions are met in the industrial or natural processes where density gradients take place in a gravity field.

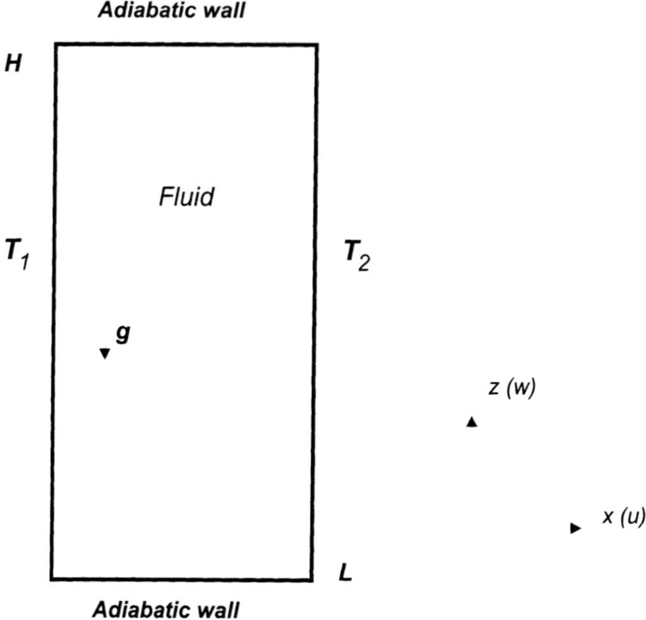

Figure 1. Schematic view of the enclosure

The purpose of the present section is to give a general insight into the main mechanisms and to yield the basic orders of magnitude. This is the reason why we will consider he simplest situation of a vertical wall, heated at a uniform temperature, and immersed in a newtonian fluid.

Equations and Parameters. Let us consider a two-dimensional rectangular vertical enclosure. The main notations are presented in Fig. 1. The governing equations for natural convection are the usual conservation equations for mass, energy and momentum. Under the classical hypotheses of incompressible flow, and Newtonian Boussinesq fluid, the following set of dimensional equations is obtained:

1. Mass conservation (continuity equation):

$$\frac{\partial u^*}{\partial x^*} + \frac{\partial w^*}{\partial z^*} = 0 \tag{1.1}$$

2. Energy conservation:

$$\rho C_P \left(\frac{\partial T}{\partial t^*} + u^* \frac{\partial T}{\partial x^*} + w^* \frac{\partial T}{\partial z^*} \right) = \lambda \left(\frac{\partial^2 T}{\partial x^{*2}} + \frac{\partial^2 T}{\partial z^{*2}} \right) \tag{1.2}$$

3. Momentum conservation (Navier-Stokes equations):

a/ along the horizontal x-axis (u-component):

$$\left(\frac{\partial u^*}{\partial t^*}+u^*\frac{\partial u^*}{\partial x^*}+w^*\frac{\partial u^*}{\partial z^*}\right)=\nu\left(\frac{\partial^2 u^*}{\partial x^{*2}}+\frac{\partial^2 u^*}{\partial z^{*2}}\right)-\frac{1}{\rho}\frac{\partial p^*}{\partial x^*} \tag{1.3}$$

b/ along the vertical z-axis (w-component)::

$$\left(\frac{\partial w^*}{\partial t^*}+u^*\frac{\partial w^*}{\partial x^*}+w^*\frac{\partial w^*}{\partial z^*}\right)=\nu\left(\frac{\partial^2 w^*}{\partial x^{*2}}+\frac{\partial^2 w^*}{\partial z^{*2}}\right)-\frac{1}{\rho}\frac{\partial p^*}{\partial z^*}+g\beta(T-T_0) \tag{1.4}$$

The associated boundary conditions are adiabatic horizontal walls and imposed uniform and constant temperatures at the opposing vertical walls.

In order to give more generality to the solutions of this set of equations, they are usually presented in a non-dimensional form, using reduced variables with respect to some reference quantities which depend upon the problem under study. A possible choice, leading to the dimensionless equations (1.6) to (1.8), considers the height H of the enclosure as the reference length and uses the kinematic viscosity ν for the time and velocity references:

$$x(z)=x^*(z^*)/H\,,\ V=\left(\frac{V^*}{\nu/H}\right)\,,\ t=\left(t^*\frac{\nu}{H^2}\right)\,,\ \theta=\frac{(T-T_0)}{\Delta T} \tag{1.5}$$

Dimensionless Equations:

$$\nabla.\mathbf{V}=0 \tag{1.6}$$

$$\frac{\partial\theta}{\partial t}+\mathbf{V}.\nabla\theta=\frac{1}{Pr}\nabla^2\theta \tag{1.7}$$

$$\frac{\partial\mathbf{V}}{\partial t}+(\mathbf{V}.\nabla)\mathbf{V}=\nabla^2\mathbf{V}-\nabla P+Gr_H\theta\,\mathbf{k} \tag{1.8}$$

This allows to identify the dimensionless parameters of the problem: here, the Prandtl and Grashof numbers, and the aspect ratio of the enclosure:

$$Pr=\frac{\nu}{\alpha}\,,\ Gr_H=\frac{g\,\beta\,\Delta T\,H^3}{\nu^2}\,,\ A=H/L\,. \tag{1.9}$$

Remark: A different non-dimensionalization (i.e. a different choice of the velocity or time scales) would have led to the same parameters or to a combination of those, namely the Rayleigh number: $Ra_H=Pr\,Gr_H$. However it is useful to insist here on the relevance of the height H of the enclosure as the reference length of the problem: it is indeed along the z-direction that the gravity field exerts its influence on the fluid: as a consequence, as it will be assessed later, the Rayleigh number *based on H* is the relevant dimensionless parameter of this problem, even if the temperature difference is along the x-direction. A very common practice, arising from the earlier works on Rayleigh-Bénard convection in an horizontal fluid layer, consists in using the distance L between the active walls as the length scale: in this case both the temperature gradient and the gravity field are along the same direction. As a consequence, the correct scale for the heat flux is:

$$\phi_R=\frac{k\,\Delta T}{H}$$

in such a way that the Nusselt number - which is nothing but the *dimensionless heat flux* - is not 1, but H/L = A, when conduction prevails in the fluid.

It is well known (Patankar (1980)) that the above equations have the same general form:

$$\frac{\partial \phi}{\partial t} + \mathbf{V}.\nabla \phi = \nabla(\Gamma.\nabla \phi) + \mathcal{S}(\phi). \tag{1.10}$$

where ϕ represents one of the scalar variables of the problem and Γ the corresponding diffusivity. The first term is an accumulation term, the second term is an advection term, the third the diffusion term and the last one is an eventual source term, depending on the scalar quantity ϕ upon consideration.

The complete resolution of the above equations, associated to the relevant initial and boundary conditions, makes use of complex numerical methods the detail of which is not within the scope of the present text. We will however provide a short analysis of the main scales, particularly concerning the heat transfer rates.

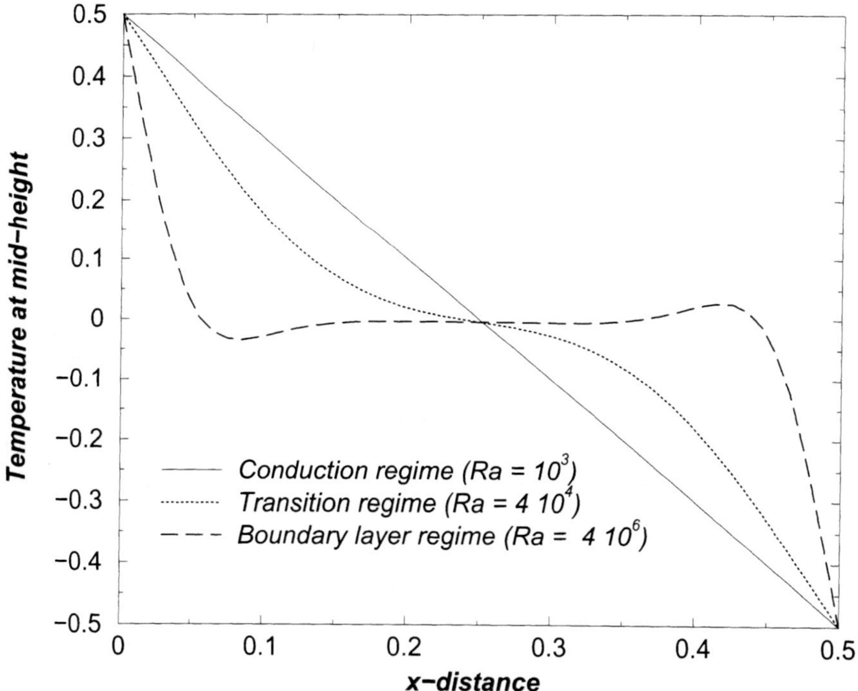

Figure 2. The regimes of laminar natural convection (Pr = 1 ; A = 2)

Flow Regimes for Natural Convection in Enclosures. The dimensionless number characterizing the intensity of the buoyancy force, and thus of the natural convective flow, is the Rayleigh number. For a given fluid and geometry, this number essentially depends on the temperature difference between the "active" walls of the enclosure. According to the particular problem under study, non uniform temperatures or variable heat fluxes may be imposed at these walls, but still this analysis is presented for the standard academic situation of an imposed uniform temperature at the vertical walls of the enclosure.

A classification of natural convective flows in the laminar regime has been proposed for air, based on the observation of the horizontal temperature profiles in the fluid (Eckert and Carlson (1961); Gill (1966); Ostrach (1972)). Depending on the cell aspect ratio (A = H/L) and the temperature difference between the walls, three regimes may be defined (Figure 2):

1. in the range of low Rayleigh (or Grashof) numbers, it is clear that the temperature profiles are almost linear in the central part of the enclosure. This so-called *conduction regime*, corresponds to extremely low flow velocities with almost no effect on the temperature field and heat transfer.
2. in the range of higher Rayleigh numbers, the temperature profile is characterized by strong gradients close to the walls and by a central stratified zone where temperature is roughly constant in the x-direction. This regime is called *separate boundary layer regime*, where the temperature gradient - the heat transfer - takes place in a thin layer close to the wall. At a given H/L, these gradients increase with the Ra number.
3. Between these two clearly identified regimes, there is a *transition regime*, where the forming boundary layers still interact in the central part of the cavity.

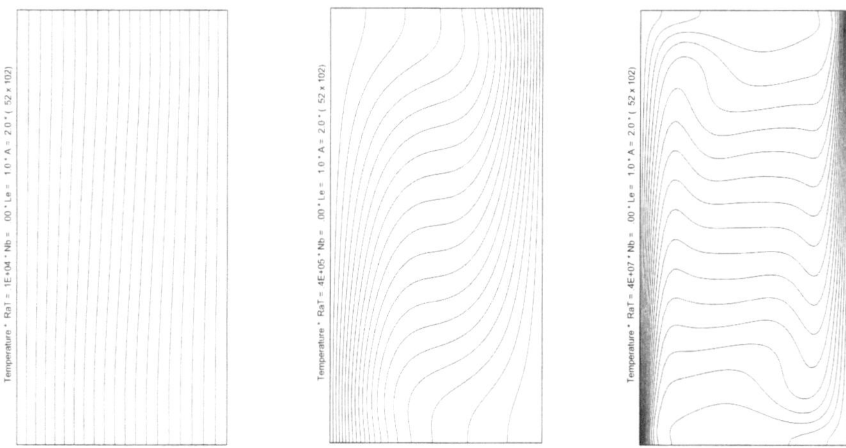

Figure 3. a - Conduction ; b - Transition ; c - Boundary layer

Many studies have been dedicated to the assessment of criteria for defining the limits between these regimes. For instance, for air, Grondin and Roux (1979) have numerically

established that, for large aspect ratios ($A \geq 4$) and using a Rayleigh number built on the width L of the enclosure:
1. Conduction Regime for: $Ra_L < 250$ A ,
2. Boundary Layer Regime for: $Ra_L > 2500$ A .

Scaling Laws. In order to provide the main tools for the analysis of natural convection in the phase change processes, we will recall the classical orders of magnitude for steady state natural convection in the frame of the boundary layer approximation. This presentation is widely inspired from the scaling law analysis developed by Adrian Bejan (1995), and the interested reader may refer to the more complete presentation given there.
Let us call δ_T the thickness of the fluid layer adjacent to the wall where the temperature gradients are present: we will focus our interest to the region of width δ_T and height H along the hot wall. The boundary layer approximation assumes that $\delta_T << H$, so that the dominating terms are due to the gradients normal to the wall (here, in the horizontal x-direction). In the steady state, the energy equation (1.2) expresses the balance between the convective term and the diffusive term, whose orders of magnitude are respectively:

$$u\frac{\Delta T}{\delta_T}\,,\; w\frac{\Delta T}{H} \sim \alpha\frac{\Delta T}{\delta_T^2}\,. \tag{1.11}$$

After the continuity equation (1.1), the orders of magnitude of the u and w components of the velocity at the δ_T scale are related by :

$$u \sim w\,\frac{\delta_T}{H}\,. \tag{1.12}$$

So the scale of the vertical component of the velocity along the wall writes:

$$w \sim \frac{\alpha H}{\delta_T^2}\,. \tag{1.13}$$

The upstream flow along the hot wall is due to the buoyancy force created by the dilatation of the fluid layer. The balance of the forces exerted on a fluid particle is expressed by a compact form of the Navier-Stokes equations (1.4), where the pressure term may be eliminated by cross differentiation of the two components of the equation. Assuming that the gradients along the vertical z-direction and the horizontal component u may be neglected compared to the gradients in the x-direction and the vertical velocity w, it is easy to show that the dominating terms in the Navier-Stokes equation (1.4) are:
- the inertia term:

$$u\frac{\partial^2 w}{\partial x^2} \sim u\frac{w}{\delta_T^2}\,,$$

- the viscous friction term:

$$\nu\frac{\partial^3 w}{\partial x^3} \sim \frac{\nu w}{\delta_T^3}\,,$$

- the buoyancy term:

$$g\beta\frac{\partial T}{\partial x} \sim \frac{g\beta\Delta T}{\delta_T}\,.$$

In dimensionless form, using equation (1.12), these terms are written:

$$\left(\frac{H}{\delta_T}\right)^4 Ra_H^{-1} Pr^{-1}\,,\ \left(\frac{H}{\delta_T}\right)^4 Ra_H^{-1}\,,\ 1 \tag{1.14}$$

The dominating term among the two first terms clearly depends on the order of magnitude of Pr. For the case of fluids with a high Prandtl number ($Pr > 1$), the buoyancy term is balanced by the viscous force:

$$\left(\frac{H}{\delta_T}\right)^4 Ra_H^{-1} \sim 1\,. \tag{1.15}$$

where the Rayleigh number, built on the height of the "active" wall is:

$$Ra_H = \frac{g\,\beta\,\Delta T\,H^3}{\alpha.\nu}\,,$$

and the thermal boundary layer thickness is:

$$\delta_T \sim H\,Ra_H^{-1/4} \tag{1.16}$$

while the vertical velocity scales as:

$$w \sim \frac{\alpha}{H} Ra^{1/2} \tag{1.17}$$

Thus, the Nusselt number (dimensionless) wall heat transfer, proportional to the heat flux in the δ_T layer, scales as H / δ_T:

$$Nu \sim Ra^{1/4} \tag{1.18}$$

The same kind of scale analysis in the transient regime (Patterson and Imberger (1980)) shows that the time scale τ_f for the thermal boundary layer development, given by the conduction-convection balance is:

$$\tau_f \sim \frac{H^2}{\alpha} Ra^{-1/2} \tag{1.19}$$

Outside the thermal boundary layer, the buoyancy force effects vanish, since the thermal gradient is negligible, but the fluid is entrained on a certain distance by the boundary layer flow. The dominating terms in the momentum equation are the friction and the inertia terms. Let δ_V be the thickness of this layer, the balance writes:

$$w\,\frac{w}{H} \sim \nu\,\frac{w}{\delta_V^2} \tag{1.20}$$

leading to:

$$\frac{\delta_V}{\delta_T} \sim Pr^{1/2} > 1 \tag{1.21}$$

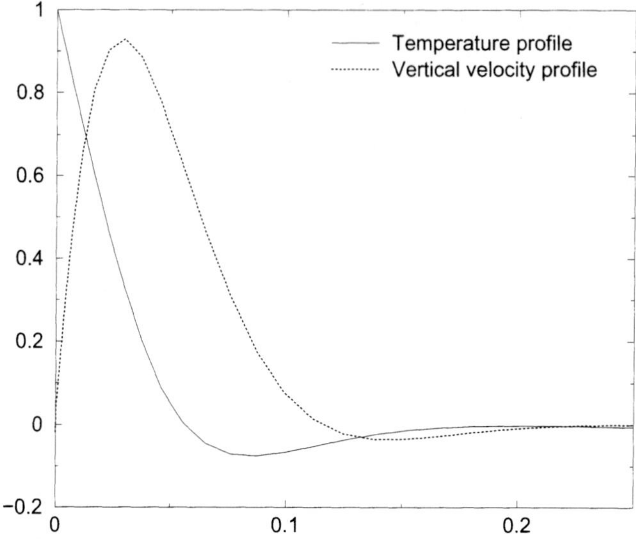

Figure 4. The thermal and viscous boundary layers for Pr >> 1

(Pr = 10 ; Ra = 4×10^6 ; A = 2)

This is illustrated in figure 4, where the thermal and viscous boundary layers are displayed (Pr = 10 ; Ra = 4×10^6 ; A = 2). This allows finally to provide a criterion for the separation of the boundary layers: the condition is that the boundary layer thickness δ_T is smaller than the cavity width, leading to:

$$\frac{H}{L} < Ra_H^{1/4} \tag{1.22}$$

Real situations may be much more complex from the hydrodynamic viewpoint (multicellular regimes, oscillatory solutions, ...) and many theoretical studies are concerned with the associated stability problems of such flows. This is out of the scope of this presentation whose interest is in the estimation of the main orders of magnitude of the dominating terms in such phase change problems where laminar natural convection is the driving mechanism. The coupling with phase change will be examined in a forthcoming section.

1.2 Double Diffusive Natural Convection

This section is concerned with a specific feature of natural convection in multi-component systems, commonly met in solid-liquid phase change processes where mixtures are involved: *double diffusive* convection. This expression defines a certain kind of natural convection where the fluid flow is driven both by concentration and temperature gradients. This kind of situation is very frequent in solidification of binary mixtures, for instance, due to the segregation of components during phase change (Figure 5). Indeed the solid equilibrium composition at a given temperature is different from the equilibrium composition of the liquid (partition coefficient). As a consequence, solute redistribution at the interface induces concentration gradients in the liquid phase which combine to the temperature gradients necessarily imposed to the system and modify the local density of the fluid. In a gravity field, the buoyancy forces thus created induce the so-called *double diffusive* (or thermohaline, or thermosolutal) convection, due to the thermal and solutal origin of the density variation. The resulting convective flow structures may significantly modify the energy and species balances at the interface which govern the phase change process.

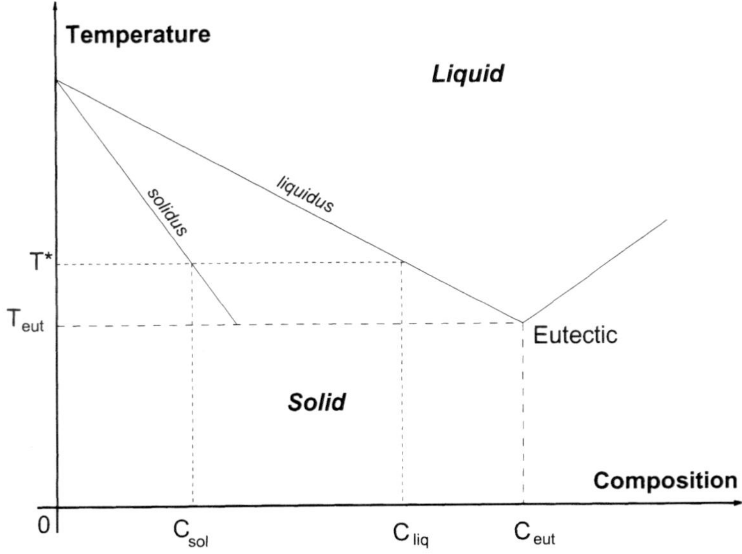

Figure 5. Simplified phase diagram of a binary system

Equations and Parameters. The classical mathematical model, based on the usual conservation equations, leads to a set of equations very similar to the previous one, except for the species conservation equation and for the source term in the Navier-Stokes equation. For a newtonian Boussinesq fluid in an incompressible flow, and neglecting the Soret and Dufour effects, this results in the following set of dimensionless equations:

$$\nabla . \mathbf{V} = 0 \qquad (1.23)$$

$$\frac{\partial \theta}{\partial t} + \mathbf{V}.\nabla\theta = \frac{1}{Pr}\nabla^2\theta \tag{1.24}$$

$$\frac{\partial \phi}{\partial t} + \mathbf{V}.\nabla\phi = \frac{1}{Sc}\nabla^2\phi \tag{1.25}$$

$$\frac{\partial \mathbf{V}}{\partial t} + (\mathbf{V}.\nabla)\mathbf{V} = \nabla^2\mathbf{V} - \nabla P + [Gr_T\,\theta + Gr_S\,\phi]\,\mathbf{k} \tag{1.26}$$

The thermosolutal convection problem is thus characterized by two more parameters: the Schmidt number and the solutal Grashof number:

$$Sc = \frac{\nu}{\mathcal{D}}\,, \; Gr_S = \frac{g\,\beta_S\,\Delta C\,H^3}{\nu^2}\,. \tag{1.27}$$

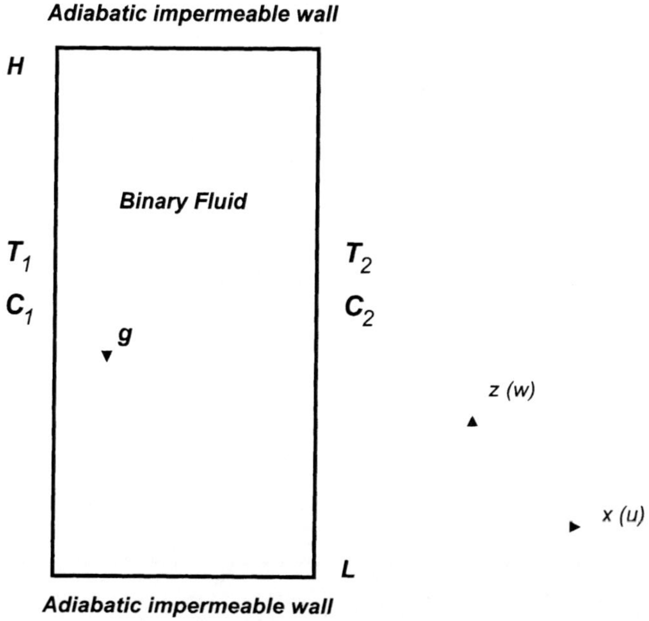

Figure 6. Schematic view of the enclosure

In this text, we will rather use combinations of these parameters: first, the ratio of the thermal and molecular diffusivities, the Lewis number,

$$Le = \frac{\alpha}{\mathcal{D}} = \frac{Sc}{Pr}$$

then, the buoyancy number, ratio of the solutal and thermal Grashof numbers,

$$N = \frac{Gr_S}{Gr_T} = \frac{\beta_S\,\Delta C}{\beta_T\,\Delta T}\,.$$

Heat and Mass Transfer The scaling laws for double diffusive convection may be presented for all the possible ranges of parameters (Bejan, 1995). Here the scope will be limited to the "reasonable ranges" of Lewis numbers. The main orders of magnitude characterizing the most usual binary fluids are presented in Table 1. It appears clearly that, if the Prandtl number may be much larger or smaller than 1, the Lewis number is always larger than 1. Moreover the forthcoming results will essentially concern the $Pr >> 1$ range.

Fluid	Pr	Le
Liquid Alloys	10^{-2}	10^{4}
Gases	1	1
Aqueous Solutions	10	10^{2}
Organic Liquids	10^{2}	10^{3}

Table 1. Usual ranges for the Prandtl and Lewis numbers

After the first boundary layer approximation analysis proposed by Bejan (1985; 1995), the scaling laws for thermosolutal convective heat and mass transfer yield, for Pr $>$ 1 and Le $>$ 1:

1. in the "mass transfer driven" regime (defined as $|N| >> 1$):

$$Nu \approx \left(\frac{N}{Le}\, Ra_{\mathrm{T}}\right)^{1/4} = Le^{-1/2}\, Ra_{\mathrm{S}}^{1/4}$$

$$Sh \approx (N\, Le\, Ra_{\mathrm{T}})^{1/4} = Ra_{\mathrm{S}}^{1/4} \tag{1.28}$$

2. in the "heat transfer driven" regime (defined as $|N| \leq 1$):

$$Nu \approx Ra_{\mathrm{T}}^{1/4}$$

$$Sh \approx Le^{1/3}\, Ra_{\mathrm{T}}^{1/4} \tag{1.29}$$

This estimation of the orders of magnitude has been extended over the N $>$ 0 domain (Bennacer and Gobin (1996)): it may be shown (Figure 7) that the solutally dominated regime, eq. (1.28), depends on the ratio N/Le:

$$\frac{N}{Le} >> 1$$

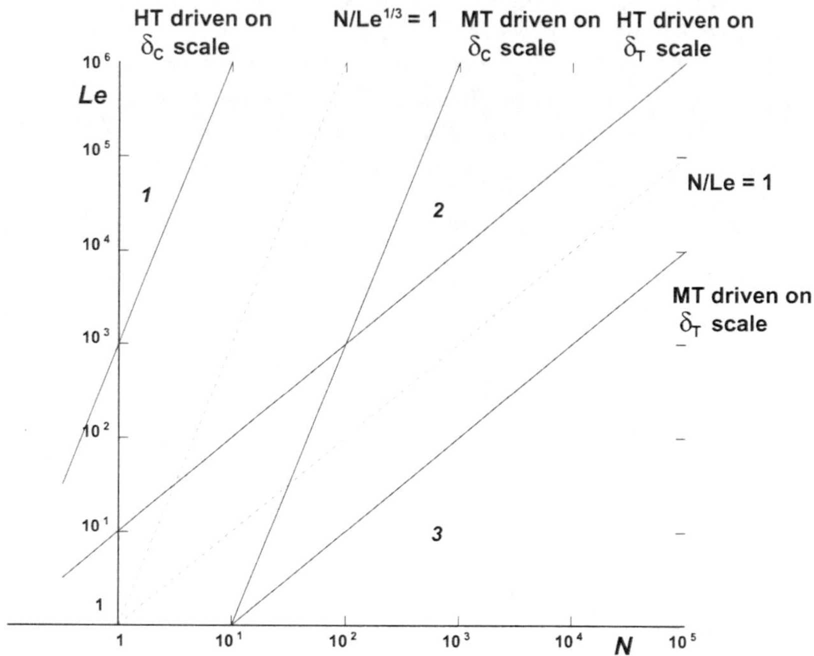

Figure 7. Thermally or solutally dominated regimes in the (*Le*-*N*) plane.

and that the domain where the transfers are thermally dominated (1.29) depends on $N/Le^{1/3}$:

$$\frac{N}{Le^{1/3}} << 1 \, .$$

The combination of these two behaviors leads to a more general correlation for mass transfer (the Sherwood number):

$$Sh \approx \left[Ra_{\mathrm{T}} \left(1 + \frac{N}{Le^{1/3}} \right) \right]^{1/4} Le^{1/3} \tag{1.30}$$

Figures 8 and 9 display the numerical results obtained in the following range of parameters:

$$Gr_{\mathrm{T}} \in [10^3 - 10^6] \quad Pr \in [7 - 100] \quad Le \in [1 - 100] \quad N \in [0 - 100] \quad A \in [0.5 - 2]$$

A very good agreement is seen with the power law defined in Eq. (7), with larger discrepancies at the lowest values of the dimensionless group where the boundary layer

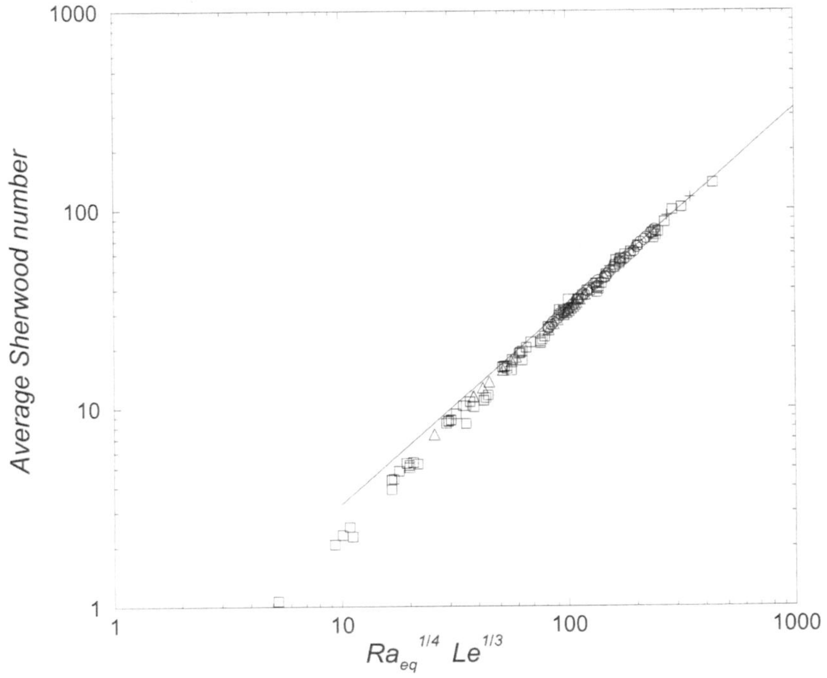

Figure 8. Mass transfer by thermosolutal convection: correlation

approximation is less relevant. A linear regression on the set of solutions gives a 0.33 coefficient for the correlation (solid line in the figure).

On the contrary, in the same range of parameters, no power law correlation fitting the Nusselt number behavior in the extremes of the parameter range may be found for heat transfer (Figure 9). It may be noted in the figure that the results obtained at constant Ra_T, e.g. for Le = 100, are showing a power law at low N values, but that increasing N yields a fast decrease in Nu, in contradiction with an intuitive behavior which would be expected that the fluid flow increases at higher N and thus the heat transfer. The same observation is made when Le is increased at a constant value of N.

Actually the transition from a flow regime dominated by *thermal* buoyancy to a flow regime dominated by *solutal* buoyancy is linked to a significant modification of the flow structure, which evolves from a monocellular regime at low buoyancy ratio towards two different regimes (for increasing N or Le):
1. the first one, close to the walls, is controlled by solutal buoyancy at the scale of the concentration boundary layer,

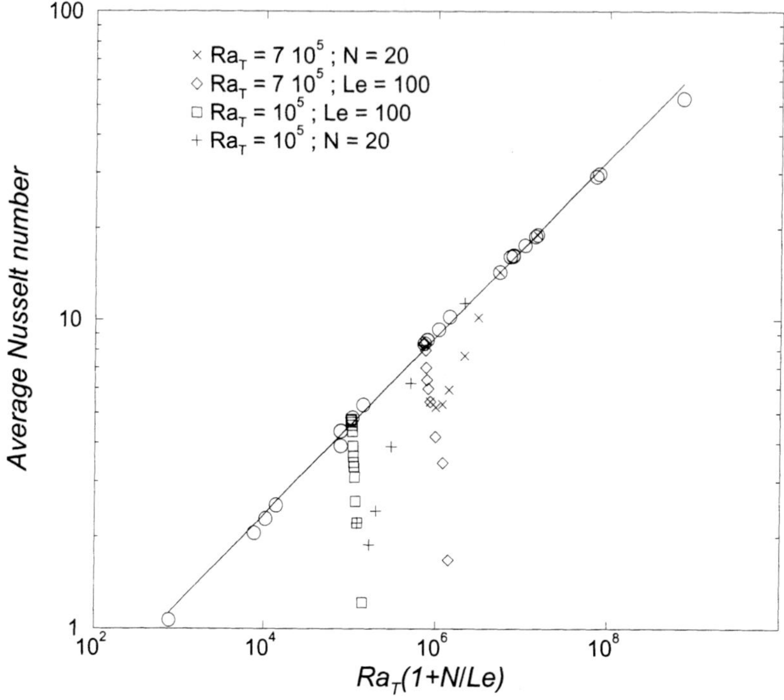

Figure 9. Heat transfer by thermosolutal convection

2. the other one, in the central part of the enclosure, is induced by the thermal buoyancy force.
As the solutal force becomes more important, the size of the central "thermal" cell decreases, the flow in the top and bottom regions of the enclosure become stagnant.

In these low velocity zones, vertical stratification in composition takes place and heat transfer is mainly conductive (vertical parallel isotherms): this explains the significant reduction in heat transfer and the lack of validity of the boundary layer approximation on the whole height of the enclosure. Before the thermal cell complete vanishes, the stratified zones may be destabilized by the lateral temperature gradient, in a similar mechanism to the process initially observed by Chen et al. (1971)and give rise to a multicellular regime constituted of co-rotative thermosolutal cells at a homogeneous concentration.

This evolution through such multicellular regime obviously depends on the aspect ratio and only exists at high Lewis numbers, for intermediate values of N and relatively high thermal Rayleigh number. These qualitative considerations reveal that a precise criterion relating these various parameters has still to be assessed, as it has been done in sideways heating of stratified layers both from experimental (Jeevaraj and Imberger (1991)) and stability (Thangam et al. (1981)) analysis.

The main characteristics of laminar natural convection in enclosures presented in this section allow for understanding the governing mechanisms and the associated scaling laws. In the next chapter, these elements are used to study the influence of such flows in phase change processes.

2 Melting driven by laminar natural convection

In this chapter, the problem of phase change driven by natural convection flows is studied in the case of melting, where the interface may be considered as a surface between the solid and the liquid phases. First we will consider the problem of a pure substance, and the main features of the purely diffusive problem, the *Stefan problem* are briefly recalled, before analyzing the scaling laws of the coupling with natural convection. Then the case of mixtures is presented, and the role of double diffusive convection is tackled.

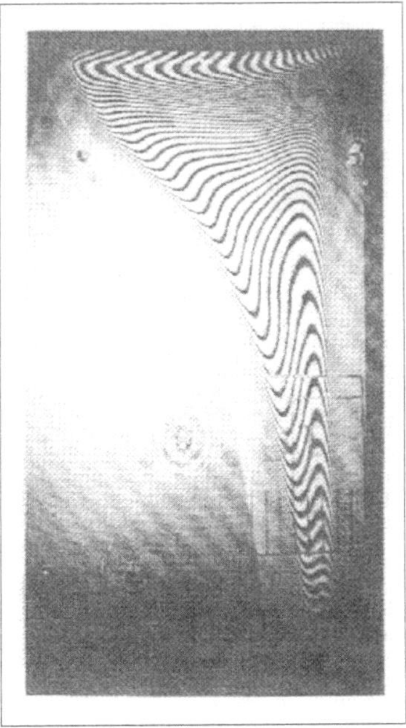

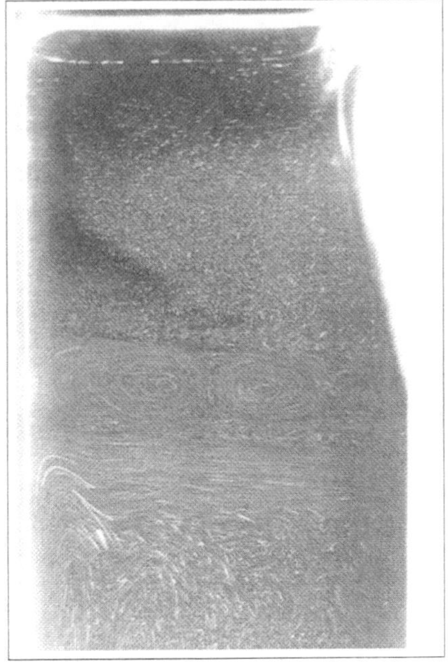

Figure 10. Flow structures during convection driven melting: a - Pure substance: melting of paraffin from a vertical hot wall (Van Buren & Viskanta, 1980) ; b - Binary fluid: melting of ice in a Na_2CO_3 aqueous solution (photo: Sophes Merged)

2.1 Melting of a pure material

The Stefan Problem. The so-called "Stefan problem" consists in describing heat transfer during phase change of a pure substance by pure heat diffusion. The typical feature of this problem is that the time-depending position of the interface between both phases (the melting - or solidification - front) is unknown: this non-linear situation is the archetypical problem of *moving boundary* problems (Crank (1984); Hill (1987); Rubinstein (1971)). This section gives a presentation in the simple one-dimensional case, in order to characterize the problem and present the most popular solutions.

Formulation of the two-phase Stefan problem : Let us consider a pure material with a known phase change temperature T_F. The domain is a semi-infinite medium initially in phase 1 at temperature T_0, eventually different from T_F. At $t^* = 0$, the wall temperature at $x^* = 0$ is set at T_P. The solution of the problem is to describe the time evolution of the interface position $s^*(t^*)$ between the initial phase (1) and the new phase (2).

Phase in formation (phase 2):

$$\frac{\partial T_2}{\partial t^*} = \alpha_2 \frac{\partial^2 T_2}{\partial x^{*2}} \tag{2.1}$$

Initial phase (phase 1):

$$\frac{\partial T_1}{\partial t^*} = \alpha_1 \frac{\partial^2 T_1}{\partial x^{*2}} \tag{2.2}$$

Energy balance at the interface (Stefan condition):

$$-k_2 \left[\frac{\partial T_2}{\partial x^*}\right]_i + k_1 \left[\frac{\partial T_1}{\partial x^*}\right]_i = \rho_S L_F \frac{ds^*}{dt^*} \tag{2.3}$$

Boundary conditions:
Phase 2: $T_2(0,t^*) = T_P$, $T_2(s^*,t^*) = T_F$,
Phase 1: $T_1(\infty,t^*) = T_0$, $T_1(s^*,t^*) = T_F$.

Dimensionless equations: Distances are reduced using a reference length L ($x = x^* / L$) and time using the conduction time in the forming phase ($t = t^*/t_{ref}$), with $t_{ref} = L^2/\alpha_2$
- for melting: $t_{ref} = L^2/\alpha_L$
- for solidification: $t_{ref} = L^2/\alpha_S$.

The reference temperature differences in each phase are:

$$\Delta T_2 = T_P - T_F \ , \ \Delta T_1 = T_0 - T_F$$

$$\alpha^* = \frac{\alpha_1}{\alpha_2} \ , \ k^* = \frac{k_1}{k_2} \ , \ \rho^* = \frac{\rho_S}{\rho_2} \ , \ \Delta T^* = \frac{\Delta T_1}{\Delta T_2}$$

$$\theta_1 = \frac{T_1 - T_F}{\Delta T_1} \ , \ \theta_2 = \frac{T_2 - T_F}{\Delta T_2}$$

Phase in formation:

$$\frac{\partial \theta_2}{\partial t} = \frac{\partial^2 \theta_2}{\partial x^2} \tag{2.4}$$

Initial phase:

$$\frac{\partial \theta_1}{\partial t} = \alpha^* \frac{\partial^2 \theta_1}{\partial x^2} \tag{2.5}$$

Boundary Conditions:
Phase 2: $\theta_2(0,t) = 1\ ,\ \ \theta_2(s,t) = 0\ ,$
Phase 1: $\theta_1(\infty,t) = 1\ ,\ \ \theta_1(s,t) = 0\ .$
Interface energy balance (Stefan condition):

$$- \left[\frac{\partial\theta_2}{\partial x}\right]_i + \Delta T^*\ k^* \left[\frac{\partial\theta_1}{\partial x}\right]_i = \frac{\rho^*}{\mathrm{Ste}}\frac{ds}{dt} \tag{2.6}$$

The energy balance equation at the moving front (eq. 2.6), introduces the governing parameter of the phase change problem: the *Stefan number*, defined as:

$$Ste = \frac{\mathrm{C}_{P2}\,\Delta T_2}{\mathrm{L}_F} \tag{2.7}$$

which corresponds to the ratio between the sensible heat and the latent heat involved in the process. Note that the temperature difference used in the Stefan number definition is the ΔT driving the phase change process.
Analysis: In terms of orders of magnitude, let us consider the simplified problem where $\Delta T^* = 0$ (initial phase at T_F). The interface energy balance (eq. (2.3)) gives:

$$-k_2\frac{T_P - T_F}{s^*(t^*)} \sim \rho_S L_F\frac{ds^*}{dt^*}$$

leading to:

$$s^*\ ds^* \sim \alpha_2\ \mathrm{Ste}\ dt^*$$

and

$$s \sim 2\sqrt{\mathrm{Ste}/2}\ \sqrt{t}\ .$$

Analytical solution: The above system of equations has a similarity solution, where the time evolution of the front position is assumed to be of the form:

$$s = 2\lambda\sqrt{t}\ .$$

Expressing the system of equations (2.4 to 2.6) in terms of the similarity variable (see any textbook on conduction, e.g. Crank (1984)): $\eta = x/\sqrt{t}$, one gets the so-called Neumann solution where λ is obtained by solving:

$$\frac{e^{-\lambda^2}}{erf\lambda} - \Delta T^*\frac{k^*}{\alpha^{*1/2}}\frac{e^{-\lambda^2/\alpha^*}}{erfc(\lambda/\alpha^{*1/2})} = \frac{\lambda\pi^{1/2}}{\mathrm{Ste}} \tag{2.8}$$

where:

$$erfx = \frac{2}{\sqrt{\pi}}\int_0^x e^{-u^2}du\ .$$

This equation has to be solved numerically.

The corresponding temperature fields in both phases are (recalling that $\alpha^* = \alpha_1/\alpha_2$):

$$\theta_1 = 1 - \frac{erfc(\eta/\sqrt{\alpha^*})}{erfc(\lambda/\sqrt{\alpha^*})} = 1 - \frac{1}{erfc(\lambda/\alpha^{*1/2})}erfc\left(\frac{x}{2\sqrt{\alpha^* t}}\right)$$

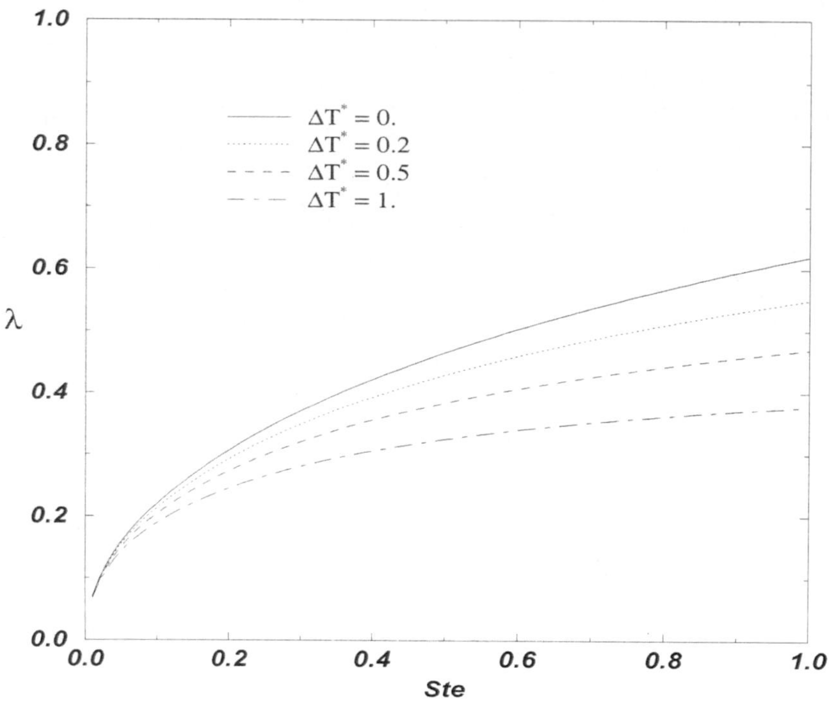

Figure 11. Variation of λ with Ste and ΔT^*

$$\theta_2 = 1 - \frac{erf\eta}{erf\lambda} = 1 - \frac{1}{erf(\lambda)} erf\left(\frac{x}{2\sqrt{t}}\right) \tag{2.9}$$

Figure 11 displays typical values of λ over a range of Stefan numbers and of the ratio ΔT^* values. Obvious features of the Stefan problem may be observed on the picture: the fact that the interface velocity (proportional to λ) increases with Ste (smaller latent heat) and decreases with ΔT^* (more resistance to phase change due to conduction in the initial phase). Although of a very limited interest, this analytical solution maybe used as a reference solution for testing numerical procedures in phase change problems. A similarity solution for the two-dimensional "corner problem" maybe found in the bibliography (Rathjen and Jiji (1971)).

Remarks: 1. In the particular case where there is no conduction in the initial phase ($\Delta T^* = 0$):

$$\lambda \, erf\lambda \, e^{\lambda^2} = \frac{Ste}{\sqrt{\pi}} \, .$$

2. In the limit case where Ste $\rightarrow$ 0, the specific heat is negligible compared to latent heat:

$$\lambda = \sqrt{\frac{Ste}{2}} \, .$$

This approximate solution may be used whenever the heat diffusion time in phase 2 is small compared to the characteristic time of interface displacement. A number of approximate methods have been developed in the literature, using integral methods (Goodman (1964)) or perturbation methods. Although these methods may be relevant to the simplified analysis of some phase change problem, they have few applications to phase change driven by convective heat transfer, and they will not be developed here. The interested reader may refer to the corresponding bibliography.

Numerical solutions: In many situations, the phase change problems have no analytical nor approximate solutions. The complete solution of the problem then requires the help of numerical simulation. This section is dedicated to the" description of the two main alternatives met in the bibliography: the "front-tracking methods" and the "one-domain methods". These methods will be developed in other parts of this lecture, and we will present their main features in the frame of the most simple 1D conduction with phase change problem.

Let us consider the previous two phase Stefan problem in a finite domain of extension L in the x-direction. The initial solid phase is isothermal at a temperature $T_0 < T_F$. As previously, at t* = 0, the temperature of the wall at x* = 0 is imposed at $T_P > T_F$, while the wall at x* = L is maintained at T_0. The reference length of this problem is naturally the thickness of the plate, L and the reference time is the diffusion time in the liquid phase:

$$z = \frac{x}{L}\,,\ t_{ref} = L^2/\alpha_L\,.$$

Coordinate transformation: the Landau method : In each phase, a coordinate transformation is applied, such that the interface position is a fixed surface in the transformed space (Landau (1950)). Thus, one defines respectively, for the liquid phase:

$$x_L = \frac{z}{s}$$

and, in the solid phase:

$$x_S = \frac{1-z}{1-s}\,.$$

The transformation is such that, in the liquid:

$$df = \left(\frac{\partial f}{\partial z}\right)_t dz + \left(\frac{\partial f}{\partial t}\right)_z dt = \left(\frac{\partial f}{\partial x_L}\right)_t dx_L + \left(\frac{\partial f}{\partial t}\right)_{x_L} dt\,.$$

Then:

$$\frac{\partial f}{\partial z} = \frac{1}{s}\frac{\partial f}{\partial x_L}$$

$$\left(\frac{\partial f}{\partial t}\right)_z = \left(\frac{\partial f}{\partial t}\right)_{x_L} - \frac{x_L}{s}\frac{ds}{dt}\frac{\partial f}{\partial x_L}\,.$$

The transformed conservation equation is then:

$$\frac{\partial \theta_L}{\partial t} - \frac{x_L}{s}\frac{ds}{dt}\frac{\partial \theta_L}{\partial x_L} = \frac{1}{s^2}\frac{\partial^2 \theta_L}{\partial x_L^2}\,, \tag{2.10}$$

with the corresponding boundary conditions:

$$z = 0\,,\, x_L = 0\,,\, \theta_L = 1\,,$$

$$z = s\,,\, x_L = 1\,,\, \theta_L = 0\,.$$

In the solid phase, a similar transformation is used:

$$x_S = \frac{1-z}{1-s}\,,$$

and the heat conservation equation becomes:

$$\frac{\partial \theta_S}{\partial t} + \frac{x_S}{1-s}\frac{ds}{dt}\frac{\partial \theta_S}{\partial x_S} = \frac{\alpha^*}{(1-s)^2}\frac{\partial^2 \theta_S}{\partial x_S^2}\,, \tag{2.11}$$

with the corresponding boundary conditions:

$$z = s\,,\, x_S = 1\,,\, \theta_S = 0\,,$$

$$z = 1\,,\, x_S = 0\,,\, \theta_S = 1\,.$$

The interface condition writes:

$$-\frac{1}{s}\left[\frac{\partial \theta_L}{\partial x_L}\right]_i - \Delta T^*\, k^*\, \frac{1}{1-s}\left[\frac{\partial \theta_S}{\partial x_S}\right]_i = \frac{\rho^*}{\mathrm{Ste}}\frac{ds}{dt}\,. \tag{2.12}$$

Resolution: The mathematical problem, formulated by the previous system of equations, clearly shows the non-linearity of the problem. Actually the coefficients of the equations depend upon the solution of the problem through the instantaneous position and the velocity of the front. On the other hand, the simplification of the formulation due to the "front immobilization" is compensated by a new "pseudo-convective" term which arises from the deformation of the physical space and from the fact that the mass of material included in the *fixed* computational space changes with time with the displacement of the interface.

The solution procedure is thus iterative, and a possible algorithm is as follows:

- Estimation of the front position and velocity over the considered time-step,
- Resolution of the conservation equations in each phase using the corresponding estimation of the coefficients,
- Resolution of the interface equation with the resulting temperature fields,
- The calculated front velocity is compared to the estimation at the beginning of the cycle, and the iteration loop proceeds up to a predetermined convergence criterion is satisfied.

Melting driven by thermal natural convection: equations and parameters
The problem under consideration deals with melting of a pure substance controlled by natural convection in the melt. One considers a 2D square cavity (height H = width L) initially filled with a solid material uniformly at the melting temperature ($T_0 = T_F$). At

$t^* = 0$, the temperature of one of the vertical walls (the left wall in Figure 12) is raised at a value $T_1 > T_F$, while the other vertical wall is maintained at the initial temperature. The horizontal walls are assumed to be adiabatic and no-slip. The fluid flow is supposed to be in the laminar regime, and the thermophysical properties of the material to be constant.

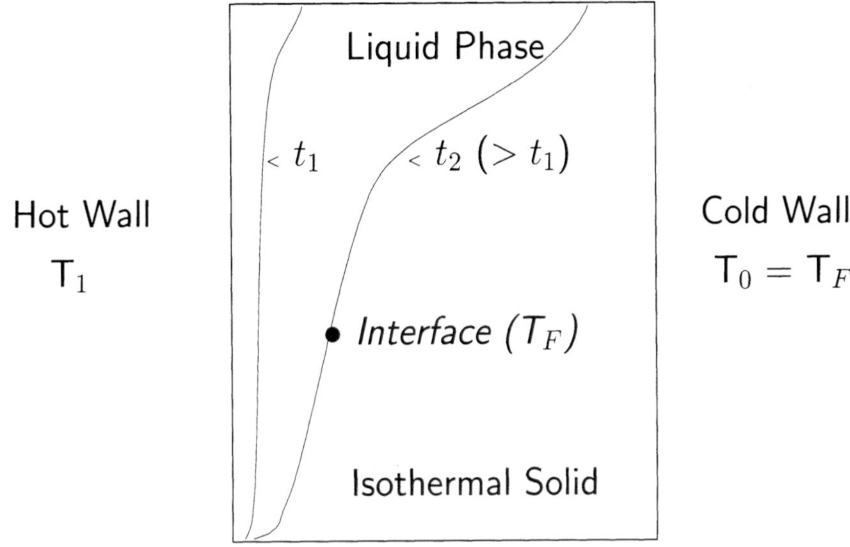

Figure 12. melting driven by natural convection: schematic description

After a pure conduction stage, thermal convection develops in the liquid phase, causing a non-uniform distribution of the heat flux at the interface and a non-uniform displacement of the melting front. The set of equations governing the problem is the classical set of equations for natural convection (1.1) to (1.4). Let $s^*(z^*, t*)$ be the position of the melting front, specific boundary conditions have to be written at the interface:

$$T(s^*(z^*, t*), t^*) = T_F \ ,$$

$$-k_L(\nabla T_L)_i.\vec{n} = \rho_S \ L_F \ \vec{v_F}.\vec{n} \ .$$

The problem is characterized by a set of four main dimensionless parameters. The fluid phase is defined by its Prandtl number: $Pr = \nu/\alpha$ and the intensity of natural convection is given by the thermal Rayleigh number: $Ra = g\beta(T_1 - T_F)H^3/(\alpha\nu)$. Given the temperature conditions, the Stefan number defines the relative importance of the latent heat in the overall energy balance, $Ste = C_{PL}(T_1 - T_F)/L_F$. Finally the geometrical parameter is the global aspect ratio of the enclosure ($A = H/L$).

Scaling Laws and Correlations. The analysis of the melting process based on the preceding numerical and experimental results allows for proposing a simplified modelling leading to a number of correlations. The model is generally concerned with the overall

heat transfer at the interface in order to predict the time evolution of the melted fraction in a simple way.
Limit Regimes: We will first focus on the ideal case where the solid phase is initially at the melting temperature and the adjacent wall is maintained at this temperature. Consequently, there is no heat conduction in the solid phase and all the energy transferred at the interface by natural convection is used for phase change. The melting process in the presence of natural convection is characterized by the existence of two limit regimes:
1. at short times, the process is dominated by heat conduction, the melting front remains parallel to the active (hot) wall, and the time evolution of the liquid layer thickness is given by the Neumann solution or, in the limit of small Stefan numbers, by the approximate solution (in the rectangular geometry):

$$s \sim (2\tau)^{1/2} , \tag{2.13}$$

where $s = s^*/H$ is the dimensionless thickness of the liquid layer, and the dimensionless time $\tau = Ste \times Fo$ is built on the Fourier number $Fo = \alpha_L/H^2$. The corresponding expression of the dimensionless heat transfer at the interface (the Nusselt number):

$$Nu_{cd} = \frac{ds}{d\tau} \sim (2\tau)^{-1/2} . \tag{2.14}$$

2. at higher times, after the convective regime is established, natural convection dominates the melting process and when a separate boundary layer regime is reached, the average Nusselt number at the interface does not depend on time. Nu is only a function of the dimensionless numbers characterizing natural convection (the Prandtl and Rayleigh numbers):

$$Nu_{cv} \sim Ra^{1/4} , \tag{2.15}$$

and the dimensionless front velocity is a constant (as far as the geometry - cold wall - does not interfere with the front displacement).
Global Correlations: From this description of the limit regimes, one can establish an analytical expression of the time evolution of heat transfer: the aim is to find a good description of the transition between the two limits, taking in consideration eq. (2.14) at small times and eq. (2.15) in the developed convection regime.

Scaling Laws for high Prandtl number materials. The foregoing scale analysis is taken from the work by Bejan (1988), which takes into account the whole melting process. It consists in analyzing the orders of magnitude in the two regimes we have identified and also in the transition regime. This allows for an estimation of the average heat transfer, and thus of the instantaneous average melting front velocity during the whole process.

1. Conduction regime (Fig. 13-a)

Heat transfer is dominated by conduction during the initial step of the melting process. We can thus use relations (2.13) and (2.14) to describe the time evolution of the front position and of the average heat transfer at the beginning of the melting process. We thus consider the rectangular liquid domain of height H and width s^*, to which we apply the classical scaling laws of natural convection in enclosures. From the previous chapter, we know that natural convection is present even for high aspect ratios ($A = H/s^*$) and

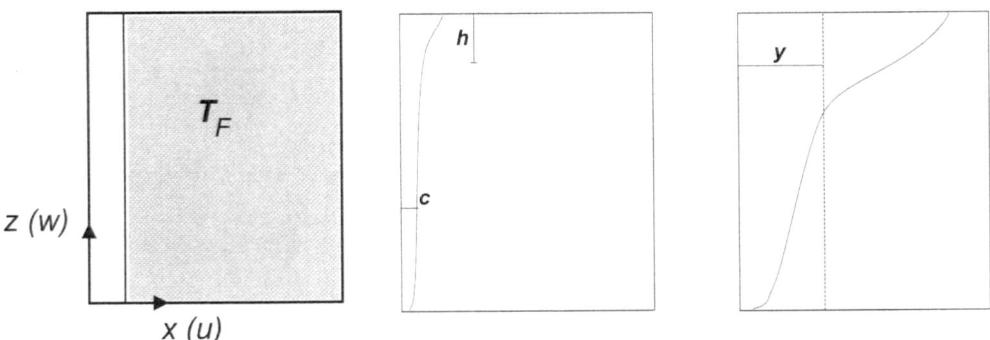

Figure 13. Evolution of the fluid domain: a. Conduction - b. Transition - c. Convection

our purpose is to determine when the influence of convection on the heat transfer is significant.

Assuming that the buoyancy term is balanced by the diffusion term in the momentum equation, we know that the vertical velocity scales as :

$$w \sim s^{*2} \frac{g\beta(T_1 - T_F)}{\nu}$$

The energy transported by convection along the hot wall is thus:

$$\Phi_{\mathrm{cv}} =\sim s^* w \, \rho C_p \, (T_1 - T_F) \tag{2.16}$$

and the total dimensionless heat transfer (conduction + convection) scales as:

$$Nu \sim \tau^{-1/2} + Ra \, \tau^{3/2} \, . \tag{2.17}$$

Let us recall that this is an order of magnitude and that we are interested in the dominating mechanisms of the processus: the determination of the proportionality constants will be treated further. Expression (2.17) clearly emphasizes that the relative importance of convection ($Ra \, \tau^{3/2}$ term) increases with time while the conductive contribution term ($\tau^{-1/2}$) actually decreases.

2. Transition regime (Fig. 13-b)

Here *transition* means the stage of the process where convection develops in the fluid, as long as the boundary layer regime is still not established. During this stage, the convective contribution to the overall heat transfer mainly concerns the top part of the narrow liquid enclosure, which thus melts faster and gets progressively irregular. Let h be the scale of the height and width of this zone. It may be shown (Jany and Bejan (1988)) that eq. (2.17) is also valid in the transition regime when the contribution of convection cannot be neglected any more before conduction. This expression is valid up to the limit $h \sim H$. This limit is reached when the value of the Nusselt number is minimum, and corresponds to τ on the order of:

$$\tau_1 \sim Ra^{-1/2} \, . \tag{2.18}$$

and to a minimum value of the Nusselt number on the order of:

$$Nu_{\min} \sim Ra^{1/4} . \tag{2.19}$$

3. Convective regime (Fig. 13-c)

When $\tau > \tau_1$, the separate boundary layer regime prevails over the whole height of the liquid cavity and it is well known that the Nusselt number scale for $Pr > 1$ is given by:

$$Nu_1 \sim Ra^{1/4} . \tag{2.20}$$

This order of magnitude, assessed for rectangular cavities, remains valid if the inclination of the melting front on the vertical is moderate.

During this stage of the process, the local heat transfer rate is not uniform along the interface and that the front position $s(z, \tau)$ depend on the height z. We will thus consider average values, and particularly the mean position of the interface defined as:

$$y(\tau) = \frac{1}{H} \int_0^H s^*(z^*, \tau)\, dz^*$$

which leads to the melted fraction ($f = y/L$), is such that:

$$Nu_1 = \frac{1}{H} \frac{dy}{d\tau} , \tag{2.21}$$

and

$$y(\tau) \sim H\, Ra^{1/4}\, \tau . \tag{2.22}$$

The width L of the enclosure has been ignored so far, assuming that the melting front does not reach the cold wall. The contact with this wall ($s^* = L$) happens (eq. (2.22)) at a time τ_2 on the order of:

$$\tau_2 \sim \frac{L}{H} Ra^{-1/4} , \tag{2.23}$$

and it may be assessed that the fully developed convection regime will establish only if $\tau_2 > \tau_1$, that is:

$$Ra^{1/4} > \frac{H}{L} . \tag{2.24}$$

4. General Correlation for the whole melting process.

The preceding analysis leads to general scaling laws for the global heat transfer at the interface and for the average melting front position as a function of the governing parameters of the problem and of the dimensionless time $\tau = SteFo$ only. It allows for proposing a general correlation describing the different regimes:

1. Nu $\sim (2\tau)^{-1/2}$ for $\tau \simeq 0$,
2. Nu shows a minimum $Nu_{\min} \sim Ra^{1/4}$ for $\tau_1 \sim Ra^{-1/2}$,
3. Nu $\sim Ra^{1/4}$ for $\tau \to \infty$

in one single expression:

$$Nu(\tau) = \frac{1}{\sqrt{2\tau}} + \left[Nu_\infty - 1/\sqrt{2\tau}\right] \left[1 + (c_1\, Ra^{3/4}\, \tau^{3/2})^n\right]^{1/n} .$$

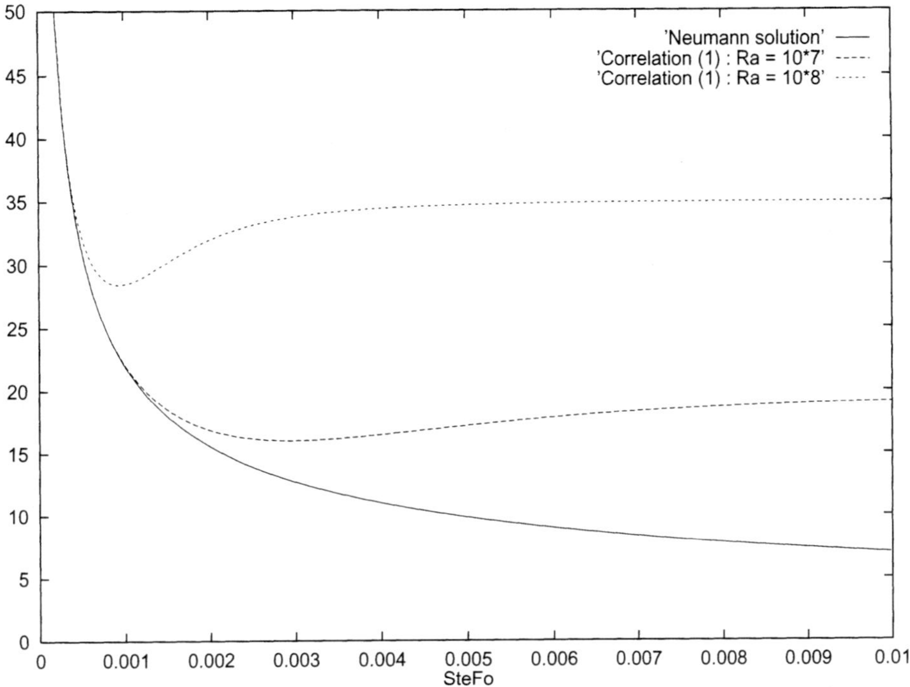

Figure 14. Interface heat transfer evolution for high Pr number (Pr = 50.)

The coefficients c_1 and n of this relation have been determined by adjusting numerical simulations performed in the $[10^5 - 10^8]$ Ra range for $Pr = 50$: $c_1 = 0.0175$, $n = -2$. Thus, for $Pr \geq 1$, the correlation writes:

$$Nu(\tau) = \frac{1}{\sqrt{2\tau}} + \frac{Nu_\infty - 1/\sqrt{2\tau}}{\sqrt{1 + \frac{1}{(0.0175\, Ra^{3/4}\, \tau^{3/2})^2}}} . \tag{2.25}$$

Nu_∞ is given by $Nu_\infty = 0.33\, Ra^{0.25}$ for $Pr >> 1$ (Bénard et al. (1986)). This correlation gives access to the time evolution of the liquid fraction by integrating (2.21) between 0 and τ, but a fairly good approximation may be obtained from the asymptotic limits at $\tau \to 0$ and $\tau \to \infty$:

$$f(\tau) = \frac{H}{L} \left[(\sqrt{2\,\tau})^5 + (0.33\, Ra^{1/4}\, \tau)^5 \right]^{1/5} \tag{2.26}$$

These correlations are valid on the time interval $0 < \tau < \tau_2$, provided that the criterion (2.24) is satisfied.

Scaling Laws for low Prandtl number materials. The present analysis may be extended to the range of low Prandtl number fluids ($Pr << 1$), using $Ra \times Pr$ as the

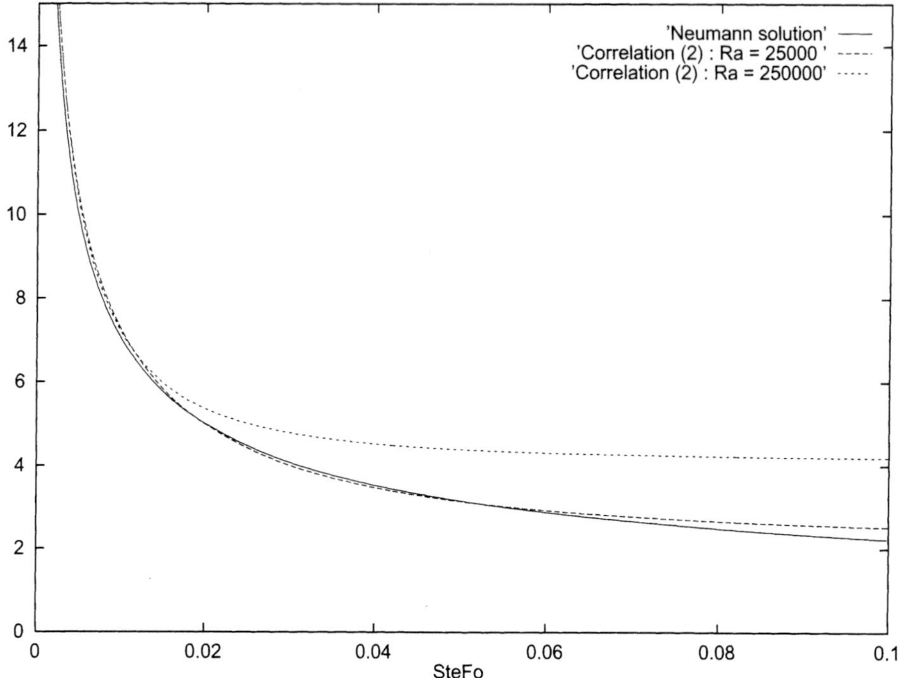

Figure 15. Interface heat transfer evolution for low Pr number (Pr = 0.02)

governing parameter (Bejan (1995)). Using this kind of approach, we could show a separate influence of the Rayleigh and Prandtl numbers (Gobin and Bénard, 1992) in the convection regime, but the influence of the $Ra \times Pr$ group seems to be determining in the transition regime. The general features of the Nusselt number evolution are roughly the same except that the simulations for the melting of metals do not exhibit a minimum.

The expressions of the Nusselt number in the limit regimes lead to a correlation for the whole melting process of the following form:

$$Nu(\tau) = Nu_\infty + \frac{1}{\sqrt{2\tau}} \left[1 - \frac{1}{\sqrt{1 + \frac{1}{((Ra.Pr)^{0.36}\,\tau^{0.75})^2}}} \right], \tag{2.27}$$

where the numerical parameters have been identified from numerical simulations for Ra between 10^4 and 10^6 and Pr between 0.01 and 0.1, corresponding to $Ra \times Pr$ values ranging from 200 to 20000.

The existing correlations for Nu_∞ are:
$Nu_\infty = 0.29\, Ra^{0.27}\, Pr^{0.18}$ (Gobin and Bénard, 1992) for the $Pr << 1$ range

$$Nu_\infty = \frac{0.35\, Ra^{1/4}}{\left[1+(0.143/Pr)^{9/16}\right]^{4/9}}$$ (Lim and Bejan, 1992) for any Pr value.

Note that these correlations are not based on a complete scale analysis, and that there validity may be questioned. The physics of natural convective flows in low Pr number fluids is still a challenging issue and the quasi-steady boundary layer analysis may be not fully relevant in this range (see LeQuéré and Gobin (1999)).

2.2 Melting in a binary mixture

The problem of solid-liquid phase change in a binary mixture is relevant to model most of the solidification processes. The pure diffusive problem corresponding to the Stefan problem (*i.e.* with a smooth interface, see section 2.1) involves the conservation equations for energy and species in both phases, energy and species balance at the interface and the liquidus relation between the interface concentrations and temperature. The interested reader may refer to Caroli et al. (1992) where the complete formulation and the underlying assumptions are given in detail. Simple analytical solutions in some limiting cases are analyzed in Flemings (1974) or Kurz and Fisher (1998). A complete analytical solution in the quasi stationary regime is developed by Rubinstein (1971), but these solutions find little applications since the experiments are either affected by the existence of a "mushy" dendritic growth zone for solidification and/or by strong natural convection effects.

This is why the scope of this section is limited to the interaction between thermosolutal natural convection in the liquid phase and phase change phenomena, in the case of melting of a pure solid in a binary mixture. This situation has been extensively studied at the experimental level (Bénard et al., 1996): it consists in melting of a vertical layer of pure ice in an aqueous solution. The 2D cavity (height H, width L) is initially filled with a solution at a uniform temperature ($T_0 = T_F$) and concentration (C_0). The cold wall is made of pure ice at T_F. At $t^* = 0$, the temperature of the opposite vertical wall is raised at a value $T_1 > T_F$, while the ice wall is maintained at the initial temperature. The horizontal walls are assumed to be adiabatic and no-slip. The fluid flow is supposed to be in the laminar regime, and the thermophysical properties of the material to be constant. Heat transfer by thermal convection induces ice melting and thus a composition gradient at the interface between the solution and pure water. The solutal buoyancy force thus created combines with or opposes to the initial thermal buoyancy force.

Experimental Observations. In agreement with previous experiments (Huppert and Turner (1978); Beckermann and Viskanta (1988)) the observation show the existence of an initial thermoconvective cell which induces melting of the ice at the "cold wall". This creates a significant concentration gradient between the pure water from the melt and the initial solution, and a strong ascending flow along the interface due to solutal buoyancy. This results in the formation of a quasi-motionless "stagnant" zone in the top part of the enclosure accompanying the recession of the initial thermal cell. The stagnant zone may then be destabilized by lateral heating, depending on the experimental conditions, and give rise to a sequential formation of thermosolutal cells which get formed in the shear layer between the bottom of the stagnant zone and the last cell. The figure displays a zoom of the visualization of corotative rolls formation at the interface between the initial thermal cell and the stagnant zone in the top of the cavity. This destabilization mechanism is essentially of double diffusive origin.

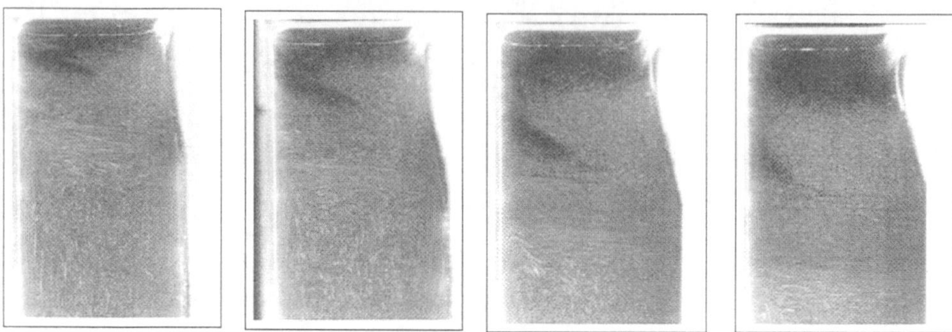

Figure 16. Evolution of the flow structures

Figure 17. Destabilization of the stagnant zone and formation of a thermosolutal cell

Equations and Parameters. The basic conservation equations for mass, heat, species and momentum are the same as presented in section 1.2 (1.23) to (1.26) and the convective problem is characterized by the four dimensionless parameters characterizing double diffusion previously defined. The complete analysis of the problem would involve the phase change conditions at the moving interface (energy and species balance), plus the phase diagram defining the local equilibrium temperature and concentration conditions at the interface. Here we are essentially interested by the specific structure of convective flows in such a system, and some simplifying hypotheses may be made, concerning the concentration and temperature boundary conditions at the melting front. If we assume

that the front movement does not disturb the flow in the liquid phase, one may accept to simulate the process in a fixed cavity with a non-moving melting front. This hypothesis is consistent with the assumption of constant concentration and temperature at the interface. In dimensionless terms, the boundary conditions of the problem are :
– at the interface (fixed cold wall - $x = L/H$) :

$$\theta(1, z, t) = 0 \; ; \; \phi(1, z, t) = 0 \; . \tag{2.28}$$

– on the hot exchanger ($x = 0$) :

$$u = w = 0 \; ; \; \theta(0, z, t) = 1 \; ; \; \frac{\partial \phi}{\partial y}(0, z, t) = 0 \; . \tag{2.29}$$

– at the other walls :

$$u = w = 0 \text{ (at the bottom wall) } ; \; \nabla v.\vec{n} = 0 \; ; \; w = 0 \text{ (at the top wall) } , \tag{2.30}$$

$$\nabla \theta.\vec{n} = 0 \; ; \; \nabla \phi.\vec{n} = 0 \; . \tag{2.31}$$

The initial conditions are :

$$\theta(x, z, 0) = 0 \; ; \; \phi(x, z, 0) = 1 \; . \tag{2.32}$$

Simulation results. Numerical simulations in a *fixed* cavity, that is neglecting the effects of melting in the calculation, including the interface motion and shape, confirm that the driving mechanism for the building-up of the cells is only due to thermosolutal convection.

The numerical simulation represents the transient evolution of a thermosolutal convective cell in the binary solution initially at a concentration C_0 and at the temperature T_0 of the right wall. The streaklines show :
- on one hand the filling process of a stagnant zone in the top part of the enclosure, on top of the initial "thermal" cell (the particles driven upwards by the right solutal boundary layer remain on their iso-density level, and a vertical stratification builds up in this zone),
- on the other hand, the destabilization in the bottom part of the stagnant zone which leads to the formation of mixing cells, called "thermosolutal" cells.

But although the results are in good qualitative agreement, the transient behavior of the system is not well represented by the computations (growth velocity of the stagnant zone, formation time of the cells, etc.). The main reason for this discrepancy is probably in the description of the thermodynamic equilibrium conditions at the interface : the assumptions made in the model are only correct if convection is strong enough to evacuate the melted (pure) ice ; in the domains where the melting rate is small, solid-liquid phase change is dominated by dissolution (Woods, 1992) and the interface conditions correspond to an equilibrium point on the liquidus line which is between pure ice (zero concentration) and the initial liquid concentration. This leads to an interface temperature below 0^oC and by an interface concentration higher than 0. Consequently, the actual temperature difference between the hot wall and the interface (governing the

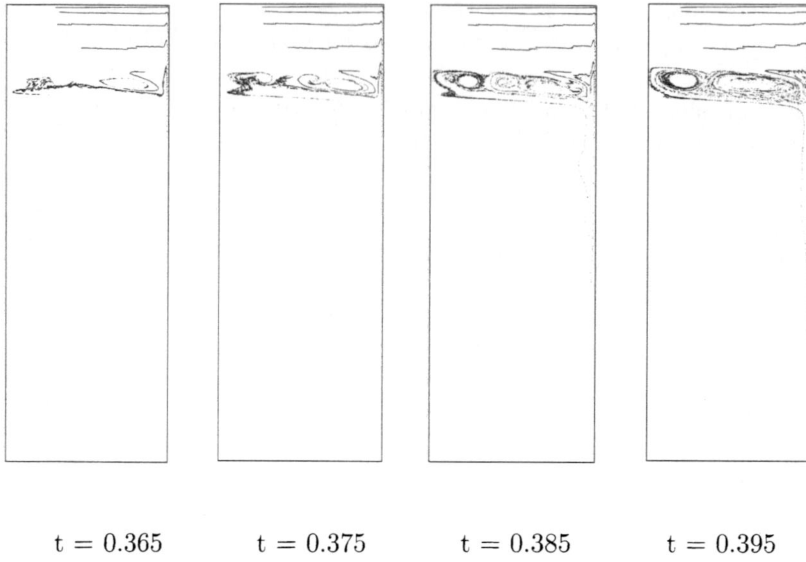

Figure 18. Numerical simulations : flow structures

(A = 3 ; Pr = 10 ; Le = 210 ; $\mathrm{Ra}_T = 2.2 \times 10^8; N = -22$)

thermal Rayleigh number) is higher than the reference difference and the actual interface concentration (defining the solutal Rayleigh number) is smaller.

These few indications are intended to stress the importance of convective phenomena in the description of solid-liquid phase change processes. Double diffusive effects are to be expected in most solidification processes, and it is not sure that the generally dominating solutal buoyancy force allows to neglect the thermal body force and consider mere conduction of the temperature field. Oscillatory behaviors or multicellular flows may occur and significantly affect the heat and species transfer distribution in the mushy zones. Accurate description of these highly convective flows is still a challenge in many situations, and this difficulty should not be underestimated when accounting for convection in solidification codes.

3 The melting problem : a comparison exercise

In the heat and mass transfer bibliography, many studies have bee devoted to the problem of melting driven by natural convection in the melt. Different numerical procedures have been proposed and many experiments referring to the melting or solidification problem have been performed, using either low melting temperature metals (Ga or Sn) or pure paraffin waxes (octadecane, eicosane). In spite of a relatively rich literature, very few attempts have been made to systematically compare these solutions or to provide a reference solution to such a classical problem.

In this chapter, the results of a test problem for comparing the different models and numerical procedures are presented. The conclusions of an international comparison exercise are summarized (Bertrand et al. (1999); Gobin and LeQuéré (2000)), leading to a general reflection on benchmark solutions of phase change processes.

3.1 The benchmark problem.

The situation considered for the test problem is the one described in section 1 of Part II. The analysis of the bibliography shows that :
1. in the absence of reference experiments, purely numerical comparisons must be performed,
2. two groups of numerical tests have to be proposed, corresponding to the distinct Prandtl number ranges : the high Prandtl number liquids ($Pr \sim 10^2$), typically paraffin waxes, and the low Pr domain ($Pr \sim 10^{-2}$), for the melting of metals.

Proposed test cases. The governing parameters have been estimated using approximate values of the thermophysical properties of tin and octadecane : in the low Prandtl number range, the Pr value is taken to be 0.02, and in the high Prandtl number range $Pr = 50$. For a given geometry ($A = 1$), the values of the Rayleigh and Stefan numbers correspond to a dimensional height of the enclosure H = 0.10 m and a reference temperature difference $T_1 - T_F = 3$ °C for tin (Case 2) and 10 °C for octadecane (Case 4), leading to the values displayed in Table 1. The description of the melting process resulting from the scale analysis presented in Part II for these two cases is displayed in Figs. 14 and 15. For each Pr range, a 10 times smaller Rayleigh number (Cases 1 and 3) is also considered.

$Pr = 0.02$ Ste = 0.01	Case #1 Ra = 2.5 10^4	Case #2 Ra = 2.5 10^5
$Pr = 50$ Ste = 0.1	Case #3 Ra = 10^7	Case #4 Ra = 10^8

Table 2. Parameters of the test cases.

In order to limit the number of outputs, the following results are requested :
1. the time evolution of the melted volume and of the average Nusselt number at the hot wall,
2. the position of the melting front and the local Nusselt number distribution at four different times (expressed in the dimensionless form $\tau = Fo \times Ste = \alpha t^* \times Ste/H^2$) :

- at $Pr = 0.02$: $t_1 = 4 \times 10^{-3}$ $t_2 = 10^{-2}$ $t_3 = 4 \times 10^{-2}$ $t_4 = 10^{-1}$,
- at $Pr = 50$: $t_1 = 5 \times 10^{-4}$ $t_2 = 2 \times 10^{-3}$ $t_3 = 6 \times 10^{-3}$ $t_4 = 10^{-2}$.

Contributions. Among the contributions to the comparison exercise (see Gobin & Le Quéré (2000) for a more complete description), 10 authors have presented results for Case #2 (melting of tin) and 6 for the high Prandtl number range (Case #4). These contributions have presented a selection of the most popular models and of the main numerical procedures. Actually, a larger number of contributions have used fixed grid or "enthalpy" methods (FG), while three (referred to as FVM5, FVM6 and CVFEM) which have used a front-tracking or two-domain procedures (FT).

One-domain methods (FVM1-4 - FDM1 - FVM7 - FEM1)

The common features of these models are the use of the enthalpy formulation for energy conservation and of the primitive variable Navier-Stokes equation for momentum conservation.

The transition from the solid to liquid phase is treated by a Darcy-like penalization term in the momentum equation depending on the local solid fraction, in all contributions but one, for which this is handled by a strong viscosity variation in two-phase volumes. The discretization technique mainly uses the finite volume approach, except one contribution using a finite element technique (FEM1). Structured fixed grids are used in most cases, with the exception of FVM7 which uses a moving structured grid to increase the accuracy in the calculation of the flow field and FEM1 where the grid is fixed but eventually non structured.

Front-tracking methods (FVM5 - FVM6 - CVFEM)

The common features of these models are the explicit calculation of the front movement and the use of some kind of upwinding for discretizing the convective terms. A finite volume procedure with coordinate transformation is used by (FVM5 and 6), while CVFEM uses a CVFEM approach with an adaptive unstructured grid. The streamfunction-vorticity formulation is used by FVM6, while both FVM5 and CVFEM solve the flow problem in terms of the U-V-P primitive variables. The last main difference lies in the use of the quasi-stationarity assumption by FVM5, while FVM6 and CVFEM solve the full transient flow problem.

Results. Only the main trends of the results are recalled in this section, and the outputs presented for Cases #2 and #4 are : the time evolution of the average Nusselt number at the hot wall, and the position of the melting front at a given time of the process.

The high Prandtl number range (Case #4). The results at $Pr = 50$ and $\mathrm{Ra} = 10^8$ (Case #4) are displayed in Figure 19 in terms of the time evolution of the Nusselt number. Although a number of solutions are in a reasonably good agreement in the average, it is clear that some simulations fail to predict the process, and show unrealistic behaviors : four solutions out of the eight contributions do not display the average behavior and the dispersion on the melted fraction at time t_4 is almost $\pm$ 20 %. If we arbitrarily discard the extreme behaviors, the remaining results are relatively close to each other, but the relative differences are nevertheless significant, resulting in a $\pm$ 4.5 % dispersion on the melted fraction at time t_4 . In the absence of any reference solution, it is difficult to further quantitatively classify the different solutions.

The observation of the melting front position at $t_4 = 0.010$ shows a large discrepancy between the computed shapes of the interface (Figure 19-a). This indicates that, although it corresponds to easily reached experimental conditions, this value of the Rayleigh num-

Contributor	Formulation (1/2 domain)	Method Space scheme	Front description	Misc.
FVM1-FG	U-V-H 1-domain	FVM (UDS) 1st order		Source Term (KC)
FVM2-FG	U-V-H 1-domain	FVM (HDS) 1st order		Source Term Penalization
FVM3-FG	U-V-H 1-domain	FVM (HDS) 1st order	$f_S = 0.5$	Source Term (KC)
FVM4-FG	U-V-H 1-domain	FVM (PLS) 1st order	PCI : 0.1 K $f_L = 0.99$	Source Term (KC)
FDM1-FG	U-V-H 1-domain	Dif. Approx. 1st order		Source Term Penalization
FVM5-FT	U-V-T 2-domain	FVM (HDS) 1st order	Coordinate Transformation	Quasi-stationarity
FVM6-FT	ψ-ω-T 2-domain	FVM 1st order	Coordinate Transformation	
FVM7-FG	U-V-H 1-domain	FVM (CDS) 2nd order	Isotherm 0.001 PCI : 0.001	Multigrid Expanding grid
FEM1-FG	U-V-H 1-domain	FEM 2nd order	PCI : 0.001	Source Term
CVFEM-FT	U-V-T 2-domain	CVFEM (EDS) 1st order	Explicit	Adaptive grid Triangular

Table 3. Contributions to the numerical test.

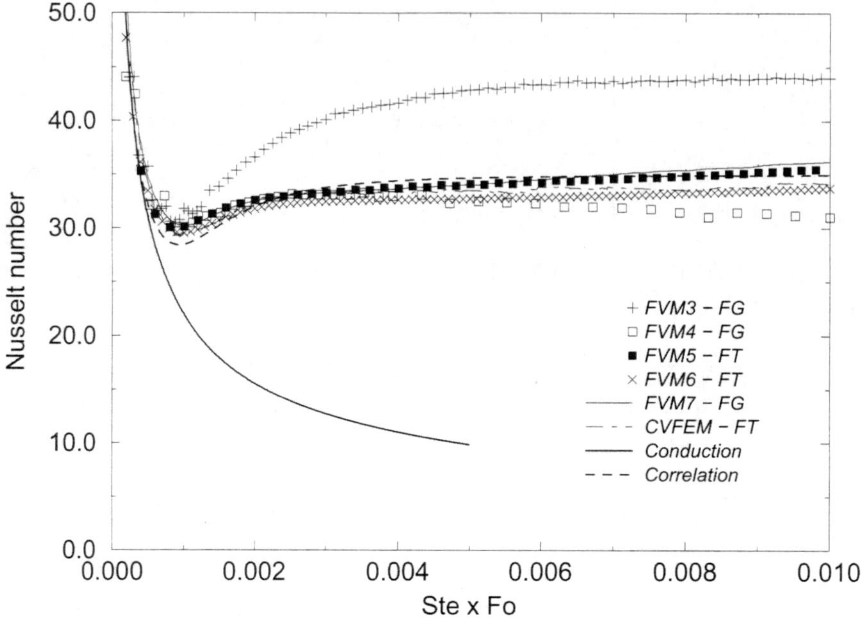

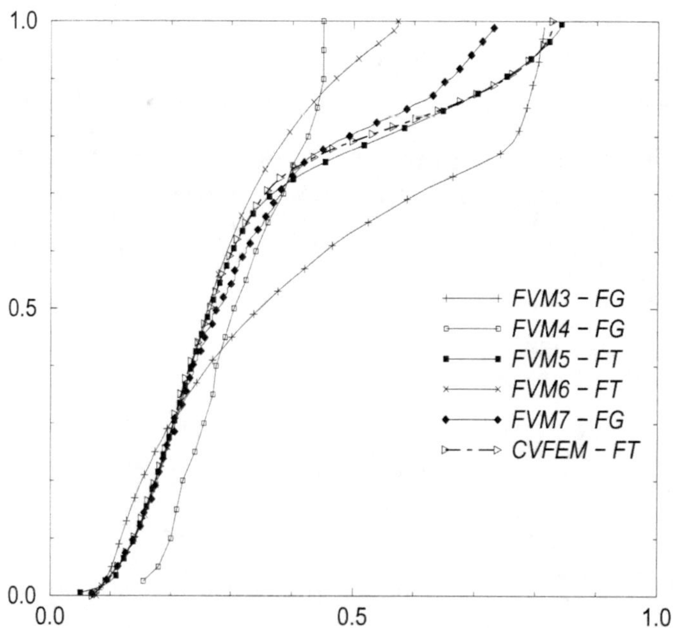

Figure 19. Melting of octadecane : Nusselt number evolution and front position

ber is a very demanding numerical test. In this case, the high velocities (strong convective contribution in the transport equations) and the very thin thermal boundary layers (of the order of 10^{-2} in dimensionless terms) may lead to unrealistic results. In this test also, the front position predicted by four simulations out of 6, represented in Figure 19-b, present a very good agreement on the 50 % bottom half of the domain. These correspond to the contributions already selected on the basis of the Nusselt number evolution, giving one more argument for this otherwise arbitrary choice. We must underline that the agreement between these four solutions on the overall shape of the interface is rather qualitative yet, since the dispersion in the front position at the top of the enclosure is still more than $\pm$ 15 % .

Interestingly, there is no clear distinction between FT and FG methods in the results of the comparison at high Prandtl and Rayleigh numbers. It may be noted however that a similar level of accuracy is obtained with rather lighter grids when FT methods are used. This confirms that the adaptive grids or coordinate transformations allow for an easier description of the thin boundary layers at the interface, where the fixed grid techniques require a much finer space discretization.

The low Prandtl number range (Case #2). Case # 2 was solved by 10 participants, at least up to t_3. The qualitative difference in the time behavior of the average Nusselt number is clearly shown in Figure 20-a. Two different classes of time evolutions are observed. A majority of simulations (FVM1 to FVM6 and CVFEM) find that the Nusselt number evolution is very close to correlation (2.27). Another behavior, displayed in continuous lines in the figure, gives a significantly higher value of the Nusselt number in a given time range and presents some oscillations.

This is confirmed by noticeable differences in the interface positions, shown at time $t_3 = 0.04$ in Figure 20-b. The melting front shapes predicted by methods FDM1, FVM7, FEM1-FG and CVFEM-FT are in fairly good agreement, indicating the presence of two recirculation rolls where the local heat transfer at the interface, and thus the local front velocity are higher. These findings bring a confirmation to the original numerical results by Dantzig (1989), and the origin of this multicellular regime and of the time oscillations have found an explanation by the stability analysis provided in a next section. Note however that this behavior has not still been reported experimentally.

Again, there is no clear-cut superiority of any model or procedure over the others : 4 of them are using the enthalpy model, 3 out of five are based on finite volume approximations and structured or non structured meshes, fixed or moving grids are equally used in these solutions.

These results show that the use of a quasi-steady assumption is not always justified, indicating the limits of the boundary layer analysis presented in a previous section. Extremely small time step is required to capture the flow pulsations. As a conclusion of this comparison exercise, it must be emphasized that the accurate description of the coupled dynamics of convection and phase change for metals should require major attention and the use of performing numerical tools.

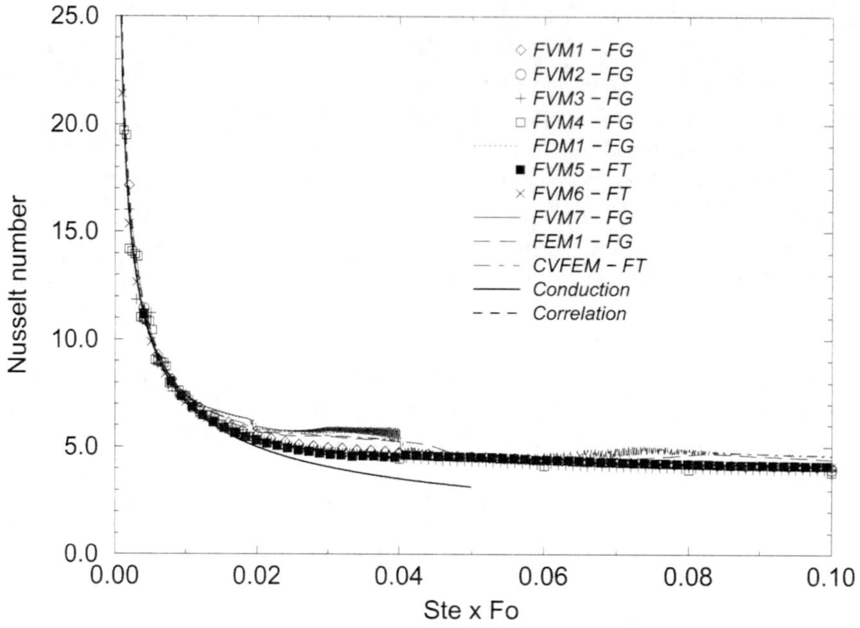

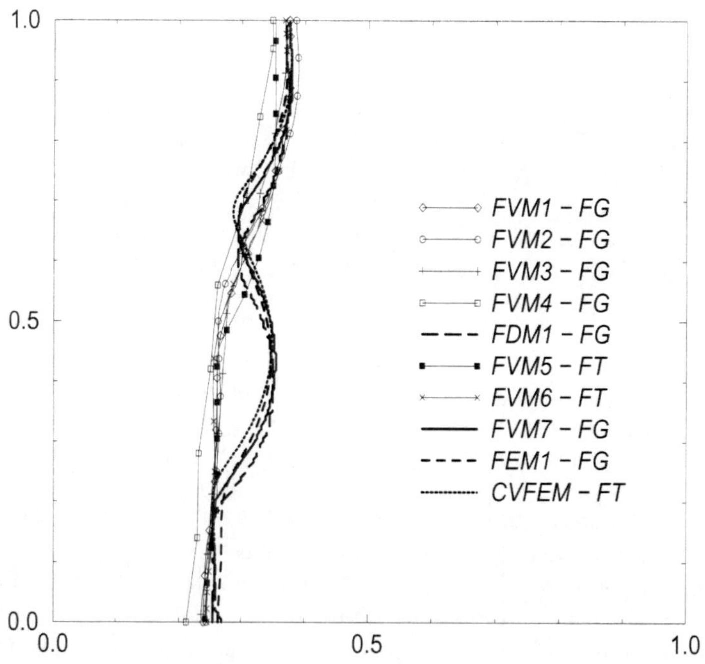

Figure 20. Melting of tin : Nusselt number evolution and front position

3.2 The stability problem

The flow instability and the oscillatory behavior of the solution in the case of melting of metals for the simple case of thermal convection indicates that phase change processes in the low Prandtl number range may be severely affected by the interaction with complex flow structures which require major attention when developing solidification codes. The present section provides some details on the stability of the solution in the present problem.

Theoretical approach. At early times of the melting process, the fluid motion develops in a high aspect ratio cavity. It is known that fluid circulation in such conditions may give rise to hydrodynamic instabilities (instability of the conduction regime). The onset of such instabilities is known to depend on the Grashof ($g\beta\Delta T L^3/\nu^2$) or Rayleigh ($Ra = Gr\,Pr$) numbers, based on the width L of the rectangular fluid enclosure. For the conduction solution, the simple fluid velocity expression is

$$w = \frac{1}{6}\frac{g\beta\Delta T L^2}{\nu} x(x-0.5)(x+0.5) \tag{3.1}$$

where x represents the horizontal length scaled with the width L of the cavity and ranges from -0.5 to 0.5. The corresponding maximum fluid velocity is approximately equal to $0.008 g\beta\Delta T L^2/\nu$. The Grashof number may be viewed as a Reynolds number based on the width of the layer and on the characteristic velocity scale. The Reynolds number based on the maximum velocity reads $Re \simeq Gr/125$. The fluid layer will then become linearly unstable when its characteristic Reynolds number reaches a critical value Re_c depending on the Prandtl number of the fluid.

Assuming that the Stefan number is very small, the temperature field across the slot is in the conduction regime and the temperature gradient is $\Delta T/s^*$ (s^* = width of the liquid layer). A conduction-phase change balance at the fluid-solid interface classically yields

$$s^{*2} \sim 2\alpha\, Ste\; t. \tag{3.2}$$

In the vertical-momentum equation, the dominant terms are the unsteady, viscous and buoyancy terms (see e.g. Bejan (1995)), whose respective order, using (3.2), are

$$\frac{w}{t}, \frac{Pr}{Ste}\frac{w}{t}, g\beta\Delta T\,.$$

In most situations of practical interest $Pr/Ste \gg 1$ (in the present cases, Pr/Ste is respectively 2 and 500) and consequently

$$w \sim \frac{g\beta\Delta T\, Ste}{Pr} t \tag{3.3}$$

The scaling factor C may be determined from the computations for cases # 2 and # 4 (LeQuéré and Gobin (1999)) and is found to be equal to 0.015. The instantaneous characteristic Reynolds number, based on the instantaneous velocity maximum and layer width, thus reads :

$$Re(t) \simeq 0.02 \frac{Ra_H}{Pr}\left(\frac{t}{H^2/(\alpha\, Ste)}\right)^{3/2} \tag{3.4}$$

Instabilities may be expected when time increases and the critical time at which the layer will become linearly unstable is :

$$t_c \simeq 0.5 \frac{H^2}{\alpha\, Ste} \left(\frac{Gr_c}{Gr_H} \right)^{2/3} \tag{3.5}$$

and the corresponding cavity aspect ratio is :

$$A_c \simeq \left(\frac{Gr_H}{Gr_c} \right)^{1/3} \tag{3.6}$$

The stability of the conduction solution for each of the Prandtl numbers considered in the benchmark problem has been investigated by various authors (Korpela et al. (1973); Gershuni and Zhukhovvitskii (1976); Bergholz (1978)). The critical Grashof number for $Pr = 0.02$ is 8000 and the most dangerous modes are steady, while for $Pr = 50$ the critical Grashof number is on the order of 1000 and the instability is oscillatory. The corresponding critical times and aspect ratios are listed in Table 1. The fact that the numerical values vary very little for each Prandtl number is purely accidental.

A first limitation of this analysis arises from the fact that the base flow solution is quasi-steady and that the instability develops rapidly compared to the slow time scale evolution. The time given by (3.5) is thus an underestimation of the actual time at which the instability should become visible. Secondly, the instability will develop provided the flow is still in the conduction regime : actually, if the aspect ratio is too small, the fluid cavity will already be highly distorted which may prevent the instability. On the other hand, stratification may develop, in particular for large Prandtl numbers, and make the flow switch from the conduction to the improperly named 'transition' or separated boundary layer regimes. The stability of the transition and boundary layer regimes may be quite different from those of the conduction regime, as evidenced by Bergholz (1978) who showed that the critical Grashof number is an increasing function of the stratification parameter.

	Ra_H	A_c	τ_c
Pr = 0.02	2.5×10^4 2.5×10^5	5 11	0.02 0.004
Pr = 50	1×10^7 1×10^8	5.5 12	0.016 0.0035

Table 4. Critical dimensionless times and aspect ratios

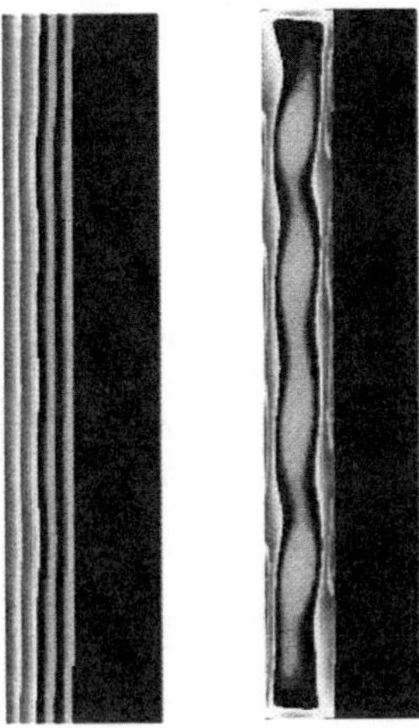

Figure 21. Temperature and vorticity fields at $\tau = 0.0048$

Numerical verification. Low Prandtl number range. In view of the above analysis, the numerical results were checked and it was indeed found that at time $\tau = 0.0042$, the solution corresponding to $Pr = 0.02$ and $Ra_H = 2.5 \times 10^5$ showed the formation of four cells, regularly spaced throughout the cavity height (see Figure 21). No cells were found neither for $Ra_H = 2.5 \times 10^4$, nor for$Pr = 50$ whatever the value of Ra_H. The reason why no cell was found for $Ra_H = 2.5 \times 10^4$ is very likely due to the fact that the instability would occur at a time of 0.02 when the cavity aspect ratio is 5. At this late time, non parallelism is already strong and likely prevents the instability to appear. Concluding that the instability can only develop when the cavity aspect ratio is still larger than 10 leads to a general criterion governing the onset of instabilities for small Prandtl number fluids

$$Gr_H \geq 5 \times 10^6 \sim 10^7.$$

Large Prandtl number range. For $Pr = 50$ no instability was found whatever the value the Rayleigh number. The explanation is that in large Prandtl number fluids, a slight vertical stratification develops very soon as the Grashof number is increased (due to the low thermal diffusivity), and the flow switches to the transition or boundary layer regime. At large Prandtl numbers, the situation is actually very complex as far as possible instabilities are considered. In infinite slots, as shown by Bergholz (1978) amongst others, the conduction regime is prone to a multicellular stationary instability if $Pr \leq 12.7$ and to an oscillatory instability if $Pr \geq 12.7$. In the latter case, this oscillatory mode persists for small stratifications until, for large enough Prandtl number (50 seems to be the

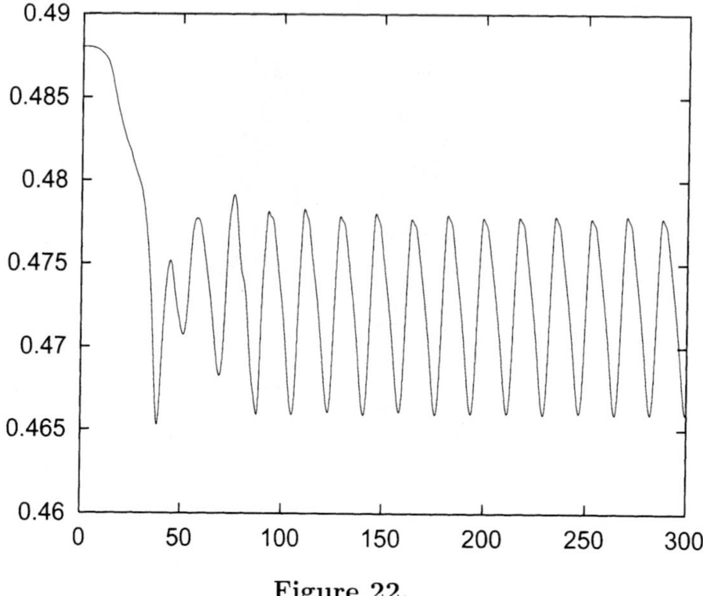

Figure 22.

Time evolution of temperature θ at $(x/H, z/H) = (0.025, 0.5)$ for $Ra_H = 2.5 \times 10^5$, $H/W = 3$.

The reference time for this computation is $H^2/\alpha Ra_H^{1/2}$.

critical value), stationary modes become more dangerous (see Bergholz (1978), figure 11). However such a transition to this oscillatory regime in finite cavities of large aspect ratio has never been observed. On the other hand numerical computations and experimental visualizations of very large Pr ($Pr \simeq 1000$) flows performed in moderately large aspect ratio cavities ($A \simeq 20$) have long shown the existence of multicellular stationary patterns (Elder (1965); de Vahl Davis and Mallinson (1975)). We can thus think that the critical aspect ratio needed to see a multicellular stationary instability increases rapidly as the Prandtl number decreases. Bergholz estimated that the aspect ratio should be on the order of 100 to see the appearance of the steady multicellular regime in water. As a conclusion, it is still not clear which is type of instability that could be seen in the case of high Prandtl number fluids.

It is however possible to show that the instability of the conduction regime is very unlikely. The end of the conduction regime was quantified to be $Ra_H \leq 300A^4$ ($Pr > 1$) (Elder (1965); Gill and Davey (1969)). Inserting the aspect ratio given by (3.6) in this expression yields a critical value

$$Ra_H = \frac{1}{300^3} \left(Pr\, Gr_c\right)^4 .$$

With a Pr value 50, the nominal Rayleigh number is larger than 10^{11}. Since $Pr\, Gr_c$ is an increasing function of Pr over the range $20 - 1000$ (Bergholz (1978)), it is clear that high

Prandtl number flows will very unlikely display the conduction regime instability in this material processing context, although possibly in some geophysical configurations. There is still one chance that it could however display the steady multicellular instability at large enough Prandtl number ($Pr \simeq 1000$) provided the Stefan number is small enough for the parallel flow approximation to remain valid long enough to let the instability develop.

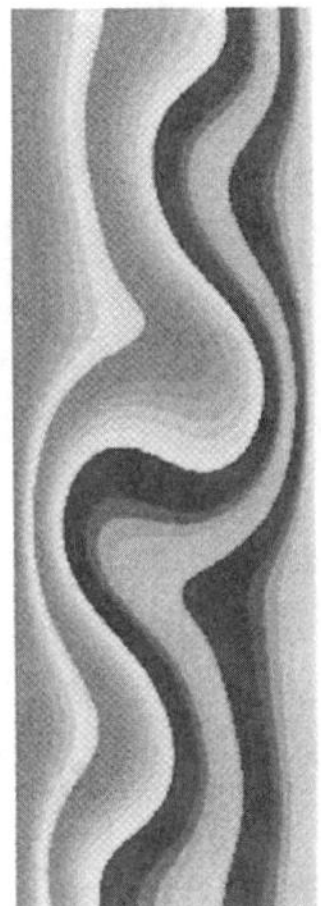

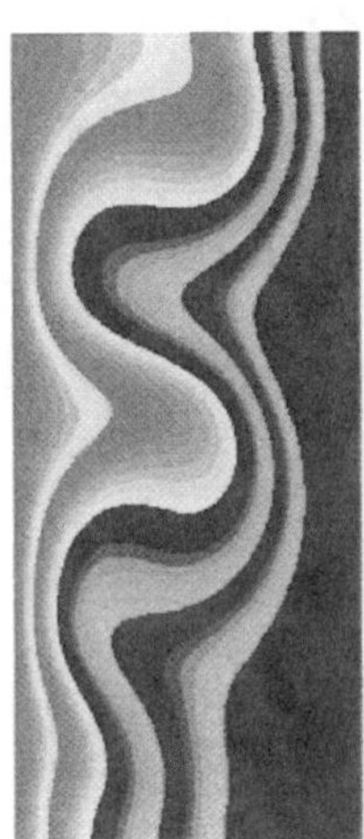

Figure 23.

Left plot : Instantaneous temperature field in the time periodic asymptotic regime for pure natural convection.

Right plot : Instantaneous temperature field of coupled phase change natural convection configuration at time $\tau_3 = 0.04$. The rightmost contour line is the front position.

Secondary oscillatory instability. It is clearly visible on the average Nusselt number evolution Figure 20 that the numerical solution becomes unsteady at a dimensionless time close to 0.03. This instability seems to originate in the shear between the two main rolls. It may be shown that this unsteadiness is not due to the coupling between natural convection and phase change, but simply intrinsic to the natural convection flow regime alone. The numerical integration of the Boussinesq equations in a differentially heated cavity of vertical aspect ratio equal to 3, starting from rest, gives the time evolution of the temperature at a given monitoring point displayed in figure 22. This clearly shows that the asymptotic solution is time periodic, very likely due to a Hopf bifurcation of the steady solution. In the convective time unit ($(H^2/\alpha)\, Ra_H^{-1/2}$) used for the computation the period of these oscillations is 18.1, which compares favorably with that of 10.8 characterizing the oscillations of the coupled problem. (Note that in a cavity of aspect ratio 4, the time asymptotic solution is also periodic with a period of 13). Snapshots of instantaneous temperature fields corresponding to the pure natural convection and to the coupled problem are presented in figure 23, showing the qualitative resemblance and thereby establishing that the oscillatory instability is intrinsic to the natural convection alone.

Conclusion. The simulation of coupled phase change processes in the presence of natural convection flows still requires much care in the simple problem of melting couples to purely thermal convection. The case of melting of pure low Prandtl number substances heated from the side are prone to the classical multicellular instability for sufficiently large nominal Rayleigh numbers. This instability strongly influences the rate of heat transfer, the melting front velocity and the shape of the liquid/solid interface. The multi-roll regime resulting from this initial steady instability becomes later also prone to an oscillatory instability, which is intrinsic to the natural convection flow alone. Although the analysis has been conducted on the simple melting problem, these stability problems may have some relevance in solidification processes, and the need for a thoroughly accurate solution of the convective flow in such problems may be emphasized. This may be of particular relevance when simulating a solidification problem at the very beginning of the process. This shows that, beyond the necessary discussion of the solidification model itself, the fluid flow problem has to be modelled with extreme care on rather small time scales compared to the overall scale of the process.

4 Closure

The purpose of this chapter is to give the basic features of natural convective phenomena involved in solid-liquid phase change processes. For the sake of simplicity, only problems involving laminar natural convection and clear-cut interfaces have been tackled in those pages. Although the presentation has a limited scope, the author is indebted to many coworkers and colleagues. [1]

As can be seen in other chapters of this course, the modelling and numerical simulation of solidification processes deal with more complex situations. The convective flows concern not only the liquid phase, but also the mushy zone where fluid flow, and heat and species transfer also take place. The macroscopic modelling of the transport mechanisms makes use of different possible upscaling methods, similar to the techniques used in the field of porous media. This aspect has not been detailed in the present chapter, but the particular heat, mass and fluid flow problems in such porous structures and their coupling with the bulk flow are of major relevance to this field.

Another aspect of this class of problems is the dynamic interaction between the solidification process and convection. As a consequence the characteristic temperature and

[1]The author wishes to thank his colleague Sophie Mergui from Laboratoire FAST for the experimental results. He also wants to acknowledge the many ideas and developments borrowed from the papers by Patrick Le Quéré and Adrian Bejan.

concentration differences leading to the dimensionless parameters to be used in the scaling laws are time dependent and often difficult to estimate. However, these estimates are necessary to get some approximate idea of characteristic velocities or boundary layer thickness. The definition of reasonable computational grids leading to sufficiently accurate solutions of the flow problem, and thus of the solidification process, depends on the knowledge of the corresponding orders of magnitude.

Bibliography

C. Beckermann and R. Viskanta. Double-diffusive convection due to melting. *Int. J. Heat Mass Transfer*, 31:2077–2089, 1988.

A. Bejan. Mass and heat transfer by natural convection in a vertical cavity. *Int. J. Heat Fluid Flow*, 6(3):149–159, 1985.

A. Bejan. *Convection heat transfer.* (2nd Edition) Wiley, 1995.

C. Bénard, R. Bénard, R. Bennacer, and D. Gobin. Melting driven thermosolutal convection. *Physics of Fluids*, 8(1):112–130, 1996.

C. Bénard, D. Gobin, and A. Zanoli. Moving boundary problem : heat conduction in the solid phase of a phase-change material during melting driven by natural convection in the liquid. *Int. J. Heat Mass Transfer*, 29:1669–1681, 1986.

R. Bennacer and D. Gobin. Cooperating thermosolutal convection in enclosures : 1. scale analysis and mass transfer. *Int. J. Heat Mass Transfer*, 39(13):2671–2681, 1996.

R. F. Bergholz. Instability of steady natural convection in a vertical fluid layer. *J. Fluid Mech.*, 84:743–768, 1978.

O. Bertrand, B. Binet, H. Combeau, S. Couturier, Y. Delannoy, D. Gobin, M. Lacroix, P. LeQuéré, M. Medale, J. Mencinger, H. Sadat, and G. Vieira. Melting driven by natural convection - a comparison exercise : first results. *Int. J. Thermal Sc.*, 38: 5–26, 1999.

B. Caroli, C. Caroli, and B. Roulet. *Instabilities of Planar Solidification Fronts*, chapter 2 of Solids far from Equilibrium, pages 155–296. Cambridge University Press, 1992.

C. F. Chen, D. G. Briggs, and R. A. Wirtz. Stability of thermal convection in a salinity gradient due to lateral heating. *Int. J. Heat Mass Transfer*, 14:57–65, 1971.

J. Crank. *Free and Moving Boundary Problems.* Clarendon Press, 1984.

J. A. Dantzig. Modelling liquid-solid phase-changes with melt convection. *Int. J. Num. Methods Engin.*, 28:1769–1785, 1989.

G. de Vahl Davis and G. D. Mallinson. A note on natural convection in a vertical slot. *J. Fluid Mech.*, 72:87–93, 1975.

E.R.G. Eckert and W.O. Carlson. Natural convection in an air layer enclosed between two vertical plates with different temperatures. *Int. J. Heat Mass Transfer*, 2:106–120, 1961.

J. W. Elder. Laminar free convection in a vertical slot. *J. Fluid Mech.*, 23:77–98, 1965.

M. C. Flemings. *Solidification Processing.* McGraw-Hill, New York, 1974.

G. Z. Gershuni and E. M. Zhukhovvitskii. Convective stability of incompressible fluids. *Israel Program for Scientific Translations*, 1976.

A. E. Gill. The boundary-layer regime for convection in a rectangular cavity. *J. Fluid Mech.*, 26:515–536, 1966.

A. E. Gill and A. Davey. Instabilities in a buoyancy system. *J. Fluid Mech.*, 35:775–798, 1969.

D. Gobin and C. Bénard. Melting of metals driven by natural convection in the melt : influence of the prandtl and rayleigh numbers. *J. Heat Transfer*, 114:521–524, 1992.

D. Gobin and P. LeQuéré. Melting from an isothermal vertical wall. synthesis of a numerical comparison exercise. *Comp. Assist. Mech. Eng. Sc.*, 7–3:289–306, 2000.

T. R. Goodman. Applications of integral methods to transient non linear heat transfer. *Adv. in Heat Transfer*, pages 51–122, 1964.

J.C. Grondin and B. Roux. Recherche de corrélations simples exprimant les pertes convectives dans une cavité bidimensionnelle, inclinée, chauffée différentiellement. *Rev. Phys. Appl.*, 14:49–56, 1979.

J. M. Hill. *One-dimensional Stefan Problems : an Introduction.* Longman Scientific and Technical, 1987.

H. E. Huppert and J. S. Turner. Melting icebergs. *Nature*, 271:46–48, 1978.

P. Jany and A. Bejan. Scaling theory of melting with natural convection in an enclosure. *Int. J. Heat Mass Transfer*, 31:1221–1235, 1988.

C.G. Jeevaraj and J. Imberger. Experimental study of double-diffusive instability in sidewall heating. *J. Fluid Mech.*, 222:565–586, 1991.

S. A. Korpela, D. Gözüm, and C. B. Baxi. On the stability of the conduction regime of natural convection in a vertical slot. *Int. J. Heat Mass Transfer*, 16:1683–1690, 1973.

W. Kurz and D. J. Fisher. *Fundamentals of Solidification.* Trans Tech. Publications, 4th edition, 1998.

H. G. Landau. Heat conduction in a melting solid. *Quart. Applied Math.*, 8:81–94, 1950.

P. LeQuéré and D. Gobin. A note on possible flow instabilities in melting from the side. *Int. J. Thermal Sc.*, 38:595–600, 1999.

J. S. Lim and A. Bejan. The prandtl number effect on melting dominated by natural convection. *J. Heat Transfer*, 114:784–787, 1992.

S. Ostrach. Natural convection in enclosures. *Adv. in Heat Transfer*, 8:161–227, 1972.

S. V. Patankar. *Numerical heat transfer and fluid flow.* Hemisphere, 1980.

J. Patterson and J. Imberger. Unsteady natural convection in a rectangular cavity. *J. Fluid Mech.*, 100:65–86, 1980.

K. A. Rathjen and L. M. Jiji. Heat conduction with melting or freezing in a corner. *J. Heat Transfer*, 93:101–109, 1971.

L. I. Rubinstein. The stefan problem. *Transl. Math. Monographs*, 27:231–338, 1971.

S. Thangam, A. Zebib, and C. F. Chen. Transition from shear to sideways diffusive instability in a vertical slot. *J. Fluid Mech.*, 112:151, 1981.

A. W. Woods. Melting and dissolving. *J. Fluid Mechanics*, 239:429–448, 1992.

Experimental Methods for Quantitative Analysis of Thermally Driven Flows

Tomasz A. Kowalewski

Department of Mechanics and Physics of Fluids, IPPT PAN, Polish Academy of Sciences
Warsaw, Poland

Abstract. Properly designed validation experiments are necessary to establish a satisfactory level of confidence in simulation algorithms. In this review recent achievements in the measurement techniques used for monitoring macroscopic flow field features are presented. In particular, optical and electro-optical methods, for example thermography, tomography or particle image velocimetry, are reviewed and their application to simple solidification experiments demonstrated. Computer supported experimentation combined with digital data recording and processing allows for the acquisition of a considerable amount of information on flow structures. This data can be used to establish experimental benchmarks for the validation of numerical models employed in solidification problems. Three experimental benchmarks based on water freezing in small containers are proposed to model flow configurations typically associated with crystal growth and mould-filling processes.

1 Introduction

1.1 Why do we need to measure?

Modern computational fluid dynamics (CFD) began with the arrival of computers in the early 1950s. The field of computational modelling of flow with heat and mass transfer has subsequently matured to the level it has. After half a century this is evidenced of a multitude of commercial codes purporting to solve almost every problem imaginable and suggesting the époque of expensive and complicated laboratory experimentation to have passed. Some foresights even profess construction of universal Navier-Stokes solvers, which implemented in *a black box* will be used in the predictable future almost the same way as pocket calculators now. Although all would welcome such a development, the validation of numerical results remains a concern tempering some of this optimism (Roache 1997). Typical difficulties in obtaining credible predictions for industrial problems lead to the often-encountered dilemma: Do we trust numerical simulations? This question seems especially pertinent when modelling solid-liquid phase change problems (see Gobin & Le Quere 2000), due to the complexity of the physical phenomena and the difficulties implied by the multi-scale nature.

One of the basic problems encountered by any model attempting to simulate physics involved in solidification is the broad diversity of length scales. The basic length scales fundamental to solidification processes arise from capillary forces, heat conduction, solutal diffusion, and convection. Different mechanisms of convection, including forced, natural, and Marangoni convection can all contribute with a different strength. Hydrodynamic interactions with solid intrusions may lead to redistribution of species and local agglomeration. A wealth of flow patterns can result at the interface, e.g. cells, fingers and dendrites, depending on the parameters pertinent to the system. The complexity of the physical phenomena and the difficulties implied by its multi-scale nature create a plethora of non-dimensional parameters which attempt to merge these different scales into reasonable sets of equations.

In many cases coarse numerical results, resulting from simplified and idealised models, are accepted and successfully applied in engineering applications. A broad class of practical problems however exists, where knowledge of general flow behaviour alone is not sufficient to obtain a full quantitative elucidation of the phenomena. Examples include the distribution of fuel or soot in a combustion chamber, the transport of impurities in crystal growth and a considerable number of solidification problems, particularly where complex geometries and fluid compositions are concerned. The growing demand for high-quality alloys and semiconductors calls for means to predict and control the distribution of impurities and additives in grown crystals. Instabilities at the solid – liquid interface can lead to micro-segregation patterns in the deposition of impurities, which later may be detrimental to the material produced. Convective influence on cellular or dendritic growth remains an open topic for research. The interface grows in zones of instability; repeated branching results in a highly convoluted shape and topological changes may result. The behaviour of the intricate interface is very sensitive to boundary conditions applied to the interface and the flow field in its vicinity.

Strong nonlinearity of the governing equations combined with a moving boundary make *a priori* prediction of the consequences of inaccuracy or simplifications used in the numerical model almost impossible. The mechanism by which nonlinear phenomena induce reorganization of the flow patterns in unstable systems cannot be elucidated by small perturbation analyses. General methods to describe the above-mentioned phenomena in their highly nonlinear states are inevitably required. These regimes are obviously the most appealing to laboratory and computational experiments. Most industrial problems unfortunately involve configurations and substances, which are very difficult to investigate experimentally. Metals and metal alloys are opaque, their melting temperature is extremely high and their physical properties are not known precisely enough (Viskanta 1988, Incropera 1997). Hence, collected data is usually insufficiently accurate to provide a definitive answer on code reliability. One solution is to use so called *analog* fluids. These are transparent and have a low melting point. Such materials are most commonly aqueous solutions of salts, which crystallize with a dendritic morphology. Some organic liquids also lend themselves favourable to this purpose

It is evident that improvement in accuracy of both theoretical and numerical models and their validation using experimental data are imperative to solve such problems. Besides this very important task, there is still place for the *classical* laboratory experiment, enabling the discovery of new flow features based on *real life* observations. Solving complicated problems is easier when experimental feedback is present. For example, the simplification of thermal boundary conditions at apparently *passive* sidewalls was responsible for difficulties in numerical modelling of the flow structure in relatively simple cases of natural convection analysed in our laboratory. Only the use of *both* experimental *and* numerical methods allowed the fine structures of the thermal flow to be comprehensively understood. Difficulties in predicting changes in the flow pattern were triggered by a minor modification of the thermal conditions at traditionally idealised „passive" side walls. It is impractical and usually impossible to include all possible factors when modelling the environment numerically. Properly planned experimental benchmarks alert one to the sensitivity of the flow to such conditions, which would otherwise be hard to predict. A brief review of experimental techniques for the study of heat and mass transfer flow problems is given with this objective in view in this work.

Velocity and temperature are the primary fields, which characterize a thermally driven flow. Point measurements were the preferred method to validate numerical codes in the past, however despite their high accuracy (e.g. Laser Doppler Anemometry), the limited number of simultaneously acquired values makes their use potentially questionable for the purpose of validation. It is therefore the author's opinion that 2-D or 3-D flow field acquisition methods are the only means, especially in the case of transient phenomena. One of this author's favourite 2D methods to determine both the temperature and velocity fields is based on a computational analysis of the colour and displacement of liquid crystal tracers. It combines Particle Image Thermometry (PIT) and Particle Image Velocimetry (PIV). Complete 2-D temperature and velocity fields were determined from two successive colour images taken at a selected cross-section through the flow. The 3-D flow structure can be reconstructed from a only few sequential measurements if the flow relaxation time is sufficiently long. In some cases even apparently good agreement of the measured and calculated fields does not guarantee their equivalence. Residual discrepancies may result in what are ultimately very different flow patterns and detection of these differences is a non-trivial task. Three-dimensional tracking of tracer particles suspended in the flow medium has been found to be particularly useful to this end. This is due to the strong sensitivity of the particle position to miniscule forces or numerical inaccuracy. It could be demonstrated that observed and simulated trajectories are often far from being in acceptable agreement, even for well known *standard* problems. In several cases simpler, 2-D particle tracking was found to be useful in demonstrating the basic differences between the observed and calculated flow patterns.

More generally speaking, we would also like to recover a complete, transient description of the analysed phenomena from the experiment. Only a very small fraction of the necessary data can, in reality, be extracted from the experiments to be compared with the numerical simulation. Quite often measurement of one of the parameters by a specific experimental method excludes simultaneous use of another method. Hence, planning the experiment must be proceeded by an appropriate decision on the significance of collected data for a specific problem as well as a careful selection of the parameters to be monitored. One also has to remember the importance of an accurate description of initial and boundary conditions. Without this knowledge even the best numerical code can lead to an incorrect solution. Even a small deviation from the physical state ultimately causes the code to arrive at a different solution due to the non-linearity of the governing equations. Flow sensitivity to thermal boundary conditions defined at "passive walls", observed for simple convective flow was already mentioned. Problems with initial conditions are even more serious. Starting with an incorrect initial flow configuration (e.g. an incorrect initial temperature, velocity or concentration field) can ultimately result in differences in its behaviour, leading to different solutions. The usual practice of starting simulations from "*rest*" does not solve the problem. In the physical experiment there is always some uncontrollable motion and non-uniformity in the temperature, and concentration fields. Since these conditions are unpredictable, they cannot be used as initial conditions in the simulation. One option to minimize uncertainty is to force some initial flow pattern and use it as an initial condition for the numerical and experimental runs. Such a method was implemented in our laboratory when resolving transient solidification problems.

The sources of the discrepancies between observed and numerical flow structures are plentiful. One of them is the three-dimensionality of the flow that complicates numerical modelling. Thermal boundary conditions (TBC) assumed at non-isothermal walls are in practice neither perfectly adiabatic, nor perfectly heat conducting, as is usually assumed in numerical models. The prevalent Boussinesq approximation for the physical determination of the fluid flow is not strictly valid for real fluids. In particular, viscosity dependence on temperature may generate severe discrepancies between expected and observed flow patterns in most liquids. In the work that follows we attempt to understand and explain the discrepancies between measured and calculated convective flow patterns in three instances of natural convection viz. a differentially heated cavity, a lid-cooled cavity and natural convection with a phase change (freezing of water). Finally, three simple configurations for the validation of numerical codes are proposed.

1.2 What do we need to measure?

Phase boundary. An accurate measurement of the shape and position of the phase front is of primary interest in solidification problems. Images of the solid/liquid interface are necessary for this purpose. This is an apparently simple task for transparent fluids, however detection of the interface for opaque substances needs application of specialized, and usually expensive tools. Measurements are usually of the form of digital images. Semi-automatic detection of the interface and the analysis thereof for comparison with numerical results can be significantly improved by use of image processing software. Image processing procedures considered basic here are smoothing, edge detection and the fitting of orthogonal functions to points at the interface.

Temperature. The description of transient temperature fields within the fluid, solidus and enclosure is the next most sought after goal of experimental modelling. Unfortunately, performing accurate temperature measurements is not an easy task. Point measurements (most common) usually give misleading information that can be. Differences, if present, erroneously can be interpreted as small changes of the flow structure close to the measuring point when compared with the numerical results. For the same reason agreement between numerical and experimental results can often be only superficial. Full field measurements, although not as accurate, may offer greater confidence.

Velocity. The fluid velocity field is important with regard to understanding both mould filling, as well as the complicated mass and heat transfer processes operative during solidification. Non-intrusive velocity measurements are non-trivial and measuring the velocity of a liquid metal flow is extremely challenging. Very few non-optical means are known for this purpose and even optical measurements are severely restricted. The full velocity field is still more challenging. The velocity vector has three spatial components and collecting complete three-dimensional information is both expensive and difficult. It is as a consequence of this that the flow field is usually reconstructed from two-dimensional cross-sections. Such methods, however much welcomed by those involved in numerical simulations, may lead to erroneous conclusions.

Concentration. Solutal convection is driven by species concentration gradients, yet another variable difficult to quantify experimentally. Serious discrepancies relating to the physical modelling of diffusion, the depletion of solutal by the solidus and the segregation of components in binary systems dog many numerical simulations. Some integral measurements of component concentration can be done using optical or electrical methods; however, the accuracy of such measurements is poor.

Material properties. Although it is not the intention of this paper to dwell on problems relating to the measurement of material properties, it is nonetheless worth noting that one of the difficulties in modelling industrial problems is the uncertainty in material properties. A "*comedy of errors*" is evident when comparing many published test cases in which material properties may differ from one another by an order of magnitude. Difficulties in getting data under extreme conditions of high temperature or pressure are often characteristic of many industrial problems. Hence, depending on the method used, collected data may differ substantially. One hopes to get more accurate measurements when using well known and documented *analog* fluids.

2 How do we want to measure?

In the pages to follow we describe a selection of experimental methods that can be useful for analysing thermally driven flow undergoing phase changes. Most of the described methods belong to the broad class of "*laboratory*" experimental methods, often precise and well defined, however, difficult to apply outside of the laboratory environment. Some of these methods are classical, well known standards. Some of them are rather *exotic*, and it is hoped that advancements in technology will make them more practical. Most of the full field measurements involve some kind of imaging procedure. Using optical methods, images of the flow are directly acquired. Non-optical methods such as ultrasound, X-ray, NMR, IR thermography, or tomography, convert acquired information to easily interpreted images. The images are stored as two or three-dimensional arrays in the computer memory and describe spatial or volumetric variation of a measured parameter. Using this format for the measured data it is easy to perform any basic image processing operation (see Bovik 2000, Jähne 1997). It can be used to smooth, to enhance or to extract details of the measured parameter. In the experimental methods presently to be described several image-processing operations are used to enhance information archived in the "images" of measured flow features.

2.1 Point measurements

Temperature measurements. Thermocouples and resistance temperature detectors (RTDs) are the traditional electric output devices used for measuring temperature, however, a number of semiconductor devices have recently been finding application at moderate temperatures (see Emrich 1981). Thermocouples are small sensors making use of the Seebeck effect, voltage existing at the junction of two different metals. The junction of two wires can be made extremely small and the resulting response time and spatial resolution of the thermocouple probe is therefore very good. Probes with a response time of several kHz and fractions of millimetre in size are commercially available. Very low voltage generated by thermocouples makes accurate measurement difficult, especially for low temperatures. Resistance thermometers (using Platinum wire with well-known properties) are more accurate and measuring a variation in temperature of the order of 0.001°C is possible. One drawback of the resistance thermometers is they relatively large size (1 mm or more) and the response time of the probe is therefore in the order of seconds. Point measurements of temperature are very common in industrial experiments where the application of more sophisticated methods is inhibited. The intrusive nature of the methods make point probes more useful in determining wall temperature and in the calibration of other methods, than in directly estimating the flow temperature field in laboratory experiments.

Very recently, ultrasound waves propagating in the fluid have been used to map the temperature field. One approach is based on the scattering of ultrasound from impurities in the fluid (Sielschott 1997). Another promising non-intrusive method, based on the variation of the speed of sound, has been proposed by Xu et al. (2002). The method can be used for opaque fluids, hence its application in industrial problems has revolutionary potential.

Velocity measurements. Only a few of the numerous techniques for measuring flow velocity are considered suitable for solidification problems in laboratory models. Some of these methods can be used directly or easily adopted for the bulk flow, while other more restrictive methods are useful for controlling external inflow. Monitoring flow velocity during industrial solidification is much more difficult.

Hotwire anemometry: Hot-wire and hot-film anemometry is a well-established velocity measuring method for fluids. It is based on the convective heat transfer from a small diameter heated wire, a function of the fluid velocity (see Goldstein 1983). In a hot-wire anemometer, the heat transfer assumed is that of a cooling cylinder perpendicular to the flow direction. Precise and fast measurement of the local fluid velocity is obtained by measuring the electrical power necessary to keep the temperature of the wire constant. A typical hot-wire probe has a 2-3 millimetres support with a short tungsten wire or film spanned between two metal electrodes. The dimension of the sensing wire can be as small as 1mm in length and several micrometers in diameter. It permits a frequency response as high as 100kHz. With additional wires (three-wire probe) it is possible to measure all three components of the local velocity field. Its relative simplicity and quick response make hot-wire anemometry one of the standard tools for studying turbulent flow. It is possible to achieve very accurate measurements of the local fluid velocity (relative error less than 1%). The main drawback of the method is sensitivity to impurities, which easily damage the very thin sensing wire. Hot-film anemometers, based on the same principle, are somewhat more robust. An insulating ceramic substrate is coated with a thin layer of metal (of the order of 5 micrometers). The sensing element is much larger in size but much more durable than that of hot-wire anemometer and it may be used in liquids.

Several instances in which hot-film anemometry has been used for investigating the thermally driven flow of liquids have been highly successful. With a modified probe it is even possible to obtain reasonable data close to the phase change front during solidification. The probe itself, however, creates serious perturbation of the investigated velocity field, as well as disturbing the local temperature field. It is less sensitive and accurate for low velocity flows. Its use for solidification experiments should therefore be limited to controlling global variables (e.g. inlet velocity) or to calibrate other measuring tools.

Laser Doppler Anemometry: The laser Doppler anemometer, also called a laser Doppler velocimeter (LDV), is a highly advanced non-intrusive method for measurements of fluid velocity. It is rather sophisticated and expensive, nonetheless, widely used in fluid mechanics research, including turbulence measurements (Rinkevichius 1998). The method makes use of the well-known principle formulated by Christian Doppler, exploiting the fact that the motion of a radiating source relative to a detector results in a shift in frequency proportional to their relative velocity. The Doppler method of measuring local flow velocities is based on detecting the frequency shift of laser radiation scattered by particles moving within the flow. Since direct measurement of scattered radiation frequency in an optical range is often challenging, the method is adapted to measure instead, the frequency difference between the laser and scattered radiation. In particular, differential schemes with two crossing laser beams directed at a moving particle are widely used. The scattered radiation is recorded by a photodetector in some arbitrary position. The crossing point of the beams defines a measuring volume, typically $10^{-3}mm^3$, thought this can be decreased by as much as three orders of magnitude. By applying additional phase shift it is possible to measure velocities from several μm/s to high speed hypersonic flows. With two and three laser beams having different wavelengths the method can easily be extended to create a two or three component velocimeter. The laser Doppler anemometer is routinely used for detailed resolution of velocity fields. It is a non-intrusive and accurate method, well suited to a point diagnostic of the flow. There are, however, several drawbacks to the method:

- the method does not measure actual fluid velocity but the velocity of particles dispersed in the flow. Pure liquids therefore require some kind of seeding
- the method is sensitive to variation in refractive index and the optical properties of the light scattering particles. For complex fluids (mixtures and dispersions) variation of the composition and presence of a dispersed phase may produce a modulation of the detected frequency shift; the same effect is produced by local temperature variations.

Very careful analysis of the measured data is therefore necessary, especially for low velocity convective flow. The response time of the method is relatively short, however its sensitivity to optical variations of the analysed area makes it difficult to change location of the flow analysis quickly. Hence, a more detailed diagnosis of the flow field is practical only for steady state conditions. The method is expensive, especially for two or three component systems. Using the equipment requires a fair degree of operational competence and needs good optical access to the flow interior. In conclusion, LDA is an excellent and advanced method for non-intrusive flow diagnosis, however, its application in routine measurements of thermal convective flow with phase change is impaired tedious and possible only in limited instances.

Ultrasound anemometry: Ultrasonic devices use the same principle of Doppler effect to measure local flow velocity. A high-frequency acoustic pulse is directed at the flow interior, where it is reflected by particles or other acoustic discontinuities. Either the same device or a separate receiver receives the back-scattered acoustic wave. The distance travelled by the wave is simply related to the time for the pulse to travel from the transmitter to the receiver and to the speed of sound for the medium in question. By measuring the time taken for the echo to return, the source of the reflection signal can be localized. The Doppler frequency shift of the reflected signal carries information about flow velocity (see Takeda 1999). By repetition of the ultrasound pulses with continuously shifting time delay, a velocity profile along the beam direction is measured. For liquids, the transmitted carrier frequency may lie anywhere in the range of 0.5MHz to 10MHz, offering spatial resolution of 0.1 to 1mm and a velocity field measured with an error less than 5%. The pulsed Doppler ultrasonic velocity meters are commercially available as devices used to monitor flow of blood in arteries. Ultrasound Doppler anemometry, like LDA, is non-intrusive method for measuring local fluid velocity; moreover it can automatically scan the flow along the direction of wave propagation. The method is consequently described as a "line-wise" method, and this description is particularly succinct in instances where the flow velocities are slow enough to neglect scanning time (typically about 1ms). The method does not require optical transparency of the flow medium and can therefore be used for opaque liquids. The method is, however, impaired by the acoustic properties of the walls and medium to be investigated. Uncertainties present due to the variation of the speed of sound with temperature and composition have to be accounted for to obtain accurate velocity data. At low flow velocities additional errors can be generated by dynamic interaction of acoustic waves with flow medium (particles), an effect usually called *acoustic streaming*. Ultrasound anemometry, nevertheless, seems to be an interesting alternative for penetrating solidifying flow fields of non-transparent media, e.g. metals. It is difficult to seed a molten metal, however it appears that acoustic waves may also interact with small heterogeneities in the fluid. Back-scattered ultrasound waves are used to determine both the scattering position and frequency shift. Successful tests have been performed using mercury, and more recently, liquid gallium (Brito et al. 2001). Preliminary tests with liquid sodium have also been preformed. The experimental apparatus used by Brito et al. consists of a cylinder filled with liquid gallium, on the top of which is spinning disk. Particles 50μm in diameter were used to seed the flow. An ultrasonic Doppler velocimeter with 8mm diameter transducer was used to measure the velocity profile of the fluid along the beam at various locations in the cylinder. The spatial resolution was about 6mm and the measured velocity had a standard deviation of 5-10%.

Concentration measurements. Solutal convection is very important to the acquisition of any observational data on variability in concentration. Point methods are usually simpler to use and some of these are discussed below.

Capacitance methods: The electric capacitance method for measuring components concentration is based on the principle that the dielectric constant of the fluid flowing between two external electrodes, depends on the concentration (Plaskowski et al 1994). This method of concentration measurement has attracted great interest because of its non-intrusive character and the simple interpretation. Accurate measurement of weak electrical signals produced at the sensor electrodes is, however, a non-trivial task. To achieve acceptable sensitivity to detect a phase boundary, the electrodes must have a surface of several square millimetres. Using several electrodes and sequentially measuring the potential between different pairs it is possible to collect data for tomographic reconstruction of the volumetric variation in concentration. Such a method was successfully applied for detecting phase boundaries, however its spatial resolution is still very low, about 5-10% of the characteristic distance between the electrodes.

Optical probes: The angle and intensity of reflected light depends on the change in refractive index of a medium. An optical probe consists of an optical fibre and photo-sensor. It measures the intensity of reflected light at an end, while immersed in a liquid. Optical probes are relatively accurate and useful devices for point sampling of a transparent fluid. They measure changes in refractive indices and can be used to determine local concentration in aqueous solutions of salts. If the temperature of the fluid changes, however, independent temperature measurement is necessary to correct additional change in refractive index. Some probes come ready equipped with a thermocouple for the simultaneous temperature measurement. The size of the probe may be as small as fractions of millimetre. Concentration changes in a flow are usually continuous and relatively slow; hence response time is not an issue. The main drawback of this method is its optical basis; hence it can be applied only for transparent or semi-transparent liquids.

Electrical resistance methods: It is possible to ascertain temperature and concentration of dissolved phases by measuring electrical resistance between two electrodes, immersed in the test liquid. For solidifying liquids separate temperature information enables accurate determination of species concentrations. An accuracy of the method depends on both the size of the electrodes and the distance between them. The electrodes can be very small, of the order of millimetres or less. Conductivity of the liquid must be low in order to apply the method. It is therefore useful for aqueous solutions only, and not metals. The method can be extended to tomographic measurements if several electrodes are utilized. An example that employs such a system will be described in the section to follow.

2.2 Integral measurements

Schlieren, shadowgraphy and interferometry. These optical density measurement techniques are based on variations in the refractive index (see Mayinger & Feldmann 2001) and are very popular in aerodynamics. They have been used due to their simplicity and rapid response for many years in heat and mass transfer measurements. The sensitivity of the three optical methods are very different, the most sensitive being interferometry. This method is well suited to quantitative measurements. Hence, its use is more suited to the study of small density gradients; however, the well-tuned optics is necessary to avoid effects produced by windows and the external environment. The Schlieren and shadowgraph methods are more robust and less sensitive. It is important to note the following limitations of these techniques:

- they are essentially integral methods. They integrate the quantity measured over a length of the light beam. The information collected is a two-dimensional projection of the volumetric changes in the investigated area. These techniques cannot therefore resolve three-dimensional structures without additional information.
- they are sensitive to variation in the refractive index. Such changes are usually due to variable temperature and concentration. If both parameters change simultaneously, and this is certainly in the solidifying mixtures, an independent technique must be used to resolve the resulting ambiguity. Point or full field measurements of temperature/concentration are necessary, in addition, to evaluate concentration or temperature profiles.

The measuring technique of the Schlieren and shadowgraph technique is based on the deflection of a collimated light beam passing through a transparent medium with a variable refractive index. The shadowgraph is used to indicate the variation of the second derivative of the refractive index (normal to the light beam), which is responsible for deflecting part of the beam out of the plane of observation. The Schlieren uses optical focusing as the basis for a method strongly sensitive to rate of change in the gradient of the refractive index. A light source located at a focal point of a lens or concave mirror produces a parallel beam of light passing through the transparent medium under investigation. A second lens or mirror collects transmitted light and projects it onto the plane of observation. A diaphragm is located at the focal point of the second lens. The parallel light beam is focused on a small point at the opening of the diaphragm. Any variation of the refractive index in the media deflects light and is consequently screened by the edges of the diaphragm. Regions of light deflection show up darker at the plane of observation. Several different configurations for the Schlieren system exist. The optical arrangement in which a single mirror, double-pass system is used, improves sensitivity of the method (Lehner et al 1999). Use of only one mirror is also an important cost saving factor, especially if illuminating large areas. In the classical Schlieren system a sharp plane (called a knife), which cuts only half of the beam, is used instead of a diaphragm. It allows for the detection of both orientation and direction of beam deflection. Hence, both the direction and orientation of the refractive index changes can be deduced from the intensity of the observed image. The human eye is not a good detector of light intensity; it is better at colour detection and variation of the Schlieren method, called colour Schlieren, has consequently been developed. It replaces the Schlieren edge with a set of transparent colour filters. The direction and magnitude of light deflection can be coded at the observation plane as a variation of colours by arranging a set of colour filters.

Interferometry is the preferred method in quantitative studies. Unlike the Schlieren and shadowgraph methods, an interferometer is not dependent on the deflection of a light beam (see Lehner et al 1999). To the contrary, any deflections of the beam in interferometry are undesirable, as they introduce errors. The underlying principle on which the method is based is the interaction of two light waves. One of two parallel beams is passed through the investigated volume of interest and the other is used as a reference beam. The phase shift induced by the variation of the refractive index in the first beam produces a variation in intensity in the region of their recombination. A camera, located behind the medium of interest, records the pattern of fringes of maximum and minimum intensity. The fringe pattern is directly related to the phase change, hence to the refractive index and consequently, density variation. By analysing the shape, density and the number of fringes produced by the interfering beams precise quantitative information on the variation in refractive index can be obtained (Van Buren & Viskanta1980). A very narrow spectral width and coherence in the light source are crucial to interferometry. A laser light source is preferable, however, good results have also been reported using light emitting diodes (LEDs).

2.3 Full field direct measurements

Infrared Thermography. Infrared (IR) thermography is a two-dimensional, non-invasive technique for surface temperature measurement. IR thermography measures the long wave (thermal) radiation emitted by surfaces. *Black body* radiation increases with the fourth power of temperature and can be evaluated from a simple formula, the Stefan-Boltzmann law. General thermal radiation of the surroundings cannot be neglected and, in practice, the net measured radiation is given as a difference between the emitted and ambient radiation. Usually investigated surfaces have unknown emissivity, the extent to which a real body (surface) radiates energy compared to an ideal *black body*. Emissivity varies between 0 and 1 and must be known in order for the method to be applied. It makes quantitative measurements more difficult.

The modern infrared camera detects radiated electromagnetic energy in the infrared spectral range and converts it to an electronic video signal. This signal corresponds to a temperature map, which can be displayed or saved in the same way as an optical image. A properly calibrated infrared system allows for resolving temperature differences smaller than 0.1K in a response time of the order of 10^{-1} - 10^{-2} *s*.

Infrared thermography, like any other optical method, requires transparency of the medium in order to be applied. Most standard materials and fluids are, however, opaque to infrared radiation. In such instances it is possible to investigate external thermal conditions, such as external wall or free surface temperature, only. It remains, nonetheless, a unique method for non-intrusive, full field temperature measurement, useful in the control of thermal boundary conditions and in estimating heat fluxes (Carlomagno 1993). Prerequisites to temperature measurement by IR thermography are an accurate characterisation of the IR imaging system's performance, calibration of the IR camera, use of additional external optics to improve spatial resolution, determination of surface emissivity, identification of measured points and design of an optical access window composed of an appropriate material. Additional image processing techniques may be used to improve measurement accuracy.

Digital Holography. Holography is a natural extension of plane interferometry. Unlike classical interferometry, holography can store and reconstruct three-dimensional information. In the holographic interferometer, a reference and test light beam recombine on a camera plane to create a diffraction pattern. The test beam is created by a point source (pinhole) and it forms a spherical wave propagating through the volume under investigation. If both waves are coherent they form a very complex interference pattern when observed on the camera plane. The pattern usually consist of 1000 to 5000 fringes per millimetre, and recording therefore requires high resolution imaging systems, usually reliant on high quality photographic plates. The amplitude of the waves is stored as intensity variation. If the recorded plate is subsequently illuminated with coherent light source, the macroscopic pattern acts as a diffraction grating with variable constant. Observing the diffracted pattern from different directions facilitates the visualization of all optical details in the original medium. The main drawback of classical holography arises from problems in the digital recording of holograms. The spatial resolution of existing CCD cameras is still too low for quantitative holography. High-resolution films must therefore be used to obtain a sufficiently dense hologram, which can be used afterwards for quantitative measurements. The procedure is tedious, has limited potential for digital post-processing and is difficult to use for *online* flow analysis.

Digital holography is based on the same principle as holographic interferometry, the difference being that speckle images (a random interference pattern created by light scattered by a rough surface) are used. In holographic interferometry two object waves are recorded on a photographic plate and their interference pattern, called a hologram, is used. In digital holography the photographic plate is replaced by a video camera. The camera records the interference between an object wave and a reference wave as a *specklegram*, the relative angle being kept as small as possible so as to accommodate the low spatial resolution of the CCD sensors. The specklegram is a field of speckles whose size and intensity appears to vary randomly, however, the characteristics of the specklegram represent the characteristics of the object beam (see Fomin 1998). In digital speckle phase interferometry (DSPI), two specklegrams corresponding to two object waves are recorded separately in different frames on a CCD camera. An image processor can automatically calculate the difference between two frames during the acquisition of images. Combining two specklegrams gives rise to a speckle interferogram, the brightest fringes corresponding to the same phase shift (very much like in holographic interferometry). The most straightforward techniques for automatically measuring the phase shift from an interferogram are based on image intensity, i.e. on detecting fringe maxima and minima. Using these techniques for low-resolution CCD images produces sizable inaccuracies. One alternative is to change the phase during acquisition and collect a sequence of images. Such temporal phase shifting techniques can be done with a simple piezo driver mounted on the mirror. In this case, a known phase change is superimposed on the deformation phase between the two object waves. A different interferogram will be obtained for each phase change. A deformation phase is now obtained for each CCD pixel by a simple equation involving the pixel intensity in several interferograms. A minimum of three interferograms is needed, however, four to five are usually used. The main advantage of these phase shifting techniques is that the processing is very simple and the precision is at least an order of magnitude better than that of the intensity based techniques. Fringe numbering is, furthermore, not a problem even in the most complicated interferograms. The possibility of correlating subsequent interferograms and calculating the relative phase shift due to fluid motion also exists. Such techniques allow for complete field measurement of the velocity field in the selected plane of flow (Andres et al. 1999) and are similar to particle image velocimetry.

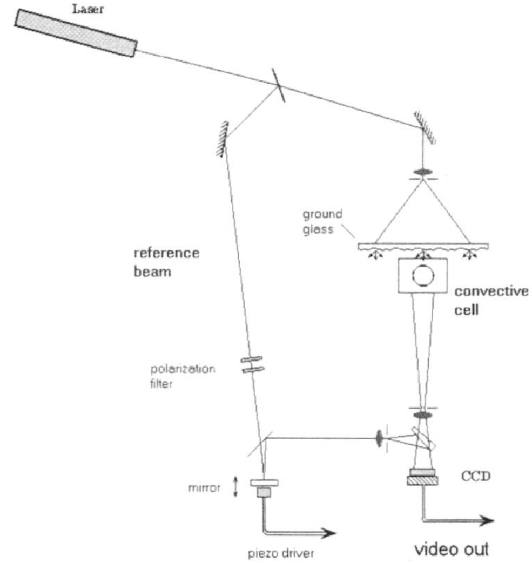

Figure 1. The DSPI arrangement used to investigate natural convection of water filled cylindrical cavity with cooled lid. A laser beam is split into a reference and illuminating beam. A piezo driver is used for the phase shift (Soeller 1994).

Attenuation methods. Light attenuation can be used to measure changes in the optical properties of a liquid. The principle on which this technique is based is the well-known exponential intensity attenuation law, which combines the absorption coefficient with properties of the subject medium. If all absorption parameters are known or the possibility for an accurate calibration procedure exists, the attenuation method allows the possibility of obtaining quantitative measurements of species concentration, during the solidification process. Optical changes arise due to natural differences in light absorption of binary mixtures or solutions. Dye is often added to one component to observe the convective flow associated with the solidification or melting. The pH indicator technique is very useful for introducing dye without disturbing a flow (McDonough & Faghri 1994). The technique involves a dilute solution of thymol blue and two electrodes located close to the region of interest. Applying voltage across the electrodes turns fluid in the vicinity of the positive electrode dark blue. The coloured fluid follows the flow perfectly as no difference in density exists. The spatial distribution of the dye along a complex interface boundary may require the use of several light sources and cameras (Ibrahim et al. 2000).

For opaque liquids attenuation of short-wavelength radiation such as γ or X-ray absorption techniques are used. In these methods the attenuation coefficient is dependent on the third power of the atomic number. They are used to resolve the concentration field or even the motion of tracers in opaque solids and liquids. These methods have the advantage of being able to analyse *in situ* industrial solidification products. The attenuation method gives integral values across a projected volume in a manner similar to optical shadowgraphy. Single beam systems have a typical accuracy of 5-10% in measuring absorption values. Accurate determination of the concentration profile depends on the absorption properties of the investigated media and the calibration procedure.

Only the phase front is detectable in most practical applications of a single beam method. A more detailed evaluation of the concentration field requires many projections and tomographic reconstruction must be used to obtain a volumetric description of the absorption coefficient. Multi-beam systems are, however, complicated and expensive. Hence, their use for detailed laboratory analysis is very rare. Good results have been reported using attenuation methods with a collimated beam of neutrons on metals (Mishima & Hibiki 1996). Thermal neutrons can easily penetrate most metals, however they are strongly attenuated by such materials like hydrogen, water, boron, gadolinium, and cadmium. Kim and Prescott (1996) used of different capture cross-sections for neutrons to measure macro-segregation associated with solidification of a gallium-indium alloy.

Nuclear resonance and electro-magnetic tomography. Nuclear magnetic resonance can be used to detect liquid composition as well as the velocity within a volume. The spin-state magnetic resonance is created using a combination of a radio frequency and a permanent magnet field. The radio-frequency field is time gated and the delay time for a change in spin is used to measure both velocity and concentration. It is also possible to follow tracers immersed in a liquid with this method, obtaining images similar to those obtained using optical tomography. NMR devices are commercially available for medical purposes and are exorbitantly expensive. It is therefore likely that many advantages of this very attractive method remain as yet undiscovered to the science of fluid mechanics.

Novel developments in tomography offer an electro-magnetic flow measuring method, a promising new technique applicable to molten metals. The method is based on the analysis of eddy-currents induced by low frequency magnetic fields within the metallic melt (Pham et al. 1999). The induced current depends on the conductivity distribution. The eddy currents produce a scattered electro-magnetic field, which can be measured in the region exterior to the flow. The eddy currents are reasonably sensitive to continuous and discontinuous changes in the metal. They can be used to detect solidification front, mushy zone, and strong temperature or composition gradients. The technique is very advantageous for large-scale industrial problems, where generated eddy currents are strong enough for accurate measurements.

Resistance tomography. Electrical impedance tomography has become a promising method for the laboratory measurement of thermosolutal convection in recent years. The method is based on the measurement of current flowing between a working electrode and a reference electrode. Chemists often use the technique to measure the concentration of substances in solution. Analysis is complicated by effects such as chemical reactions at the electrodes, adsorption of reactants, electro-deposition and erosion of the electrode. The application of an alternating potential of appropriate frequency to the working electrode and the use of chemically inert electrodes is essential to all conductivity techniques. The method can be extended to volumetric measurements of the impedance distribution (concentration) within the solution. A multi-electrode system with alternative pairs of electrodes allows for the complete tomographic reconstruction of the conductivity field, hence, the concentration field is in this way resolved. Figure 2 shows the wiring in a 38mm Plexiglas cube used for studying the concentration of salt during the freezing process in a differentially heated cavity.

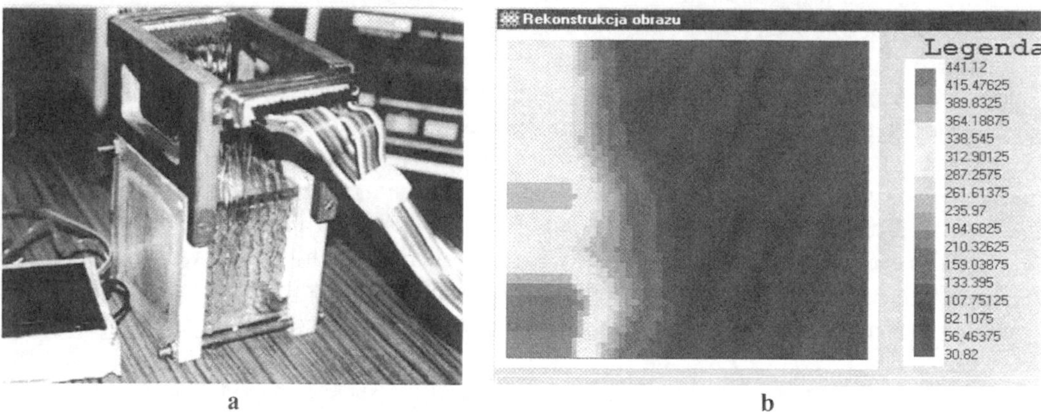

Figure 2. (a) The cavity used for concentration measurements. Two sets of 49 electrodes (7 rows x 7 columns) have been mounted on two opposite Plexiglas sidewalls of the cavity. Freezing of a salt water solution was investigated in the differentially heated cavity (Rucki et al. 2002). Ice created on the cold wall decreases electric conductivity of the investigated media. (b) The reconstructed resistance field for a salt water solution freezing from the left wall.

2.4 Tracer methods

Tracer methods are the most basic of all full field measurement techniques in fluid mechanics. Tracer methods are usually considered to be non-intrusive, however, it should be remembered that the principle involves the presence of seeding assumed neutral in the investigated fluid. Early visualization experiments based on the formation of a light sheet and tracer particles provided valuable quantitative information about the flow pattern and its basic properties. The rapid development of optical methods, lasers and finally digital imagery layered the foundation for the development of several new quantitative full field measuring techniques. For example Particle Tracking and Particle Image Velocimetry (see Grant 1994). Tracer methods arose from optical image analysis. Optical methods are usually inapplicable in most instances in volume solidification due to the opaqueness of the media e.g. metals and semiconductor crystals. However, it is noteworthy that the principles of quantitative flow image analysis are applicable to other than optical visualization methods, for example ultrasound, X-ray, NMR or infrared.

Flow Visualization using tracers. Flow visualization can be enhanced by addition of some small particles called tracers, whose paths within the flow provide basic information about flow structure. The tracers are selected on the basis of a compromise between their visibility and their ability to follow the flow. The quantitative interpretation of bulk flow observations is usually difficult. Hence, tracers are usually followed within some small "area of interest" by "cutting" the flow with a thin, well-defined plane of light ("light sheet"). Observing the plane from some angle, typically perpendicularly, only tracers within the light plane are visible. This basic visualization technique, called "light sheet" illumination, also happens to be the most convenient illumination method for particle tracking and velocimetry.

One of the biggest problems associated with flow visualization (and limitation of tracer methods) is the selection of a suitable seeding, the particles or inclusions responsible for detecting the fluid motion. A requirement of the seeding is that it has no motion relative to the fluid. This means that it has to be neutrally buoyant and its characteristic dimensions small enough to follow local flow acceleration. Using a simple models of hydrodynamic interactions, the limitation for particle diameter is dictated by $d^2 < 18\tau\mu / \rho_p$. Here τ, μ, ρ_p denote flow field relaxation time, viscosity, and particle density respectively. The buoyancy effects must be negligibly small thought sufficient to negate settling velocity, hence the particle diameter and fluid – particle density difference and must fulfil similar criterion, $V / d^2 >> \Delta\rho\, g / 18\mu$, in which V is the local flow velocity component in the vertical direction. The dimension of particles of the seeding should, on the other hand, be large enough to be detectable by optical methods. In practice this means that the diameter should be larger than several micrometers. For very bright objects it is possible to record diffraction patterns using smaller particles (e.g. small air bubbles in liquid and fluorescent tracers). Seeding concentration is another issue and to avoid any effect on the flow pattern, concentration by volume must be very low (below 1%). This condition is usually easily satisfied, as high concentrations of the seeding are also counterproductive for recording methods. Ensuring the uniform concentration of tracers throughout the medium is a requirement not easily fulfilled. Residual particle – flow interactions usually cause redistribution; agglomeration in some regions and depletion in others. This can be highly problematic when dealing with complicated flow configurations and seeding may need to be done in very careful doses. The most commonly used seeding particles for liquids are latex or polystyrene spheres of a diameter between 1 and 50 micrometers. The use of thermochromic liquid crystals was found to be useful in experiments concerned with thermally driven flows. The typical diameter of such particles is between 20 and 50 micrometers, their density is close to that of water and they are easily visible when following the flow for typical thermal convection experiments.

Selection of a light source depends on the visualization technique to be used. For classical visualization volume illumination is sometimes used. Here diffused light from a standard light source is sufficient and bright particles are observed against a dark background. Backlight illumination using a parallel light beam is sometimes more convenient for particle tracking. Tracer particles are reproduced as black dots clearly visible on bright images and easy to identify by typical image processing software. Such illumination does, however, exclude multi-exposure of a single image, a technique very useful in digital recording.

Particle image velocimetry requires a well-defined plane of illumination, in which two or three components of the tracer velocity are evaluated. The quality of the light plane, i.e. uniform illumination intensity and small depth is an important deciding factor when it comes to the accuracy of the velocity evaluation. Lasers are commonly used to create light sheets because of their ability to emit monochromatic light with high energy density. Laser light can easily be bundled into a very thin light sheet. Constant output lasers are common for flow visualization. The economic helium-neon (He-Ne) laser generating red light of a wavelength 633nm and power between 1 and 10mW can be useful for illuminating small flow fields. More powerful are argon-ion (Ar+) lasers which output green-blue light at wavelengths 514nm and 488nm and up to 100W. The intensity of light emitted by Ar+ lasers is high enough to illuminate large areas or to form short light pulses using electro-optical choppers. For better temporal resolution pulsed lasers are required, typically 10mJ to 1J per single pulse. The most widely used are ruby (Ru) and neodim (Nd:Yag) lasers. The first generates red light with wavelength of 694nm and can generate sequence of pulses over period of 1ms, the time duration of each pulse being of the order of about 30ns. The Nd:Yag lasers are available as tandems generating two 532nm wavelength light pulses within a time interval as short as 10ns. This facilitates the study of very fast flows in a large volume or micro-flows at considerably enlarged magnifications.

Monochromatic laser is an appropriate and easy solution for most visualization experiments, although there are cases where white light is required. One such case is the use of thermochromic liquid crystal tracers, where the colour of light refracted by TLC tracers indicates liquid temperature. A strong and relatively easily made light source is that produced by a linear halogen lamp with a tungsten filament spanning a 100-150mm tube. In our laboratory 1000W lamps with an additional air cooler and filament preheating circuit were used. High-energy xenon discharge lamps are employed for short illumination times. Such lamps consist of a 150mm long tube connected to a battery of condensers and can deliver as much as 1kJ of energy during a 1ms pulse. Repetition of the light pulses is relatively slow (several seconds). The total thermal load and condenser charging time are both factors limiting the repetition rate. In practice, two sets of condensers with an electronic switch can be used to allow two light pulses to be emitted from the same tube within approximately 200ms. The basic acquisition configuration in our laboratory is shown in Fig. 3.

The recording system is an important element in any tracer based measuring technique. Using a charged coupled device (CCD), an image can be stored directly in digital form, then numerically analysed. A typical CCD is a metal-oxide semiconductor composed of a two-dimensional array of closely spaced independent, light sensitive sensors, called pixels. The spatial resolution of the CCD recording is determined by the pixel dimension (usually 5 to 10μm) and the number of pixels in the sensor array. A standard CCD camera typically consists of a 768 x 572 pixel array, while high-resolution sensors may contain 5000 x 7000 pixels. By far the greatest handicap of CCD sensors is their refresh time i.e. image recording speed. This limitation is due to the charge read-out characteristics of the device. Increasing the transfer speed of photon-induced charges above a typical limit of 10 to 30Mpixels/s deprecates the signal quality, rapidly decreasing the signal to noise ratio. Hence, the frame rate of most commercially available cameras is limited to that of standard video i.e. 25 or 30 frame/s. Special CCD constructions that permit the acquisition of limited sequences of low-resolution images with frame rates up to 40000 images/s do, however, exist. In some applications, for example particle image velocimetry, two successive images are all that is necessary. Most standard CCDs fortunately have the ability to acquire two successive images within a very short time interval, a so-called straddle exposure. The first image is exposed on the CCD just before it is transferred to internal memory and the second one immediately after. Using the straddle exposure the time interval between both acquisitions can be as short as 200ns, but the next pair of images can be acquired after normal timing period.

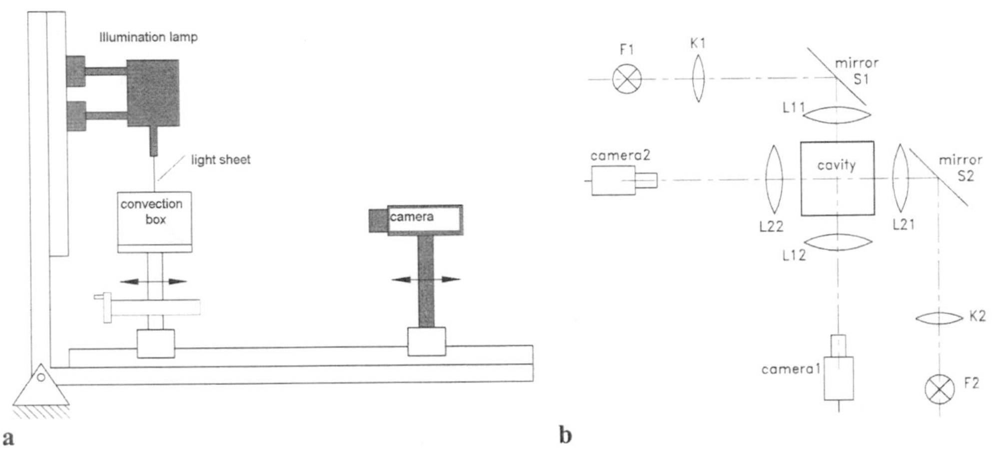

Figure 3. (a) - Acquisition system with CCD camera and light sheet illumination used for studying natural convection; (b) - two-camera system used for particle tracking in convective motion for a differentially heated, cube-shaped cavity

Electrical signals produced by each pixel of a CCD sensor are converted to a digital format and saved in the computer for future image analysis. The more professional CCD cameras, sometimes called digital CCD cameras, use an internal analogue-digital converter to provide a digital signal in 8 to 16 bit format, for monochrome images, and 24 bit format for signals containing RGB colours. The internal analogue-digital converter is matched to CCD characteristics so as to obtain optimum image quality. Special imaging boards (frame grabbers) are, however, necessary to acquire, display and archive such signals. Popular cameras, especially those having standard video format, deliver an interlaced analogue signal, which can easily be displayed on a television monitor. The analogue signal is acquired, than converted by a frame grabber to a digital format, before being stored and numerical processed. Quality of the digital image depends on the analogue-digital conversion procedure, and is impaired by low quality frame grabbers. Common causes of errors are conversion noise, conversion non-linearity and inaccurate timing (pixel position jitter). Another important property of the frame grabber is its ability to acquire images in real time (i.e. synchronic with the camera frame frequency) and to store them instantaneously to an on board memory or to a computer. The last solution is preferable as the computer memory (RAM) is easily expandable allowing an increasing number of images acquired. It is possible to transfer 1GB of data, with a speed well above 100MB/s, to memory (RAM) using current Pentium technology. This amounts to almost three hundred 16-bit, high-resolution images acquired at a rate of 30 frames/s. The image data must be transferred to the hard disk, before the next sequence can be acquired. The transfer process can take several minutes, and statistical analysis of the flow properties based on such images is therefore limited to relatively short (10s) bursts of data interrupted by save-to-disk time.

Particle Image Velocimetry. The basic concept and configuration of PIV is similar to that of a typical flow visualization (Raffel et al. 1998). The method is based on correlating parts of a sequence of flow images to establish their differences arising due to the flow movement. In contrast to the particle tracking method PIV does not search for single objects transported by the flow, instead it follows whole clusters of tracers, dyed regions within the flow, or changes in illumination. The PIV method assigns an average value to local flow velocity. Optical methods were used to correlate successive images, usually recorded on the same photographic plate (double exposure) in pioneering days of PIV. In recent times, pairs or short sequences of digital imagery, are numerically correlated by the computer to establish any local displacements caused by the motion of the flow, and this is sometimes specified by adding the adjective “digital” to the particle image velocimetry nomenclature (DPIV). The larger the amount of images in computer memory, the more detailed the study of transient flow behaviour becomes possible. It enables spatially resolved measurements of the instantaneous flow velocity in a minimal time and the detection of both large and small scale structures in the flow (Westerweel 1993). The use of PIV is very attractive in computational fluid dynamics (CFD), as the full field experimental data obtained in this way is more suitable to validating numerical simulations than the point measurements more commonly used in the past.

PIV evaluation requires two images, not necessarily two separate frames. Using single frame (double exposed) images the method can easily be adopted for high speed flows. The general trend is, nonetheless, to avoid the technical complications and ambiguity of single frame vector evaluations. In this work, we limit ourselves to PIV techniques based on pairs or longer sequences of images containing singly exposed images of the flow.

PIV analysis is based on the simple cross-correlation formula which can be stated as

$$R\left(x,y\right)=\sum_{i=0}^{M-1}\sum_{j=0}^{N-1} I_1(i,j)\cdot I_2(i+x,j+y)$$

for two arrays of pixels, where, the variables I_1 and I_2 are intensity values of the image arrays.

The whole image is divided into a regular mesh to sections, each demarcating a small interrogation area in a typical PIV evaluation procedure. Taking two small image arrays and manipulating the above summations we arrive to a cross-correlation function which describes the probability of matching the two arrays by means of an overlay. For each choice of sample shift, (x, y), the sum of the product of all overlapping pixel intensities produces one cross-correlation value *R(x,y)*. The maximum of the correlation function corresponds to the statistically most probable shift to achieve a best match. This shift, or in terms of the flow displacement, gives the statistical local velocity vector when divided by known time interval between the two images. Clearly, it is necessary to run the cross-correlation operator over the whole image to obtain a full velocity field. It requires repeating the above summations and multiplications as much as a billion times per image, a computationally expensive procedure. An alternative procedure based on Fourier transforms is therefore more commonly used. Theoretical work on periodic signal analysis reveals that the cross-correlation of two signals is equivalent to the product a Fourier transform of the template signal with the complex conjugate Fourier transform of the correlated signal. Replacement of the correlation summations with Fourier transforms does not improve evaluation efficiency in itself, however, Fourier transforms facilitate the use of Fast Fourier Transforms (FFT) giving rise to vastly more efficient algorithms. Instead of $O(N^2)$ computational operations the correlation process requires only $O[N\log_2 N]$ operations as a result of the introduction of fast Fourier transforms.

The FFT representation of the correlation function does have some drawbacks. The Fourier transform is an integral over an infinite domain and a discrete Fourier transform is, consequently, an infinite sum. Computing the transform over finite domains is justified only if the signal is periodic in all directions. Several methods are used to filter abrupt jump in the data from, artefacts arising from the use of finite image arrays. The application of Fourier transforms requires signals in the frequency domain; images of tracer particles produce a sequence of intensity peaks, which favour such analysis. The FFT gives much worse results when applied to images obtained from diffused visualization methods, like smoke, dye or intensity fluctuations. In such cases classical cross-correlation or other, similar image processing methods are more appropriate (Quenot et al. 1998, Gui & Merzkirch 1996, 2000).

Another drawback of standard PIV analysis has its origins in the interrogation window, i.e. in the principle of evaluating a cross-correlation function for a fixed sub-array of the full image. A full vector field is obtained by the use a moving step-by-step interrogation window across the whole image. Correct dimension assigned for this sub-array is crucial for the accuracy and dynamics of the evaluated velocity field (Westerweel et al. 1997). Although the spatial resolution increases with a diminished interrogation window, its minimum size is limited by two factors. Firstly, a sufficient number of tracers must be present in the interrogation window, otherwise the FFT based procedure fails to work properly due to poor statistics. Therefore, only windows larger than 16x16 pixels are used in practice. Secondly, the dimension of the interrogation window limits the maximum detectable displacement. Displacements only smaller than about half the window size can be detected. Hence, large windows are advisable for the velocity measurements when the dynamics has to be increased. Finding a compromise is not possible without some additional procedure. If the flow is relatively slow, a simple solution is to take short sequences of images acquired at suitably selected intervals. The cross-correlation analysis performed between different images of the sequence allows to preserve the accuracy for both the low and high velocity flow regions. Post-processing is necessary to combine the sequence of flow fields with the one with improved parameters. Hart (1998) proposed another solution. In this approach high-resolution analysis is achieved by the iterative use of local correlation results. The interrogation area is divided into smaller areas after standard, large window PIV correlation analysis, and each subdivided region is re-interrogated in the same manner, except that the second window is shifted by a displacement determined by the previous correlation (offset). Subdivision and re-interrogation is repeated until the size of the interrogation area attains a minimum (can be as small as the size of the individual particle-image size). The method works fine for smooth velocity fields. However, large velocity gradients lead to a false estimation of the initial window offset and subsequent evaluations with smaller windows completely degrade the vector field in this region.

Pre-processing and post-processing plays an important role for PIV as images are often less than ideal. They contain noise arising from the acquisition procedure, illumination in the image plane is never perfectly uniform, both wall reflections and the scattering properties of the seeding vary according to the illumination angle. Sever intensity differences between light pulses are also frequent when using laser or flash lamp illumination. These effects all can ruin PIV evaluation procedure and FFT based evaluation is particularly sensitive to image quality. Various image processing methods can be used to improve contrast and achieve uniform illumination by automatically analysing intensity histograms for each image of the correlated sequence and modifying them accordingly. Using direct cross-correlation (or a similar algebraic method), the adjustment of the intensity of each subsection effectively removes non-uniformity in the illumination of the image. A good contrast of images of the tracers is important for FFT based evaluation. To improve their visibility contour detection filters are sometimes used. Experience, nonetheless, shows that such procedures can sometimes result inversely. It is author's experience that the use of local thresholding filter favourable enhances PIV images, allowing for the extraction of particle images from non-uniformly illuminated backgrounds. The resulting images have almost uniform intensity, particles' shape is clearer and the bright areas which may be attributed to reflections are filtered out.

Identification and precise evaluation of the correlation peak location is crucial to the PIV process. The correlation values exist only for integral shifts since image data is discretised. By using interpolated intensity values and simply doubling the size of the analysed interrogation array, the accuracy of the evaluation improves to 0.5pixel. Irregular peak shapes with a double or triple hump form are frequently observed due to non-ideal processing conditions and a simple search for the maximum value may lead to large errors. A variety of methods for improving the estimation of the correlation peak location have been proposed. A simple and robust method to find location of the peak maximum is to fit the correlation data to some function. Parabolic peak approximation is by far the simplest and fastest procedure. The best fitting paraboloid is obtained by taking the peak value, and four or eight adjoining values of the correlation function. A Gaussian peak fit is slightly more accurate but also more tedious. Since peak approximation methods use more than one point in evaluating the peak shape, there is statistical improvement in the estimation of the peak location involving fractions of a pixel. The best values reported in the literature for the spatial resolution are close to 0.1 pixel.

PIV processing results usually need additional post processing. Correlation based PIV evaluation inevitable leads to spurious vectors due to its statistical character and the detection of spurious vectors involves a variety of approaches, or combination thereof. Some indications of the quality of the vector results can be obtained by analysing the correlation function. The detection of odd behaviour in the correlation function in some regions can lead one to critically evaluate certain vectors, either removing or interpolating them. A variety of interpolation techniques can be applied to smooth the PIV velocity field, most of them are already well-known to image processing. An adaptive procedure must normally be used as the number of spurious vectors can vary depending on the locality. It selects the degree of interpolation according to the number of removed vectors and additional constrains supplied (e.g. maximum gradients, maximum velocity). A simple median filter or two-dimensional smoothing interpolation can be applied locally and used to remove points deviating in excess of some predefined threshold, based on the assumption of a smooth velocity field. In instances where the proportion of removed vectors is high, values of the surrounding vectors can bias the interpolation and lead to false results. It is useful therefore to include some additional information pertaining to the vector field, if such information is known. If the flow can, for example, be assumed to be two-dimensional, the evaluated vector field must satisfy the continuity equation. The velocity at the walls bounding the flow field should diminish asymptotically.

Three-dimensional velocity field measurements. Fluid flow is, generally speaking, always three-dimensional. It is, nevertheless, common to analyse two-dimensional projections due to the limitation imposed both numerically and experimentally. In many cases the out-of-plane velocity component cannot, however, be neglected as it plays an important role in both overall mass and heat transfer. Real time three-dimensional velocity measurements are non-elementary. Holographic PIV could probably be deemed the most promising method for the recovery of full volumetric information about tracer locations in the flow using a single illumination shot. But development of holographic PIV is still at an infant stage and the apparatus is complicated and difficult to arrange. The stereo-PIV technique is much simpler and available commercially. It employs two standard PIV cameras to observe the flow in a illuminated plane. The third component of tracers' motion can be recovered using parallax (Raffel et al. 1998). It should be stressed, nevertheless, that the three-dimensional information gathered in this way still only pertains to the cross-section of the flow in the illuminating light sheet. A full description of the flow requires multi-plane illumination.

It is possible to make use of tomographic methods which employ ultrasound sensors, or magnetic resonance (NMR) to resolve three-dimensional flows. The accuracy of results unfortunately does not justify the effort and costs of the techniques.

Particle tracking methods are usually three-dimensional if more then one camera is used. Three-dimensional particle tracks are consequently reconstructed and the three-components of the tracer velocity evaluated (Englemann et al. 1998). Typical limitations of the method arising from particle density give rise to a very sparse velocity field. Particles also due to hydrodynamic interactions prefer some regions of the flow, whereas other are rarely visited. From statistical point of view such a description of the flow is strongly biased (Yarin et al. 1996).

Figure 3b gives an example of experimental arrangement for three-dimensional particle tracking. It employs two cameras oriented perpendicularly through the investigated cavity (Bartels-Lehnhoff 1991, Hiller et al. 1992, Mitgau et al. 1994). Bright field illumination is used with help of two light-emitting diodes. Tracers are visible in each pair of images as black dots on a white background. Evaluation of the tracers' position can be simplified by automatic subtraction of the *"zero"* image (the image without particles). Any non-homogeneities arising from the optical channel are automatically removed in this manner and only moving objects are detected.

Although three-dimensional data acquisition is challenging, most of the two-dimensional methods using a light sheet can be relatively easy adapted to obtain volumetric information, providing the flow relaxation time is sufficient. This is accomplished by scanning the flow area with a moving light sheet. In practice, a system of rotating mirrors or mechanical shift of the analysed enclosure is used. The method is the simplest to use in the case of relatively slow natural convection or solidification (typical for the laboratory scale experiments), and has been successfully applied in experiments on natural convention during the freezing of water (Kowalewski & Cybulski 1996, 1997).

Liquid Crystals Thermography. Liquid crystals are highly anisotropic fluids, existing between the boundaries of the solid and the conventional, isotropic liquid phase. Liquid crystals temperature visualisation is based on the refraction of particular colours (light wavelengths) at specific temperatures and viewing angle by some cholesteric and chiral-nematic liquid crystal materials (Straley 1994). Normally clear, or slightly milky in appearance, thermochromic liquid crystals appear to change their colour to red in response to increase of temperature, followed by yellow, green, blue, violet, finally turning colourless again at higher temperatures. In this way liquid crystals used as temperature indicators modify incident white light and display colour of a wavelength which can be associated to temperature. Thermochromic liquid crystals can be painted on a surface (Hai & Hollingsworth et al. 1996, Sabatino et al. 2000, Preisner et al. 2001) or suspended in the fluid to render the temperature distribution visible (Hiller & Kowalewski 1987, Dabiri & Gharib 1991). Application of TLC tracers facilitates instantaneous temperature and velocity fields measurements in a two-dimensional cross-section of the flow. It is a unique method of combining full field temperature and velocity measurements. The colour is red at the low temperature margin of the colour-play interval and blue at the high end. The particular colour-temperature interval depends on TLC composition. It can be selected for intervals of about 0.5°C to 20°C and associated with temperatures of -30°C to above 100°C. These colour changes are reversible providing the TLCs are not physically or chemically damaged. They can consequently be accurately calibrated and used as precision indicators of temperature. The colour play range must obviously be selected to match the temperature variation of the problem in question. Pure liquid crystal materials are thick, greasy liquids, which are difficult to handle under conventional heat transfer laboratory conditions. TLCs are also sensitive to mechanical stress. A micro-encapsulation process, enclosing small quantities of liquid crystal material in transparent, polymeric capsules, was developed to solve problems with sensitivity to stress and chemical deterioration.

In the past liquid crystals have been extensively applied in the qualitative visualisation of entire, either steady state, or transient temperature fields on solid surfaces. Since quantifying colour is a difficult and somewhat ambiguous task, application of thermochromic liquid crystals was initially largely qualitative. Application of the colour films or interference filters was tedious and inaccurate. Quantitative and fast temperature measurements were only brought about with the adoption of the CCD colour camera and digital image processing. The rapid development of hardware and software image processing techniques has made the use of inexpensive systems, capable of real-time, transient, full field temperature measurements using TLCs possible (Hiller et al. 1990, 1993, Koch 1993, Park et al. 2001). By disseminating the liquid crystal material throughout the flow, TLCs become not only classical tracers for flow visualisation, but simultaneously, minute thermometers monitoring local fluid temperature. The typical diameter of TLC tracers is 50μm. With the density close to that of water, TLC tracers are well conveyed in liquid flows. The response time of the TLC material is between 3 and 10ms, sufficiently fast for most typical thermal problems in fluids. A collimated white light source is required to illuminate selected cross-sections of the flow (light sheet technique) and colour images are acquired by a camera oriented perpendicular to it.

While tracer colour is intended to indicate temperature distribution of the liquid, the observed colour can also depend on the angle of observation. The author's investigations have shown that this relation is linear with a 10° change of angle being equivalent to a 0.07°C change in temperature (Hiller et al. 1988). It is therefore vitally important that the angle between the illuminated plane (light sheet plane) and the camera is fixed and the viewing angle of the lens is small. In a typical experiment, the flow is observed at 90° using a 50mm lens and a 1/3' sensor (i.e. the camera-viewing angle was smaller than 4°).

Temperature measurements using TLC are based on the colour analysis of images and needs a appropriate calibration. The best measure of colour would be a spectral analysis of refracted light.

In earlier investigations, sets of interferometric filters were applied to analyse the variation in light intensity, recorded by black & white video camera. The development of digital imagery has led to the acquisitions of ready pre-processed information by colour video camera. Colour video cameras split light into three basic colours components: red, green and blue (RGB). This process is known as trichromic decomposition. Each of the three colour components is usually recorded as a separate 8-bit intensity image. Numerous methods of subtracting colour information from a trichromic RGB signal exist. The most straightforward is to convert the RGB trichromic decomposition to another trichromic decomposition based on hue (colour value), saturation and intensity (HSI). Such decomposition is extremely common in image analysis and also serves as a natural means of converting colour images to their black & white representation. Classical conversion of a RGB colour space to an HSI decomposition is based on three simple relations. Light intensity (or brightness) is defined as the sum of its three primary components. The saturation represents colour purity, i.e. the relative value of the reminder after subtracting the amount of colourless (white) light. The hue relates to the dominant colour and is usually obtained from the algebraic or trigonometric relation between the two dominant primary colours. Temperature is determined by relating hue to a temperature calibration function. This is the most critical stage of TLC based thermography. Light refracted by TLC is not monochromatic, even if the observed sample has a uniform temperature. Colour depends on observation angle, scattering properties of the tracers, the colour and refractive index of the liquid and may also vary with the size of the tracers. Additional factors, like the colour of the light source, the colour transmission properties of the acquisition system, as well as reflected and ambient light. The observed colour may also depend on light intensity. Very careful calibration is therefore necessary to obtain quantitative information. The calibration procedure is performed with an identical experimental arrangement to that used for measurements. The difference is that the bulk temperature of the liquid containing the suspended tracers must be extremely uniform and well defined. This is achieved by keeping the experimental cell at a constant temperature and continuously mixing the liquid with a magnetic stirrer. The cell temperature is gradually adjusted by small increments (usually with 0.3°C steps), and several sequences of images are acquired for future processing. An area of about 50 x 50 pixels in the vicinity to the temperature sensor is extracted from each image. Hue evaluation is performed for each pixel, under constraints of minimum and maximum pixel intensity and minimum saturation. "Good" pixels are used to build a hue matrix, which is smoothed using a 5x5 median followed by a low pass filter. An average hue value is then calculated for each image and used as a reference point for the calibration procedure. The procedure is repeated for several images over the TLC's full colour range. It is accomplished by fitting polynomial with variable degree. Outliers are removed from a preliminary fit to obtain a smooth temperature-hue function. The accuracy of the measured temperature depends on the colour (hue) value as a direct consequence of the non-linearity of the curve. The relative error varies from 3% to 10%, and is based on the temperature range defined by the TLCs colour-play limits. For the TLCs used in our experiments (TM Merck), an absolute accuracy of 0.15°C results for low temperatures (red-green colour range) and 0.5°C for high temperatures (blue colour range). The most sensitive colour region is the transition from red to green, which occurs over a temperature change less than one degree Celsius. The same experiment can be repeated using different types of TLCs to improve the accuracy of temperature measurements. The combined measurements cover an extended temperature range, although the method is tedious and requires easy repetition of the experiment.

The application of TLCs to phase change problems bears additional experimental constrains. The phase change temperature must be well defined within the investigated temperature range. Hence, a proper selection of the TLC material becomes crucial. In addition, the solidus surface produces strong light scattering and reflections., which can generate unexpected colour shifts, image distortions in regions adjacent to the phase front. These must all be accounted in the calibration procedure.

In addition to being temperature indicators, TLC tracers are well suited to a seeding role in any visualization experiment, particularly particle image velocimetry. Typical PIV images, correlation-ready, are obtained by converting the colour images to intensity component and applying the usual contrast enhancement procedure. An additional advantage of TLCs is their almost complete transparency to diffused light. Hence, it is much easier to observe particles deep within the flow. Their major drawback is that the intensity of scattered light strongly decreases in regions where local fluid temperature is out of the TLCs colour variation range.

3. Examples of investigated problems

3.1 Particle tracking

Particle tracking is a extremely useful in validating computer simulations of flow. While numerically generated streamlines allow detailed analysis of flow structures, comparison with observed particle tracks provides the most reliable code verification. Some examples of the application of a particle tracking system to natural convection studies are shown below. Several images were recorded at intervals and added to the computer memory to obtain a general view of the flow pattern. The resulting images are similar to multiple exposures, showing both the flow direction as well as its structure. Only a limited number of two-dimensional projections of the particle tracks can be acquired in this way. Better elucidation of three-dimensional flow structures is possible using a stereoscopic observation of the particle paths.

A special procedure has been developed for this purpose, which allows the detection and automatic tracking of particles suspended in the flow medium. Two CCD cameras, each integrated with an identical frame grabber installed in a single personal computer, capture the images. A few, almost nearly neutrally buoyant tracers of a diameter of about 150μm are observed against a bright background from two perpendicular directions. The identification of particles followed by automatic data evaluation and storage enabled experiments to continue unattended over periods as long as several days.

Effect of thermal boundary conditions. The general approach to thermal boundary conditions (*TBC*) imposed on the limiting boundaries of a computational domain is very pragmatic. Idealizations like adiabatic, isothermal or imposed heat flux are widely accepted due to inevitable lack of detailed information. It is certainly not efficient to extend the solution domain to calculate all effects of the surrounding environment on the investigated flow. This would entail unnecessary complication. In fact some heat transfer problems appear not to suffer from *TBC* model simplifications, but it seems worth to sum up some of the possible consequences.

Our observations suggest that *TBC* imposed at the non-isothermal walls may effectively modify the three-dimensional flow structure (Hiller et al. 1990). The main flow pattern in the differentially heated cavity is predictable for most investigated case, and can be calculated by way of simple *2-D* modelling of the symmetry plane. The *3-D* modelling does, however, become more sensitive to *TBC*, away from the symmetry plane. A good example of these *TBC* effects can be found in the standard „double glazing" problem for a cube-shaped cavity. The flow pattern at the symmetry plane has one or two vortex-like rings, transporting fluid back and forth between isothermal walls. This flow is crucial to the overall heat transfer, is predictable and rather insensitive to *TBC* at the thermally passive walls. The *3-D* flow structure consists in one or two spiralling motions, responsible for cross-flow from front and back walls to the cavity centre, in addition to the main recirculation.

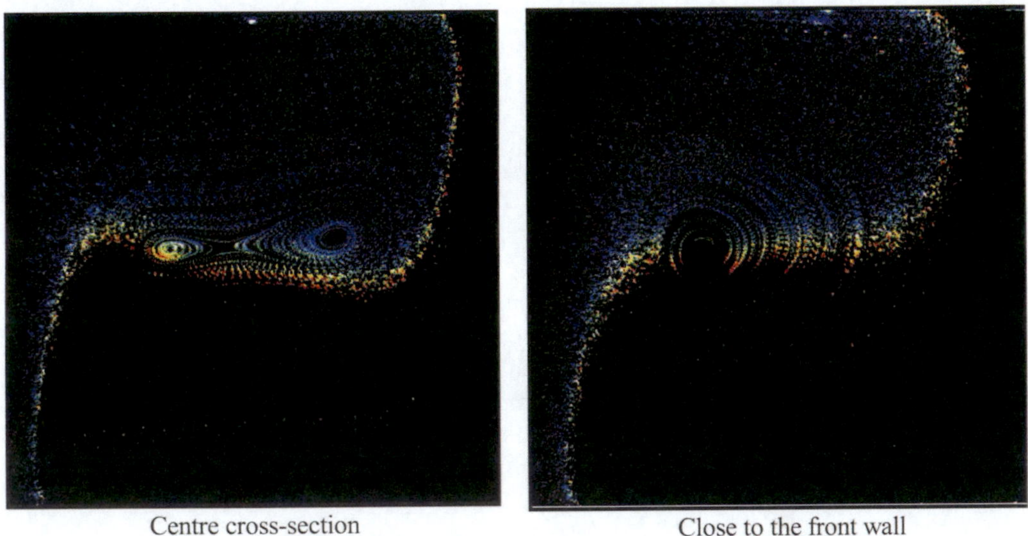

Centre cross-section Close to the front wall

Figure 4. Multiexposed photographs of liquid crystal tracers in the convective flow in the differentially heated cubic cavity; the centre plane cross-section z=0.5 (left) and the front wall z=0.95 (right); Pr=6300, $Ra=8 \cdot 10^4$. Effect of thermal boundary conditions at side-walls - merging of the spiralling structure.

It is worth noting that numerical simulation using simple adiabatic *TBC* for the sidewalls, is generally in agreement with the velocity and temperature fields measured at the symmetry plane. This agreement deteriorates progressively as the front or back wall is approached (the nominally adiabatic, vertical walls of the cavity). Both the isotherms and the flow structure differ from those predicted by simulation (Hiller et al. 1990). The temperature field at side walls is characterized by larger horizontal gradients to non-adiabatic conditions existing there. For single convective cell system (low Rayleigh number), the axis becomes shifted towards one of the isothermal walls. The straight inner spiral arising in a numerical model for an adiabatic *TBC*, in reality, has its ends curved towards the hot wall. For the two-cell system, only one spiral initially reaches the front and back wall (Fig. 4). The two-cell flow structure forms characteristic "cats eyes" in the symmetry plane, and apparently merge midway between the centre and side walls. Several numerical investigations have explored this phenomenon. It appears that the shape and location of the merging region, as well as the direction and pitch of the inner spirals are extremely sensitive to *TBC* at all non-isothermal walls. As a result the estimation of the proper *TBC* for the given experiment becomes a non-trivial task, especially for the two-cell system.

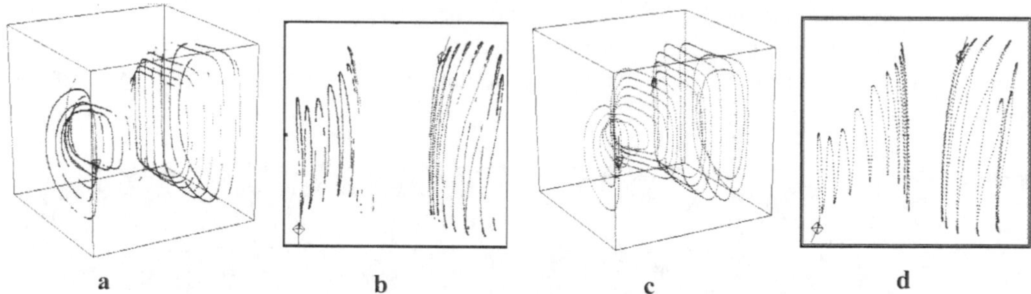

Figure 5. Particle tracks: a, b - measured, c, d - calculated using *TBC* from experiment; Ra=4·10^4, Pr=1180. Isometric and top views of the cavity.

The process of defining an explicitly measured temperature distribution for all four non-isothermal walls replaced the trial-and-error method first used to fit the TBC (Hiller et al. 1992, Leonardi et al. 1999). Figure 5 shows a comparison between observed particle tracks and streamlines calculated with the experimentally defined wall temperatures. The direction and the pitch of the calculated spirals correlate well with those measured. The improvement indicated the need to modify the numerical model, i.e. the necessity for the *3-D* modelling of heat transport through, and along, non-isothermal walls.

Particle accumulation and segregation in recirculating flow. The experimental investigation of the particle tracks usually helps us to determine whether an observed flow, is the same as one predicted by numerical simulation. New problems may also arise when particles are carried by flow. A physical particle has a finite size, mass and buoyancy. Hence, the residual hydrodynamic and gravitational forces acting on the particles may effect their distribution within the flow. Such redistribution of inclusions is also an important issue for purity in the quality of multi component melts during solidification. The experiment performed for simple convective flow in a differentially heated cavity demonstrates how apparently neutral particles aggregated after sufficient time. This occurs as a result of the cumulative effects of the periodic hydrodynamic interactions. Figure 6 shows the distribution of particles in the cavity after 100 hours and two weeks. The flow was initially uniformly seeded with an almost neutrally buoyant population of relatively large (0.35mm) particles. The recirculating convective flow field combined with the settling velocity traps most of particles in some flow regions, causing the rest of the cavity to become particle free. After about 100 hours one can identify a large ring of accumulated particles and a smaller, inner, ring structure. After two weeks almost all particles were confined to a small orbit on the left (warmer) side of the cavity.

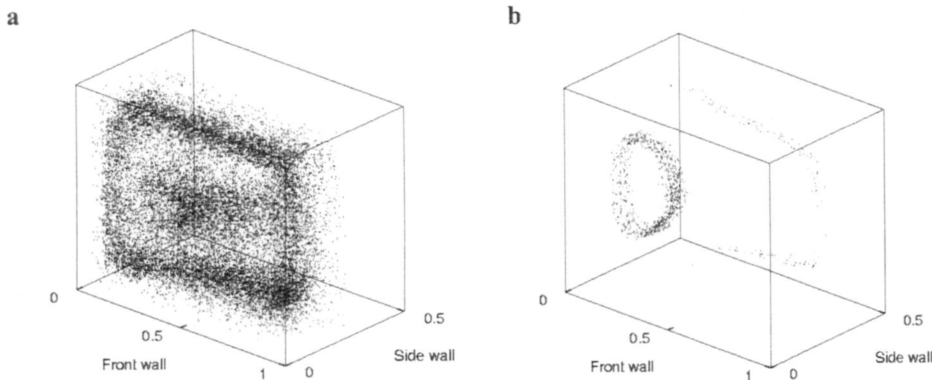

Figure 6. Isometric view of observed distribution of particles in a differentially heated cubic cavity after 100h (a) and two weeks (b); Ra=10^5. Only the front half of the cubic cavity is displayed.

Visualization of supercooling. Most of the investigations involving solidification assume isothermal conditions at the phase change boundary and temperatures above freezing point for the liquid phase. It is, however, well known that usually supercooling of the fluid usually precedes the phase change. Water of standard purity will normally, for example, supercool to about -5°C or -7°C before ice nucleation appears. It may retard the solidification process significantly. Freezing of the supercooled water results furthermore in the formation of a dendritic ice structure, which modifies the heat flux from the cooling wall. Thus, a better understanding of the role of supercooling in the solidification process seems necessary. The accurate modelling of supercooling is a non-trivial task. Supercooling depends on the purity of the fluid, concentration of the nucleation sites, the cooling rate, and sometimes even the cooling history. The theoretical prediction of the nucleation parameters is highly inaccurate. It is also not easy to extract quantitative information on the basis of empirical data only.

Our experiments with freezing water indicated that, in most cases, distilled water cools to about -7°C before phase change begins. The supercooled layer of liquid appears at the cold wall in the lid-cooled cavity. Its presence modifies the onset of convection. After cooling the lid for about 60s, the first ice layer, approximately 1mm thick, is observed. It propagates abruptly across the lid wall surface with a horizontal growth rate of approximately 40mm/s, several orders of magnitude faster than the subsequent thickening of the ice layer. In the differentially heated cavity, the observed effects of supercooling were even more dramatic, qualitatively changing the onset of freezing. In the cavity filled with warm water of 10°C, sudden cooling of the side wall to -10°C generated an instantaneous counter-clockwise vortex and an upward convection of the supercooled fluid. Figure 7 shows the supercooling using particle tracking. Colour images of liquid crystal tracers clearly indicate the presence of supercooled water along the wall (comp. Fig. 13a). During the first 40 to 60s, this supercooled water plum can cover a third of the upper surface before sudden freezing occurs. Within the next 10-15s, clockwise convective melting of excessive ice at the lid restores the regular plane propagation of the ice front. It would seem that supercooling changes the initial circulation inside the cavity. The resulting heat transfer differs from that occurring in the standard situation. The initial ice layer also has a milky dendritic structure due to the supercooling.

Thermal conductivity of this non-homogeneous layer deviates from that of the pure solidus. Neither supercooling nor imperfections on the ice block were present in the numerical model. These may be the reasons for the difficulties encountered in achieving agreement between numerical and experimental data (Kowalewski & Rebow 1999, Banaszek et al. 1999).

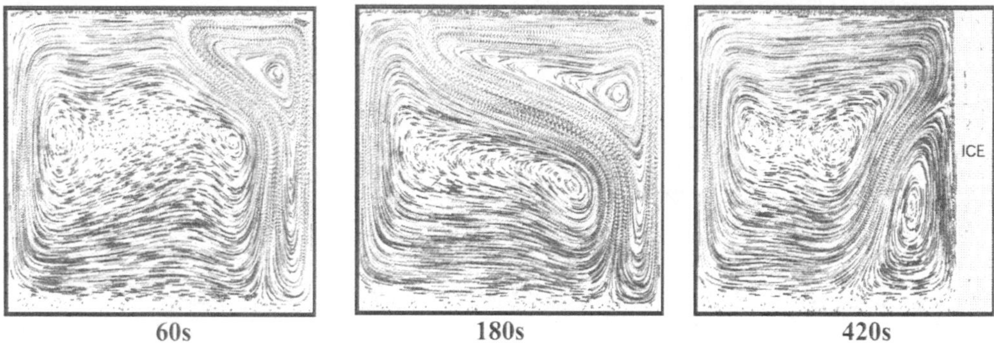

Figure 7. Supercooling of water freezing in differentially heated cubic cavity: T_c= -10°C, T_h= 10°C. Particle tracks by superposition of 20 images in the computer memory; Freezing experiment starts from developed convection.

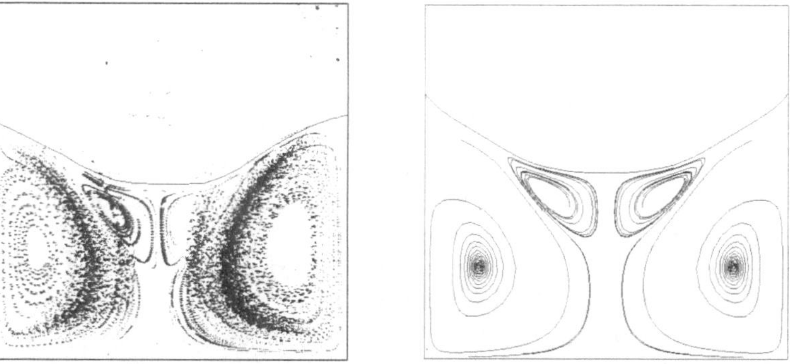

Figure 8. Natural convection beneath the ice surface. Particle tracks observed (left) and calculated (right) at the vertical centre plane. Lid-cooled cavity, T_h=20°C, T_c= -10°C.

Particle tracks under the ice crystal. A comparison of the measured and predicted ice-water interface and the associated flow structures in the lid cooled cubic cavity can be viewed in Figure 8. A primary cell carries hot fluid up the wall to the ice interface, resulting in the recession of the outer edges of this interface. The water in the centre of the cavity then flows down. Just below the ice, in the centre of the cavity, small counter-rotating secondary flows are established due to the extreme density of the water around this region. This recirculation region plays an important role in limiting heat transfer through the thermal boundary layer at the ice surface. It is noteworthy that such a small region is practically invisible, in both the experimental and numerical velocity fields. Only experimentally observed particle tracks allowed discovery of this density anomaly effects, which was confirmed afterwards by numerical tracks (Abegg et al. 1994).

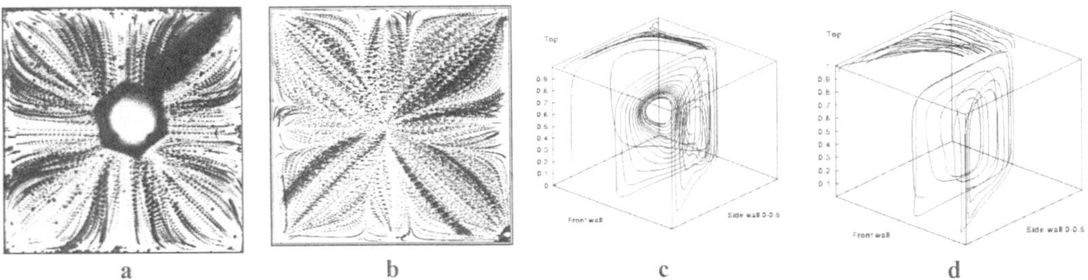

Figure 9. Lid cooled cavity, effect of *TBC* at side walls on pattern selection. (a, b) – particle tracking, top view underneath the lid; (c, d) - isometric view for numerical tracks. High conducting glass walls correspond to cases (a, c), low conducting Plexiglas walls to (b, d).

Flow pattern selection for a lid-cooled cavity. Flow visualization performed in a lid-cooled cavity using particle tracking shows the existence of a complex spiralling structure transporting fluid up along the side walls and down in a central cold plume along the cavity axis (Abegg et al. 1994). For walls of high heat conductivity (glass), eight symmetric cells formed in the flow. For Plexiglas walls, additional small regions of recirculation appeared separating the main cells. Although the computational results obtained for idealized, *1-D TBC* confirmed the eight-fold symmetry of the temperature and flow fields observed experimentally, their orientation was different (Abegg et al. 1994). Moreover, the isotherms were evidently shifted to higher values. Serious discrepancies in the temperature distribution at the horizontal cross-section were observed.

The heat flux was modified step-wise, over several simulations for the *1-D TBC* model to give better agreement with the measured temperature profiles. It was found that such agreement could be obtained by assuming nearly twice as much heat flux through the sidewalls as the nominal value calculated from the physical characteristics of the sidewalls (Leonardi et al 1999). This suggests that the heat flux along the sidewalls, neglected in the simple *1-D TBC* model, effectively increases the heat transfer from the surrounding medium. The difference between the observed and calculated flow patterns remained even after such seemingly arbitrary modification of the *TBC*. The calculated streamlines, starting in the diagonal symmetry plane, consequently spiralled in the *opposite direction* of that observed in experiments. *TBC* could not be properly modelled using an *1-D* of heat flux, especially, for 8mm thick Plexiglas walls. Physically it is indeed possible for a flow pattern with an opposite sense to develop one that spirals inwards at the central plane and outwards at the diagonal plane. Only a slight modification of the *TBC* therefore modifies the flow pattern and can be observed by replacing sidewalls of low conductivity Plexiglas with thin glass (Leonardi et al. 1999).

Numerical simulations performed of both cases, using the *3-D* model for wall heat conductivity, corroborated the *TBC* triggering mechanism for the flow pattern. The inclusion of sidewalls in the computational domain along with solving the coupled fluid-solid heat conduction problem improved the agreement with the observed flow pattern. The observed temperature distribution as well as its symmetry were fully recoverable from the numerical results. It was only as a result of the use of *both* the experimental and numerical methods that the underlying fine mechanisms of the thermal flow were fully understood (Fig. 9).

3.2 Temperature visualization

Flow pattern selection for lid-cooled cavity. Liquid crystal tracers were used to investigate the temperature field beneath the lid for a cubic cavity. Both the temperature field and the star-like grooving of the ice surface indicate the eight-fold symmetry of the flow. Colour variation of TLC-seeded flow images taken directly under the lid (Fig. 10a,b) indicate a disparity in flow structure by differences observed in temperature pattern for low conducting Plexiglas sidewalls (Fig. 10a) and for well conducting glass walls (Fig. 10b). It appears that only a slight change of thermal boundary conditions at the sidewalls can modify the flow pattern.

An interesting example of the temperature pattern observed in the cylindrical cavity is shown in Fig. 10c. Despite the cylindrical symmetry, azimuthal flow structures appear, dividing the flow domain into regular sequence of 16-18 radially spiralling cells. This phenomenon was recently confirmed by numerical simulations (Gelgat et al. 1999). Experimental evidence of this flow symmetry-breaking was only possible with the help of TLC visualisation methods. These structures remain while the phase change is taking place. Characteristic star-like grooves in the ice surface growing at the lid follow the temperature field.

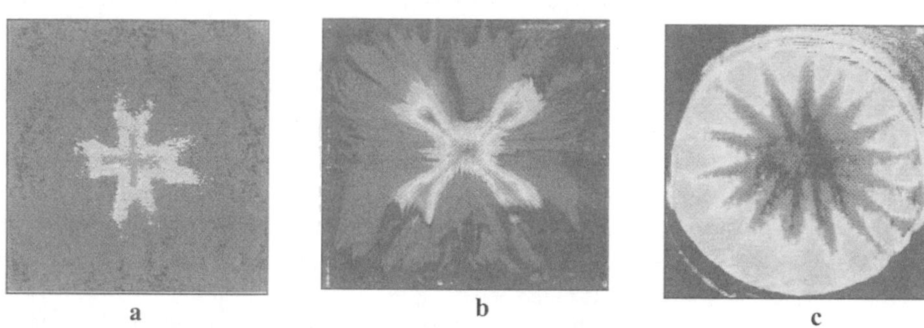

Figure 10. Natural convection in the lid cooled cavities. Temperature distribution recorded with help of TLC directly under the cooled top wall. Effect of the walls properties on the flow structure visible in the temperature fields: (a) - Plexiglas walls, (b) - glass walls, (c) - cylindrical glass cavity

Temperature distribution in lid cooled cylinder, comparing DSPI and TLC. Flow visualisation performed in lid-cooled cubic cavity shows a central cold jet along the cavity axis. Figures 11 displays temperature distribution obtained by tomographic reconstruction from DSPI images (Fig. 11b), and using thermochromic liquid crystals (Fig. 11c) (Soeller 1994).

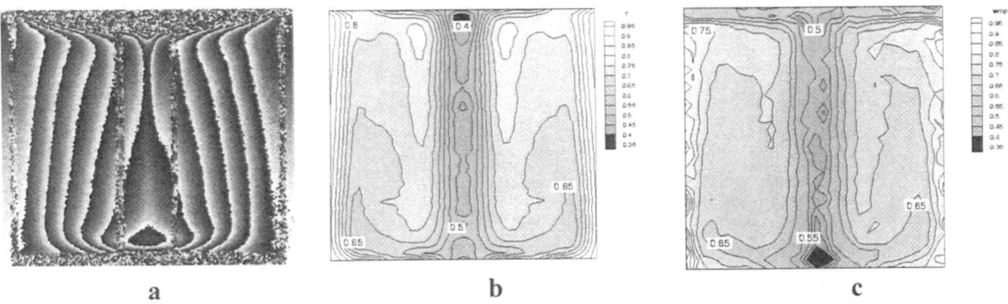

Figure 11. Temperature evaluation for lid cooled cubic cavity in vertical cross-section; (a) – DSPI speckle image for phase evaluation, (b) – DSPI evaluated temperature field, (c) – TLC evaluated temperature field.

"Hot-spots" visualization. The experimental model of the mould filling process was investigated. The experimental data were collected to create a reference database for comparison with numerical results. The method of simultaneous measurement of the flow and temperature fields using liquid crystal tracers has been successfully applied to collect transient information on the flow (Kowalewski et al. 2001). We consider forced convection in a rectangular, inclined box filled with viscous liquid. The cavity has square cross-section 38mm x 38mm and is 113mm high. The two sidewalls are made from 7.5mm tick Plexiglas. The other two isothermal sidewalls are made of copper. They are kept at low temperature Tc. Two Plexiglas plates located inside cavity are simulating shape complexity of a mould. They form three flow cavities connected only by 5mm high slits between the upper rim of the plates and the cold wall above. The cavity inclination angle α varied from 12°- 45°. The hot fluid of initial temperature T_h is forced to the cavity through a 13mm circular opening made in the bottom wall. Both forced convection and residual natural convection within the cavity are responsible for the heat transfer through the cold sidewalls.

Glycerol is used as the working fluid for its well known physical properties and very strong variation of viscosity with temperature. For the temperature range used (T_c=10°C, T_h=50°C) the fluid viscosity changes its value almost twenty times dramatically altering a flow pattern at the cold walls. The heat flux to the side cold walls is responsible for creation of the tick, viscous layer, retarding the main flow. It diminishes convective heat flux from the inner regions of the cavity. The cooling at some spots of fluid trapped in recirculation zones is delayed, leaving the "hot spots" (Fig. 12). In these places a slow conductive cooling and volumetric shrinkage of fluid can generate void bubbles, a common plague of any industrial casting. The observed interactions of free surface flow with the obstacles create an additional challenging problem for numerical simulations (Michalek & Kowalewski 2003).

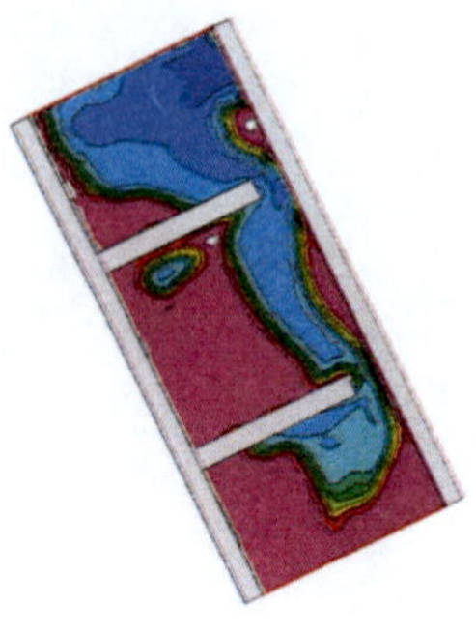

Figure 12. TLC temperature visualization (left) and evaluated isotherms (right)observed just after closing the inlet; inclination of the cavity 26°. Blue colour (dark regions) indicates higher temperature.

Supercooling, temperature visualization

Supercooling of water visualized with classical tracers (Fig. 7) can be quantitatively evaluated using thermochromic liquid crystals. It can be seen (Fig. 13a) that before freezing starts, water at temperature below null rise up covering large part of the top wall. The degree of supercooling of about –7°C can be evaluated by dominant dark-red colour of the liquid crystal tracers in the rising plum.

Infrared Thermography. Freezing of water in differentially heated cubic cavity was investigated. Discrepancies between numerical and observed results forced us to verify Thermal Boundary Conditions imposed on the "passive", non-isothermal sidewalls of the cavity. To control heat flux through these walls, measurement of both internal and external temperature fields at these walls becomes necessary. Such a possibility is offered by combined application of Infrared and TLCs based Thermography. Temperature fields obtained by IRT at external walls, and internal temperature maps from PIV & T measurements enrich our description of the physical experiment allowing better verification of numerical models (Wisniewski et al. 1998). Fig. 13b displays in false colours temperature distribution measured at the external front wall of the cavity. Temperature measured with accuracy equal to 0.3°C. The system used in our experimental studies consists of a low wavelength Agema Thermovision 900 LW infrared scanner. It uses a Mercury Cadmium Telluride (MCD) detector with liquid nitrogen cooling. The spectral band covered is from 8 to 12 μm, with some residual response outside this range. This spectral response results in low noise level measurements at room temperature. Nominal sensitivity of the scanner is equal to 0.08°C at 30°C.

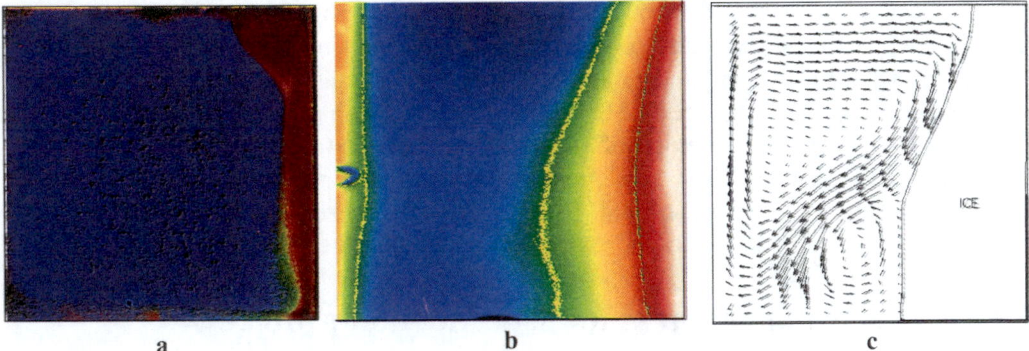

Figure 13. An example of combined application of liquid crystals and infrared thermography, and particle image velocimetry. Freezing of water in differentially heated cubical cavity (T_c=-10°C, T_h=10°C); (a) – initial supercooling, liquid crystal tracers show cold plum of water at temperature about –7°C (red colour); (b) - infrared temperature evaluation at the external front wall made of Plexiglas; (c) – velocity field measured by particle image velocimetry at the centre cross-section

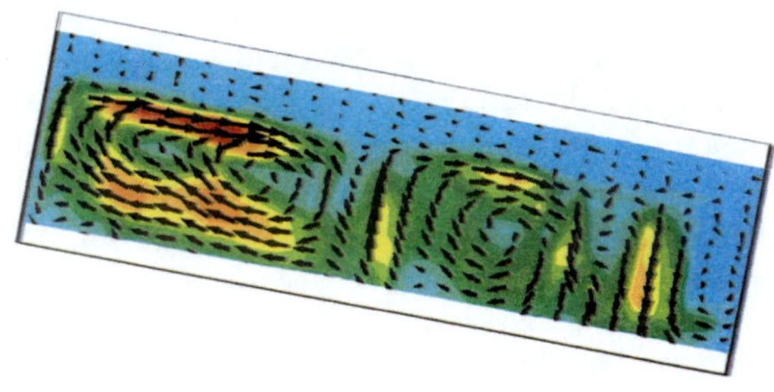

Figure 14: Water freezing in the mould model after its filling is accomplished. Ice layer is formed at both isothermal walls. Velocity vectors and contours of the velocity magnitude evaluated by PIV method. T_c=-7°C, T_h=9°C

3.3 Particle Image Velocimetry

Particle Image Velocimetry is applied to investigate flow field of water freezing in a differentially heated cubic box. The natural convection of water in the vicinity of the freezing point differs significantly from the well-known patterns of the benchmark solutions. The competing effects of positive and negative buoyancy forces result in a flow with two distinct circulations (Fig. 13c). There is a "normal" clockwise circulation, where the water density decreases with temperature (upper-left cavity region) and an "abnormal" convection with the opposite density variation and counter-clockwise rotation (lower-right region). At the upper part of the cold wall, the two circulations collide, intensifying the heat transfer and effectively decreasing the interface growth. Below, the abnormal circulation limits the convective heat transfer from the hot wall, separating it from the freezing front. Hence, the phase front is only initially flat. As time passes it deforms strongly, getting a characteristic "belly" at its lower part.

The experimental model of the mould filling process was investigated for inclined, simple rectangular cavity. The solidification process of water, the working fluid, is observed at both isothermal walls. Initially, almost uniformly and parallel to the wall, a layer of solid builds at both walls. However, as time progresses, the remaining effects of natural convection start to modify the heat transfer, resulting in a faster solidification at lower parts and an increase of asymmetry between upper and lower wall. The whole process depends on the inclination angle of the cavity. Figure 13 shows the experimental result obtained for water freezing in the cavity tilted at 11°. Initial water temperature was +9°C, the isothermal walls temperature –7°C.

4 Laboratory benchmarks for solidification problems

Since numerical simulations in fluid dynamics reached certain level of maturity an increased attention is directed to define methodology of estimating errors and uncertainties of generated solutions. The problem of verification and validation of solutions became urgent for industrial applications, where costly prototypes have to be constructed on the basis of numerical predictions only. Sources of errors and uncertainties in results from simulations can be divided into two distinct sources: numerical modelling and physical modelling. Uncertainties and errors of numerical models can be again divided into several categories, where main are due to simplifications and approximations of mathematical problems (e.g. all discretization), and errors produced by numerical procedure. Uncertainties and errors of physical modelling are due to assumption and approximations made to describe physical reality. Formulating any problem we have to limit its mathematical description to certain domain, select more important parameters and neglect apparently less important. The only possibility to find out what does it mean is to perform experimental validation of the certain approach, comparing generated results with experimental data coming from a properly designed and evaluated experiment. The second part of the above sentence elucidates already next problem. How to find criteria for selecting an appropriate experiment? There are very limited experimental possibilities to deliver complete, transient description of velocity, temperature and concentration fields for industrial problems. It is not only because of not sufficiently developed of experimental tools, but also due to inaccurate or unknown physical properties of materials used, particularly by extreme thermal conditions, in presence of chemical reactions (e.g. oxidation, solubility) or at non-equilibrium states (e.g. supercooling). Hence, taking into account that it is very difficult to create profound experimental database for problems involving industrial scales and materials, small-scale experiments using model materials are used. Of course, such approach gives possibility to validate only part of the numerical model. When laboratory models are used there are serious differences in the scale of the problem, hence flow dynamics is usually different. Even more serious differences are in material properties. So-called *analog* fluids, simulating freezing of metals differ two or three orders of magnitude in Prandtl number, representing just opposite range of heat transfer properties. Therefore, single positive validation result, particularly using laboratory experiments, is a necessary but not sufficient condition to accomplish validation procedure, and several different tests have to be performed. In the following we give short review of few solidification experiments, which can be found in the literature together with three simple configurations we proposed as benchmarks for testing solidification problems.

4.1 Selection of the solidifying media and examples

Solidification of metals. Convection pattern in molten metals are difficult to determine experimentally, especially for buoyancy-driven flows. High meting temperature for most of the metals additionally complicates collection of data in the laboratory. Therefore, experiments performed with metals concern mainly measurements of temperature using thermocouples, and phase front location using X-rays or similar technique. Physical properties of some metals used in the laboratory experiments are collected in Table 4.1. The unique property of metals, impossible to model using other liquids, is high thermal conductivity, which leads to very low Prandtl number.

Perhaps one of most frequently cited paper concerning experimental investigation of metal solidification is that by Gau and Viskanta (1986). They used Gallium, low melting temperature metal to diminish experimental problems. Gallium has relatively well known physical properties; it is also technologically important material. The Prandtl number is 0.02. Experimental data were collected for 88.8mm x 63.5mm rectangular cavity with a free surface and differentially heated walls. Transient solidification and melting was investigated by collecting temperature distribution at the centreline with 26 thermocouples. The most valuable information collected are measurements of the liquid/solid interface at pre-selected time steps during melting. The pour-out and the probing method were used to examine and measure the shape of the interface. Hence, for each time the experiment was repeated. Despite its simplicity, experiment of Gau and Viskanta defined still useful experimental benchmark for validating simulation of melting metals. Desire for more detailed information, like temperature and a velocity field is obvious.

Table 4.1. Physical properties for typical metals used to simulate industrial solidification problems

	Al	Sn	Ga
density, ρ [(kg m^{-3}]	2700	7280	6090
specific heat , c [J kg^{-1} K^{-1}]	903	220	400
thermal conductivity, k [W m^{-1} K^{-1})]	237	67	33.5
thermal expansion, β [K^{-1}]	$23 \cdot 10^{-5}$	$27 \cdot 10^{-5}$	$12 \cdot 10^{-5}$
melting temperature, [°C]	660	231	29.8
kinematic viscosity, ν [m^2 s^{-1}]	0.55×10^{-6}	$0.3 \cdot 10^{-6}$	3.1×10^{-6}
Prandtl number, Pr	$0.5\ 10^{-2}$	$1.5 \cdot 10^{-2}$	$2 \cdot 10^{-2}$

Müller et al. (1984) have performed experiments with Gallium and Gallium doped with Antimony (GaSb) in vertical Bridgman configuration. Measurements of temperature were performed using thermocouples, microphotographs of crystals has been used to analyse perturbations of the interface due to the temperature oscillations. Collected data for Gallium are difficult to use for validation numerical simulation, but the paper shows analogical experiment performed for high Prandtl number fluid (water). It gives direction how to bypass experimental difficulties and collect flow pattern information using analog fluid.

An experimental benchmark test for "real life" problem was given by Sirrell et al. at the 7th Conference on Modelling of Casting (1995). The test consisted of a simple plate connected to a narrow and tall filling system. Characteristic dimensions of the plate were 200mm length, 100mm height, and 15mm depth. It was connected to a 410mm tall filling sprue. The geometry was filled gravitationally with pure liquid aluminium. The real time X-ray radiographic unit was used to visualize the flow of liquid aluminium in the mould. The radiographic images were recorded at 50Hz on a videotape. Seven thermocouples were arranged in the mould to measure transient temperature variation. The whole filling process took about 2s, hence high speed turbulent flow was expected in the sprue. Two parameters were compared with nine numerical simulations: history of the interface and cooling rate traced by thermocouples. It appeared that despite very careful verification of thermophysical data used, and having all details of the mould geometry, it was difficult to find single code offering good agreement. Most of the codes had problems with modelling the interface instabilities. Also freezing times were not always in agreement. It is worth to note that proposed benchmark offered only global description of the process. Detailed temperature and velocity fields of the fluid were unknown, and couldn't be compared. Such comparison, if available, would be perhaps very helpful for identifications parts of the particular code that failed to simulate investigated phenomena.

Solidification of binary metal alloy. In most alloy solidification processes, a two-phase (mushy) zone forms. The zone is composed of solid dendrites and interdendritic liquid. The mushy zone separates solidified and melted regions strongly modifying heat and mass transfer trough the interface. Because dendritic zone is permeable, flow may occur in the mushy zone, as well as in the melt. Concentration gradients within the zone generate solutal convection modifying flow at the interface. Complex coupling of the melt flow and the mushy zone flow is responsible for macrosegregation of alloy components, formation of voids, entrapment of inclusions and development of residual stresses. Numerical simulations of these processes are based on several crucial simplifications. Their validation appears imperative but difficult to perform. Experimental investigations of this complex processes flow are very limited.

Perhaps most comprehensive experiment was performed by Prescott et al. (1994). They provided details of the temperature and microstructure measurements for Pb-Sn alloy solidifying in an axisymmetric cell. The mould was suspended concentrically within the cooling jacket kept at 286K. The centrally located heater had temperature 578K. Dimensions of the cell were: inner radius 15.9mm, outer radius 63.5mm and height 150mm. Six sets of radially located thermocouples, immersed at three different depths were used to monitor temperature during solidification. Post-experiment measurements were made to determine the concentration of Pb by taking samples of material from several holes drilled in the mould. Metallographic analysis of specimens taken from experimental ingots was performed to provide information regarding characteristics of the mushy zone. Measurements were used to validate authors' numerical simulations. The cooling rates obtained from the temperature probes were critically analysed to conclude presence of a strong convection flow across dendritic structures, which is probably responsible for their fracturing and detachment. Detached, free-floating crystals could not be expected in the numerical model. Detailed analysis of temperature data recorded during solidification allowed for concluding about microsegregation pattern, indirectly confirming numerical predictions. Photographs of an ingot sections were used to analyse macrosegregation pattern and to estimate the permeability coefficient for the numerical simulations. The experiment gave evidence of thermosolutal convection, however indirectly as there were no flow dynamics measurements possible. It is worth noting, that even these limited experimental data gave evidence of undercooling, recalescence (heating upon solidification), and solid particles transport. All these effects could not be predicted by numerical model.

Solidification of analog alloy models. Solutal and thermosolutal buoyancy forces play an important role in solidification process, giving rise to large-scale convection pattern. The resulting transport is termed *double-diffusive* convection. The phase change process is strongly influenced by a small-scale convective motion of the interdendritic liquid. The dendritic structure is described as porous media with some estimated permeability coefficient. Solidification process strongly depends on the complex interaction of the large-scale double-diffusive convection and filtration motion through the mushy zone. Numerical modelling of these flows depends on many unknown factors that can be only validated through experimental data. Experiments with metal alloys are not able to provide details on the mould flow. Hence, a verity of double-diffusive flows has been studied during solidification of optically transparent fluids, which freeze dendritically. These organic or aqueous solutions are known as *analog* alloys.

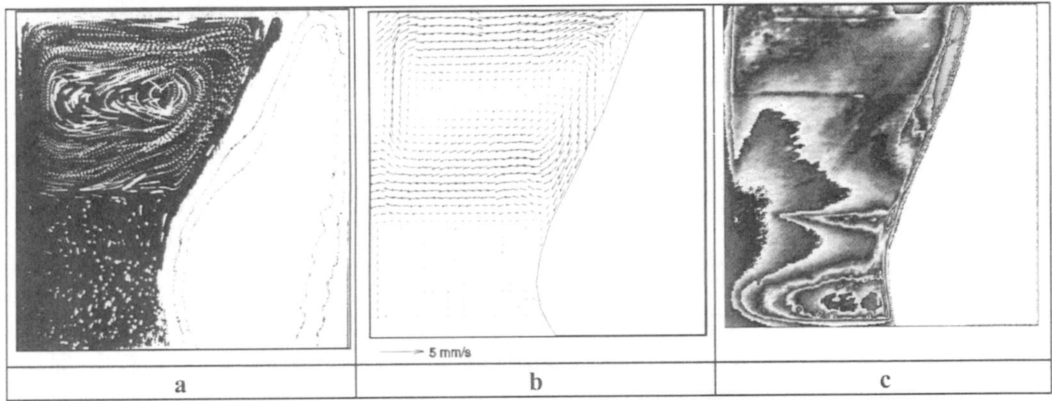

Figure 14. Solidification of NaCl aqueous solution: tracers (a), velocity field (b) and Schlieren image (c) (Kowalewski et al. 1998, Banaszek et al., IPPT 2000)

Figure 14 shows double-diffusive convection pattern during solidification of aqueous solution of NaCl in differentially heated cavity. Ice layer grows from the right sidewall. Salt rejected from the ice enriches boundary layer along the phase boundary. It induces solutal convection of denser liquid, which interacts with bulk thermal convection in the cavity. Flow stratification, which occurs along the bottom layer, is well visible in the velocity field. As a result of heating from the left wall, the second layer of fluid enriched with salt forms beneath the bottom layer. It is well visible in the schlieren image of the same flow (Fig. 4.2c). The solutal stratification may remain stable during main part of the solidification process. When the gap between hot wall and phase front decreases, the layer formed along the bottom extends vertically and becomes virtually stagnant.

Solidification of binary systems has been investigated for NH_4Cl-H_20 solutions by Chang et al (1998). The schlieren images made for different salt concentrations demonstrate evolution of multiplayer system for hypoeutectic composition to double layered system with strong stratification for hypereutectic compositions, with wave – like instability. It confirms previous observations reported by Beckerman and Viskanta (1988) for the same solution.

Water for its well-known properties, optical transparency, low melting temperature and simple handling is a best choice for laboratory experiments. The only problem, which can also be seen as a challenging advantage, is its anomaly of the density behaviour close to the freezing point. Beside water there are several, so-called phase-change-materials (PCMs) investigated intensively in the past as candidates for the energy storage. Some of them, like polyethylene glycol and n-octadekan, hexadecane, are used also for solidification in laboratory models. Another transparent organic liquid, believed to be a good analog for crystal growth problems is succinonitrile (SCN). Main physical properties of two typical *analog* fluids are collected in Table 4.2 and compared with water.

Table 4.2 Physical properties at melting point for materials used to simulate industrial solidification problems

	Water	Succinonitrile SCN	Polyethylene glycol PEG 900	Hexadecane
density, ρ [(kg m^{-3}]	999	985	1100	792
Specific heat, c [J kg^{-1} K^{-1}]	4217,8	2000	2260	2236
thermal cond., k [W m^{-1}K^{-1})]	0.552	0.223	0.188	0.18
thermal expansion, β [K^{-1}]	$-0.07 \cdot 10^{-3}$	$0.81 \cdot 10^{-3}$	$0.76 \cdot 10^{-3}$	$0.9 \cdot 10^{-3}$
melting temperature, [°C]	0	55	34	18
kinematic viscosity, ν [m^2 s^{-1}]	$1.8 \cdot 10^{-6}$	$2.6 \cdot 10^{-6}$	$9 \cdot 10^{-5}$	$3 \cdot 10^{-6}$
Prandtl number, Pr	13	23	1188	45

Selection of organic materials for investigations allows also to simulate behaviour of alloys. Frequently used in the laboratories *analog* alloy is succinonitrile with 0.2% wt acetone. This material is transparent and its solidification morphology resembles solidification of metals. It is used by Noël et al. (1997, 2000) for optical investigations of transient solidification process in vertical, cylindrical crucible. Details about interface structure, propagation of the instabilities were observed optically and detailed reported. However, despite transparency of the media, only information about the interface shape is given. It is perhaps not sufficient for detailed code validation.

Very detailed experimental study of SCN solidification in a horizontal Bridgman apparatus was reported by H. de Groh III (1994). The apparatus consists of a transparent glass ampoule with square cross-section and dimensions 6mm x 1500mm observed through the microscope. Melting is achieved by moving the heating-cooling jackets along the ampoule filled with pure SCN or SCN alloy. This experiment has been defined as an experimental benchmark at CHT'01 conference (Advances in Computational Heat Transfer, Palm Cove 2001). Experimental data available for the validation exercises consists of detailed description of the temperature history measured by thermocouples at five points on the external wall, location and shape of the interface, growth rate and flow visualization with seed particles. It is interesting to note that there was no response to this particular benchmark.

4.2 Proposed laboratory benchmark configurations

During last ten years several laboratory experiments were performed by our group to develop and improve methods of collecting quantitative data on velocity and temperature of convective motion accompanying solidification. With water as a flow media and thermochromic liquid crystals as tracers it is possible to collect transient information on temperature and velocity fields within several cross-sections of the flow. Also position and shape of the interface are retrieved form the images. The typical experimental setup we use for the flow measurements consists of a convection box, a halogen tube lamp, the 3-chip CCD colour camera and the 32-bit frame grabber. The flow field is illuminated with a 2mm thin sheet of white light from a specially constructed halogen lamp tube and observed in the perpendicular direction. It can be replaced with a Xenon flash lamp, if the flow velocity is high. The 24-bit true colour images, typically of 768x564 pixels, are acquired with a personal computer. Using PCI-bus RGB frame grabber (AM-STD ITI) the setup permits us to acquire into computer memory in real time a long sequence of true-colour images. Recording of the transient flow patterns and temperature fields is performed periodically. Typically every 10-300s, short series of images are acquired and stored on the hard disk of the computer for later evaluation. The computer controls the system of three stepping motors, switching of the light source and also records the readings from control thermocouples and the thermostats. Using three stepping motors two different configurations can be automatically arranged. First configuration, with a vertical light sheet directed by the mirror and the camera observing flow through the front wall is used to analyse flow in vertical planes. The second configuration, with a horizontal light sheet and the camera observing flow from the top through the mirror, allows to acquire horizontal cross-sections of the flow. The computer-controlled system allows for acquiring images of several cross-sections, fully automatically within several seconds. Hence, due to relatively slow variations of the flow structures, transient recording of the main three-dimensional flow features is possible.

In the experiments described below we consider natural convection of water freezing in a small cubic or rectangular cavities. A typical internal dimension of the investigated flow domain is 38mm for the cube. Either two opposite walls or one top wall are made of metal and assumed to be isothermal. Other walls are nominally insulators of finite thermal diffusivity. They are made out of 6 and 8mm thick Plexiglas or 2mm glass. The thermal conductivity of the "passive walls" seems to play an important role in the development of fine flow structures and the effect of their thickness and conductivity was investigated in different experiments.

Thermal properties of water vary with temperature. However, our numerical tests indicted (Leonardi et al. 1999) that it is appropriated to assume fluid heat capacity, thermal conductivity and viscosity constant. The values used in the numerical simulations are respectively:

$c_l = 4.202\text{kJ/kgK}$

$k_l = 0.56\text{W/mK}$

$\nu = 1.79\ 10^{-6}\ \text{Ns/m}^2$

The fluid density varies with temperature according to the given polynomial function:

$$\rho_l = 999.840281167108 + 0.0673268037314653 \cdot t - 0.00894484552601798 \cdot t^2 + \\ + 8.78462866500416 \cdot 10^{-5} \cdot t^3 - 6.62139792627547 \cdot 10^{-7} \cdot t^4,$$

where the temperature t is given in degrees Celsius.

The thermophysical properties of ice we assume to be constant and equal: $\rho_s = 916.8\text{kg/m}^3$ for the density, $k_s = 2.26\text{W/mK}$ for the thermal conductivity, and $c_s = 2.116\text{kJ/kgK}$ for the heat capacity. The latent heat of freezing water is equal $L_f = 335\text{kJ/kg}$. The thermal conductivity, heat capacity and density of the Plexiglas used were measured. When solving the energy equation for the side walls, the values of 0.195W/mK for the thermal conductivity, and $1.19 \cdot 10^{-7}\text{m}^2/\text{s}$ for the thermal diffusivity were used. The heat transfer coefficient h used for modelling the convective heat flux from the external fluid was taken to be $20\text{W/ m}^2\text{ K}$ in case of a forced air flow outside investigated cavity, and $1000\text{W/ m}^2\text{ K}$ for the forced convection in water bath surrounding the cavity. Data were collected for about 3000s of the freezing process.

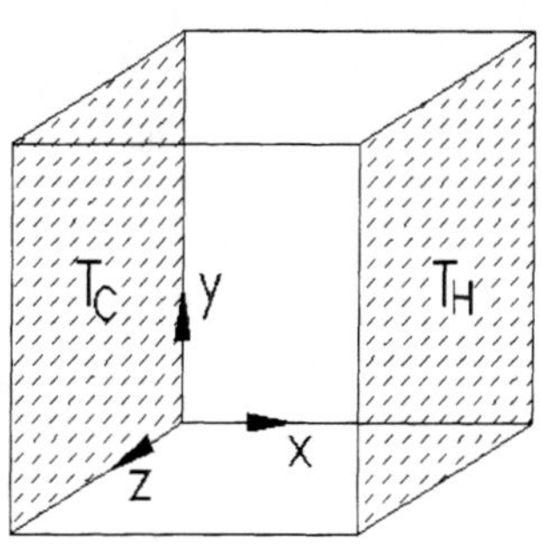

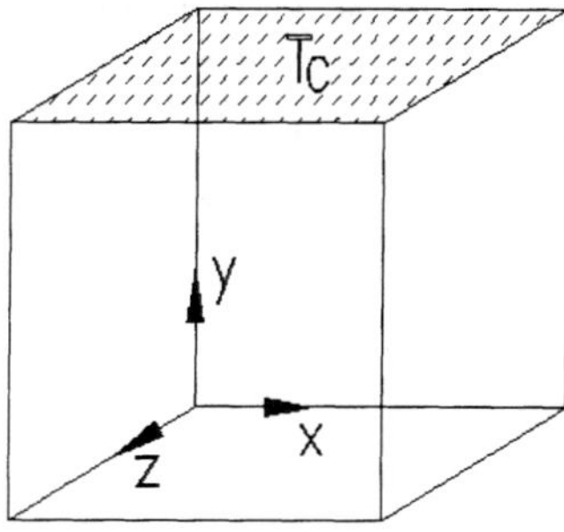

a **b**

Figure 15. (a) - benchmark 1: differentially heated cavity, cube with two metal walls and four Plexigals walls surrounded by air; (b) – benchmark 2: lid cooled cavity immersed in water bath, cube with the metal lid and with five Plexiglas walls; cavity is surrounded by an external water bath.

Benchmark problem 1, directional freezing in a differentially heated cavity: We consider convective flow in a box filled with distilled water (Fig. 15a). The flow takes place in a container with an aspect ratio of one, its two opposite vertical walls are assumed isothermal. Internal dimension of the box is 38mm. One of the vertical walls is held at temperature T_c= -10°C. It is below the freezing temperature of water, hence the solidification takes place on this wall. The opposite vertical wall is held at temperature T_h = 10°C. The isothermal walls are made of aluminium. The other four walls have low thermal conductivity, they are made of 6mm tick Plexiglas. The cavity is surrounded by a laminar stream of air at temperature T_{ext}=25°C. For transient processes uncertainty of the initial conditions may create difficulties in matching experimental and numerical results. Hence, to improve our definition of the initial condition, we start freezing after the steady convection pattern is established in the cavity. This initial flow state corresponds to natural convection without phase change in the differentially heated cavity, with the temperature of the cold wall set to T_c=0°C. The freezing experiment starts, when at time t=0, the cold wall temperature suddenly drops from null to T_c= -10°C. In the numerical runs, the solution obtained for steady state natural convection was used as the initial flow and temperature fields to start the freezing calculations. The heat transfer coefficient h used for modelling the convective heat flux from the external fluid is taken to be $20\text{W/ m}^2\text{K}$ for the forced air flow. Data are collected for about 3000s of the freezing process (about 30% volume is frozen). Below we give few examples of the data obtained.

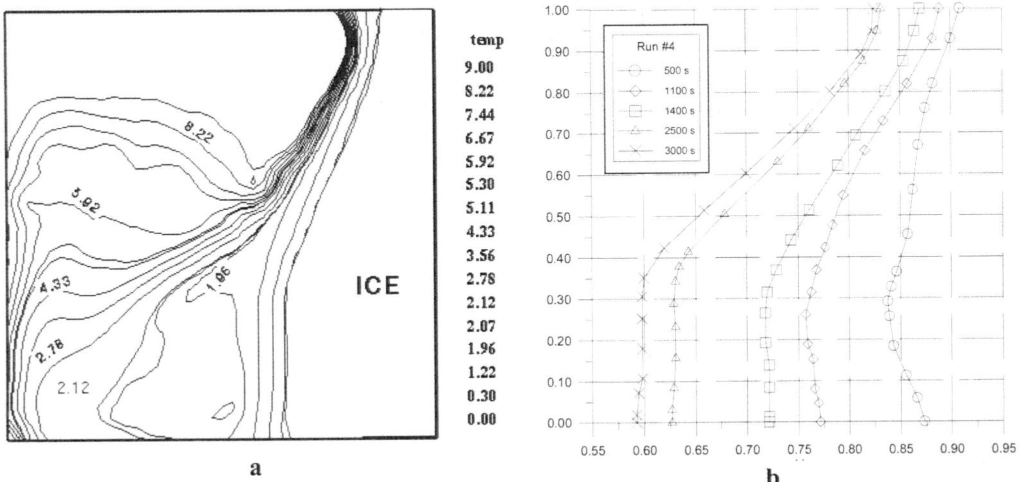

Figure 16. Freezing of water in differentially heated cube shaped cavity with four Plexiglas walls after 3000s. Left hot wall T_h=10°C, right cold wall T_c=-10°C; (a) - evaluated temperature distribution, (b) – phase front position. Compare also Fig. 12 showing evaluate velocity field and external wall temperature.

Figures 12 and 16 show examples of images showing behaviour of natural convection of water in the vicinity of the freezing point. It shows an interesting feature, two counter-rotating convection patterns, typical for the configuration with differentially heated walls. It is mainly due to the strongly non-linear temperature dependence of the density function with the extreme at 4°C. The competing effects of positive and negative buoyancy force result in a flow with two distinct circulations. There is a "normal" clockwise circulation, where the water density decreases with temperature (upper-left cavity region) and an "abnormal" convection with the opposite density variation and counter-clockwise rotation (lower-right region). At the upper part of the cold wall the two circulations collide with each other, intensifying the heat transfer and effectively decreasing the interface growth. Below, the convective heat transfer from the hot wall is limited by the abnormal circulation, separating it from the freezing front. Hence, the phase front is only initially flat. As time passes it deforms strongly, getting a characteristic "belly" at its lower part. High temperature and velocity gradients in the region of two colliding circulation patterns create challenging condition for numerical modelling. Our experimentation and numerical simulations have shown the enormous sensitivity of the flow structure and the freezing front to relatively small changes of the hot wall temperature or thermal boundary conditions at the passive walls (Leonardi et al. 1999). The simulations confirmed most of the details of the flow for the initial time steps (below 500s). However, as time proceeds the ice growth rate delays significantly in comparison to the numerical predictions (Kowalewski & Rebow 1999, Giangi et al. 1999, 2000).

Benchmark problem 2, vertical Bridgman freezing problem in lid cooled cavity: We consider convective flow in a cube filled with distilled water. The top wall of the container is assumed isothermal and is held at temperature T_c=-10°C. It is below the freezing temperature of water and ice crystal growths from the lid down into the cavity. Internal dimension of the box is 38mm, the five side walls are made of Plexiglas. The wall thickness is 8mm. The Plexiglas cube is immersed in an external water bath of temperature +20°C (see Fig. 15b). We assumed that for a forced convection in the bath the heat transfer coefficient is $h = 1000 W/m^2 K$. All other details are similar as in the benchmark 1. The isothermal metal lid has temperature -10°C, when freezing is investigated. But as described above, we start the freezing experiment (and simulations) after natural convection is developed, initially setting the top wall temperature to 0°C at the beginning.

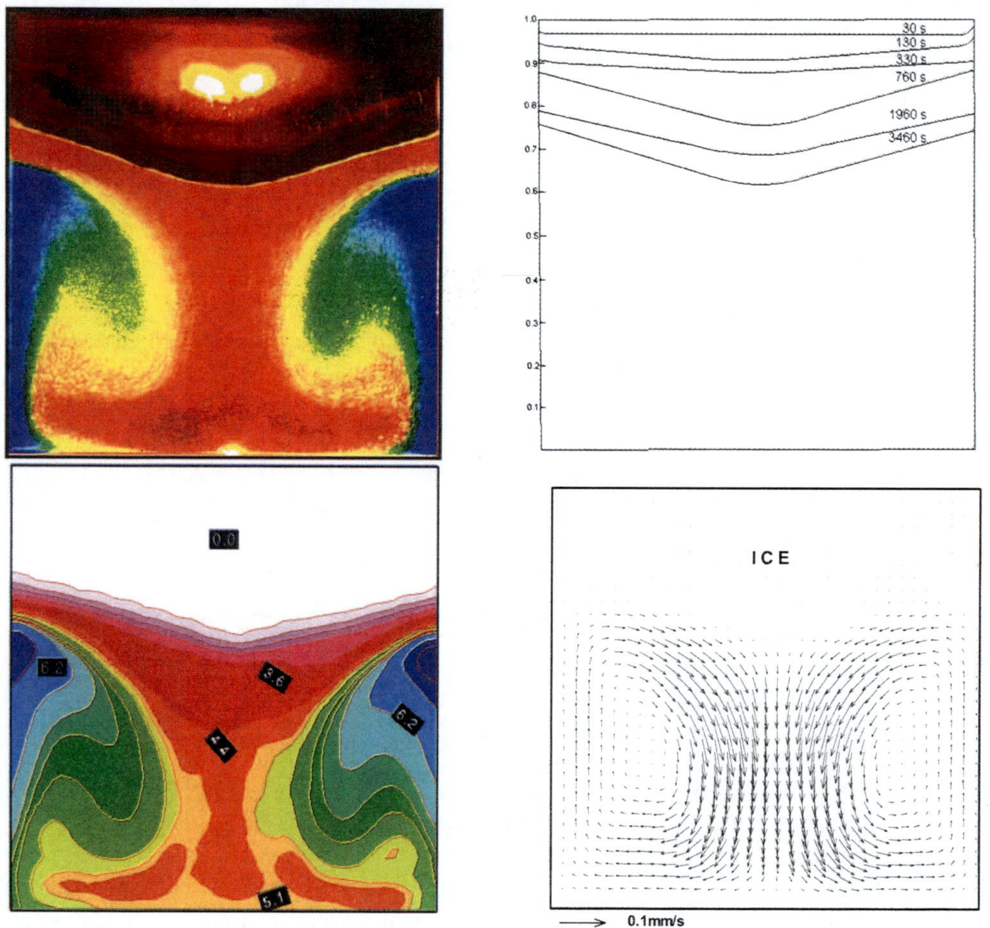

Figure 17. Freezing of water in the lid-cooled cavity. Top line: liquid crystals visualization (left), phase front profiles for selected time steps (right). Bottom line: evaluated temperature field (left) and velocity field (right) for natural convection in the centre plane at 3600s. Top wall temperature T_c=-10°C, external bath temperature T_h=20°C; Ra=$1.5 \cdot 10^6$, Pr=13.3

The data on interface position, velocity and temperature fields are collected periodically during about 3600s of the freezing process. After this time approximately 35% of the cavity volume is frozen. Figure 17 shows examples of the data acquired. Strong deformation of the initially flat ice front is well visible. The crystal has conical shape due to the recirculating flow in the cavity. The cold liquid penetrates along the cavity axis into the bottom, warms up and returns along the side walls. Detailed observations indicated presence of eight-fold symmetry of the recirculating flow field, it is reproduced in the star-like grooving present at the ice surface. These was confirmed by performed numerical simulations. However, like in the previous case, the simulated growth rate of ice seriously deviates form the observed for longer time steps (Kowalewski & Cybulski 1996, 1997).

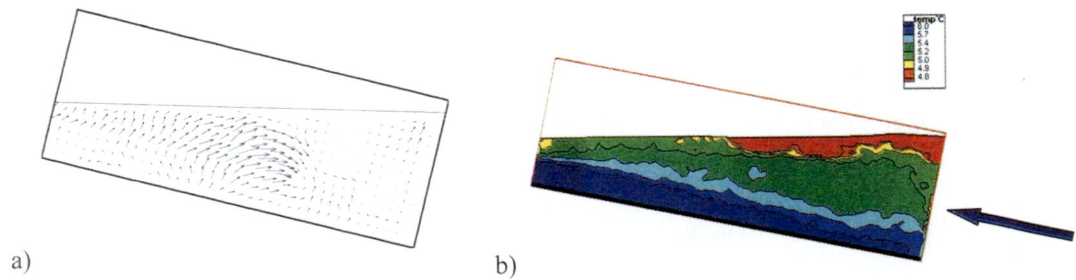

Figure 18. Flow observed in mould filling model at t=20s. Water at initial temperatutre T_H is forced through the 13mm openning in the bottom wall (arrow). It freezes at two side metal walls kept at temperature T_C. The remaining four walls (top, bottom, front and back) are made of Plexiglas. (a) - PIV evaluated velocity field; (b) - PIT evaluated isotherms. Inclination $\alpha = 78.6°$, flow rate $4.6 \cdot 10^{-6}\,m^3/s$

Benchmark problem 3, mould filling with freezing in inclined rectangular cavity: The experiments are performed in a small (113mm x 38mm x 38mm) rectangular cavity made of 7.5mm Plexiglas. The cavity is inclined, with the inclination angle from vertical α=78.6°. The two opposite side-walls made of copper are assumed to be isothermal. They are kept at low temperature T_c=-10°C. Water is used as a working fluid. The fluid of initial temperature T_h = 25°C is forced to the cavity through a 13mm circular opening made in the bottom wall. The flow rate is $4.6 \cdot 10^{-6}\,m^3/s$. Both forced convection and residual natural convection within the cavity are responsible for the heat transfer through the cold side-walls (Kowalewski et al. 2001, Cybulski et al. 2002). Due to the inclination, liquid film slips down through the opening gap and the cavity is filled (Fig. 18). After the whole cavity is completely filled the flow is immediately interrupted and fluid trapped in the cavity starts to cool down. Stable thermal stratification develops along the isothermal walls, and cooling is mainly due to the conduction. Observed residual natural convection is very weak, especially close to the freezing point. During filling process and afterwards the freezing front propagates at both isothermal walls. Initially the growth rate of ice at the lower wall is faster than at the upper one (Fig.19). But after about 4 minutes this difference vanishes completely and the uniform growth of the ice layer is observed for both walls. Transient parameters measured in the experiment are: position and shape of the free surface, position and shape of the freezing front, fluid temperature and velocity fields for centre cross-section, temperature measured by 12 thermocouples at the isothermal walls (3 points each wall), at the side wall (three points) and at the inlet and outlet. Despite of using 20mm thick copper blocks with coolant pumped through by two heavy duty thermostats, temperature of the "isothermal" walls is not constant during the filling process. It drops down about 3°C during filling process. It nearly completely recovers after filling process is finished. Hence, the transient temperature data for the metal walls must be used to simulate properly the cooling process.

Using several commercial codes to simulate this experiment we came to the conclusion that even this very simple experimental model of the mould is difficult enough to create challenging problem. Several details are difficult to reproduce numerically, like formation of the recirculating zones during filling process. Also growth rate of the solid phase differs quite seriously. It gave us indication of some weak points present in commercial codes used.

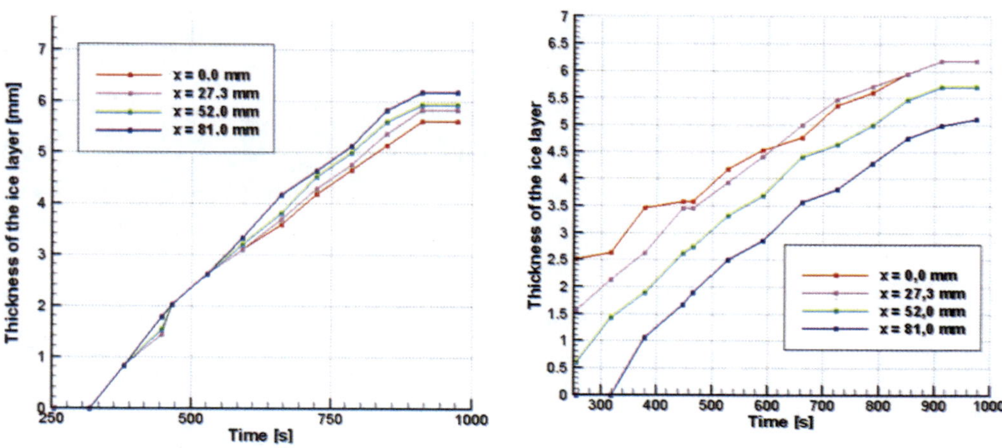

Thickness of the ice layer at the upper wall — Thickness of the ice layer at the lower wall

Figure 19. Experimental data for the ice thickness variation in time during filling and freezing process measured at 4 intervals from the bottom wall (inlet).

5. Conclusions

Convection with phase change can be analysed by variety of techniques. Some of them we mentioned in above. Assuming that the main message of the experiment is to deliver reliable and dense data about several flow parameters, full field and volumetric methods are favourable selection. Here we have in mind such methods like particle image velocimetry and thermometry (liquid crystal tracers), infrared thermography, all sets for tomographic reconstruction of temperature and concentration fields, and still useful interferometry with its holographic extension. However, trying to deliver experimental data for industrial configuration most of the modern optical methods must be excluded. Remaining methods are less accurate and still far from mature status (like industrial tomography). Hence, we have to solve typical dilemma. Is it better to test industrial code using inaccurate industrial experiment or is it better to replace materials used and try to simulate both experimentally and numerically selected physical features, typical for the industrial problem. Not all such features can be present, at least simultaneously. But if we encounter problems with simulating such simplified cases, can we have confidence in large-scale simulations with industrial materials?

Finally we do hope that experimental analysis of the simplified flow configuration, supported by numerical tests, allows for better identification of parameters playing crucial role in the specific flow problem. It offers unique possibility to perform sensitivity analysis of the problem, delivering information about tolerance span for the accuracy of defining material properties, in description of boundary conditions and flow geometry. Such analysis is essential for planning large scale industrial configuration but also very useful for determining code validation experimental procedure.

Acknowledgements. This is summary of the work which author was fortunate to share with his colleagues and students, first at the Max-Planck-Institut Goettingen and presently at his home institution. In particular contributions of W. Hiller, St. Koch, Ch. Soeller, A. Cybulski and M. Rebow are grateful acknowledged. A part of the present research was supported by Polish Scientific Committee (KBN Grant No. 8T09A00820).

References

Abegg, C., de Vahl Davis, G., Hiller, W.J., Koch, St., Kowalewski, T.A., Leonardi, E., Yeoh, G.H. (1994). Experimental and numerical study of three-dimensional natural convection and freezing in water. *Proc. of 10th Int. Heat Transfer Conf.*, edt. G.F. Hewitt, Ichem 4:1-6.

Andres, N., Arroyo, M.P., Quantilla, M. (1999). Interferometric techniques for measuring flow velocity fields. *Machine Graphics & Vision* 8: 625-636.

Banaszek, J., Jaluria, Y., Kowalewski, T.A., Rebow, M. (1999). Semi-implicit FEM analysis of natural convection in freezing water. *Num. Heat Transfer Part A*: 36:449-472.

Banaszek, J., Kowalewski, T. A., Furmanski, P., Rebow, M., Cybulski, A., Wisniewski, T.S. (2000). Natural convection with phase change in one-component and binary systems. *IPPT Reports* 3/2000, IPPT PAN Warszawa.

Bartels-Lehnhoff, H.H.(1991). *Erfassung dreidimensionaler Bahnlinien und Geschwindigkeitsfelder mittels automatischer Bildverarbeitung.* Ph.D. Thesis in Mitteilungen aus dem MPI 101, Göttingen.

Beckermann, C., Viskanta, R.(1089). Effect of solid subcooling on natural convection of a pure metal. *J. Heat Transfer (ASME)* 111: 416-424.

Bovik, A. (2000). *Handbook of Image and Video Processing.* Academic Press, San Diego.

Brito, D., Nataf, H.C., Cardin, Ph., Aubert, J., Masson, J.P.(2001). Ultrasonic Doppler velocimeter in liquid gallium. *Exp. in Fluids* 31:653-663.

Carlomagno G.M. (1993). Heat transfer measurements by means of infrared thermography, in *Measurement Techniques.* VKI Lect. Series 1993-05, Rhode-Saint-Genese. 1-114.

Chang, Y.J., Hsieh, Y.T., Chen, Y.M.(1998). Flow visualisation during a water freezing process by holographic interferometry. In *8th Int. Symp. on Flow Visualization.* 114.1-8.

Cybulski A., Michałek T., T.A. Kowalewski, M. Kowalczyk, M. Sokolnicki (2002). Experimental and numerical simulation of mould filling process, *Proceedings of Int. Conference Heat 2002*, Baranow S. June 2002. 267-272.

Dabiri, D., Gharib, M. (1991). Digital particle image thermometry: The method and implementation. *Exp. in Fluids* 97: 77-86.

Emrich, R.J. (1981). *Methods of Experimental Physics.* Vol. 18 Fluid Dynamics. Academic Press, New York.

Englemann, D., Garbe, Ch., Stöhr, M., Geißler, P., Hering, F., Jähne, B. (1988). Stereo particle tracking, in *Proc. 8th Int. Symp. Flow Visualization.* 240.1-8.

Fomin, N.A. (1998). *Speckle Photography for Fluid Mechanics Measurements.* Springer Verlag, Berlin-Heidelberg-New York.

Gau, C., Viskanta, R. (1986). Melting and solidification of a pure metal on a vertical wall. *J. Heat Transfer (ASME)* 108:174-181.

Gelfgat, A.Yu, Bar-Yoseph, P.Z., Solan, A., Kowalewski T.A. (1999). An axisymmetry breaking instability of axially symmetric natural convection *Int. J. Trans. Phenomena* 1:173-190.

Giangi, M., Kowalewski, T.A., Stella, F., Leonardi, E. (2000). Natural convection during ice formation: numerical simulation vs. experimental results. *Comp. Assisted Mech. and Eng. Scs.* 7: 321-342.

Giangi, M., Stella, F., Kowalewski, T.A. (1999). Phase-change problems with free convection: fixed grid simulation. *Comp. & Vis. in Scs.* 2:123-130.

Gobin, D., Le Quere, P. (2000). Melting from an isothermal vertical wall. Synthesis of numerical comparison exercise. *Comp. Assisted Mech. and Eng. Scs.* 7:289-306.

Goldstein, R.J.(1983). Fluid Mechanics Measurements. Hemisphere, Pub. Corp., Washington.

Grant, I., Thompson, B.J.(edts). (1994). *Selected Papers on Particle Image Velocimetry.* SPIE Milestone Series MS99, Optical Eng. Press, Bellingham.

Groh III de, H.C. (1994). *Interface shape and convection during solidification and melting of succinonitrile.* NASA Tech. Memorandum 106487.

Gui, L., Merzkirch, W. (1996). A method of tracking ensembles of particle images. *Exp. in Fluids* 21:465-468.

Gui, L., Merzkirch, W. (2000). Comparative study of the MQD method and several correlation-based PIV evaluation algorithms. *Exp. in Fluids* 28:36-44.

Hay, J.K., Hollingsworth, D.K. (1996). A comparison of trichromic systems for use in the calibration of polymer-dispersed thermochromic liquid crystals. *Exp. Thermal Fluid Scs,* 12:1-12

Hiller, W., Kowalewski, T.A.(1987). Simultaneous measurement of the temperature and velocity fields in thermal convective flows. *In Flow Visualization IV*, Ed. Claude Veret, Hemisphere, Paris. 617-622.

Hiller, W.J., Koch, St., Kowalewski, T.A. (1988). Simultane Erfassung von Temperatur- und Geschwindigkeitsfeldern in einer thermischen Konvektionsströmung mit ungekapselten Flüssigkristalltracern. *In: 2D-Meßtechnik DGLR-Workshop*, Markdorf, DGLR-Bericht 88-04, DGLR Bonn.31-39.

Hiller, W.J., Koch, St., Kowalewski, T.A. (1989). Three-dimensional structures in laminar natural convection in a cube enclosure. *Exp. Therm. and Fluid Sci.* 2:34-44.

Hiller, W.J., Koch, St., Kowalewski, T.A., de Vahl Davis, G., Behnia, M. (1990). Experimental and numerical investigation of natural convection in a cube with two heated side walls, *Proc.of IUTAM Symposium*, Cambridge UK, Aug. 13-18, 1989, (Edits. H.K. Moffat & A. Tsinober), CUP.717-726.

Hiller, W.J., Koch, St., Kowalewski, T.A., Mitgau, P., Range, K. (1992). Visualization of 3-D natural convection. *Proceedings of The Sixth International Symposium on Flow Visualization,* Eds. Tanida Y. & Miyashiro H. Springer-Verlag. 674-678.

Hiller, W.J., Koch, St., Kowalewski, T.A., Stella, F. (1993). Onset of natural convection in a cube. *Int. J. Heat Mass Transfer* 36:3251-3263.

Ibrahim, S., Green, R.G., Evans, K., Dutton K.(2000). Flow visualisation using optical tomography. *In Proc. 9th Int. Symp. Flow Visualization.* 140.1-8.

Incropera, F.P. (1997). Experimental methods for characterizing transport phenomena occurring during solidification of multi-constituent materials. In Proc. Exp. Heat Transfer, Fluid Mech. and Thermodynamics, Edizioni ETS. 1855-1865.

Jähne, B.(1997). *Digital Image Processing.* Springer Verlag, Berlin-Heidelberg-New York.

Koch, St. (1993). *Berührungslose Messung von Temperatur und Geschwindigkeit in freier Konvektion,* Ph.D. Thesis in Mitteilungen aus dem MPI 108, Göttingen.

Kowalewski, T. A., Cybulski, A. (1997). Natural convection with phase change. *IPPT Reports* 8/97, IPPT PAN Warszawa .

Kowalewski, T. A., Cybulski, A.(1996). Experimental and numerical investigations of natural convection in freezing water. *Int. Conf. on Heat Transfer with Change of Phase*, Kielce. In Mechanics 61/2:7-16.

Kowalewski, T.A. (1998). Experimental validation of numerical codes in thermally driven flows. *In Advances in Computational Heat Transfer*, G. de Vahl Davis, E. Leonardi (eds), Begel House Inc., New York. 1-15.

Kowalewski, T.A., Cybulski, A., Sobiecki, T. (2001). Experimental model for casting problems. *In Computational Methods and Experimental Measurements,* Eds. Y. V. Esteve, G.M. Carlomagno, C.A. Brebia, WIT Press, Southampton . 179-188.

Kowalewski, T.A., Rebow,M. (1999). Freezing of water in the differentially heated cubic cavity. *Int. J. Comp. Fluid Dyn.* 11:193-210.

Lehner, M., Mewes, D., Dinglreiter, U., Tauscher, R. (1999). *Applied Optical Measurements.* Springer Verlag, Berlin-Heidelberg-New York.

Leonardi, E., Kowalewski, T.A., Timchenko, V., de Vahl Davis, G.(1999). Effect of finite wall conductivity on flow structures in natural convection. *Proceedings of the International Conference on Computational Heat and Mass Transfer*, Cyprus, eds. A.A. Mohamad & I. Sezai, Eastern Mediterranean University Printinghouse. 182-188.

Mayinger, F., Feldmann O. (2001). *Optical Measurements, Techniques and Applications.* Springer-Verlag, Berlin-Heidelberg-New York.

McDonough, M.W., Faghri, A. (1994). Experimental and numerical analyses of the natural convection of water through its density maximum in a rectangular enclosure. *Int. J. Heat Mass Transfer* 37:783-801.

Michałek, T., Kowalewski, T.A. (2003). Experimental model of mould filling flow. *In Eurotherm 69 Heat and Mass Transfer in Solid-Liquid Phase Change Processes,* eds.: B. Šarler, D. Gobin

Mishima, K., Hibiki, T. (1996). Quantitative limits of thermal and fluid phenomena measurements using neutron attenuation characteristics of materials. *Exp. Thermal Fluid Scs.* 12:461-472.

Mitgau, P.M., Hiller, W.J., Kowalewski, T.A. (1994). Verfolgung von Teilchen in einer dreidimensionalen Strömung. *ZAMM* 74(5):T394-396.

Müller, G., Neumann, G., Weber, W. (1984). Natural convection in vertical bridgman configurations. *J. Crystal growth* 70:78-91.

Noël, N., Jamgotchian, H., Billa, B. (1997). *In situ* and real-time observation of the formation and dynamics of a cellular interface in a succinonitrile-0.5%wt acetone alloy directional solidified in a cylinder. *J. Crystal growth* 181:117-132.

Noël, N., Zamkotsian, F., Jamgotchian, H., Billa, B. (2000). Optical device dedicated to the non-distructive observation and characterization of the solidification of bulk transparent alloy *in situ* and real time. *Meas. Sci. Technol.* 11:66-73.

Park, H.G., Dabiri, D., Gharib, M. (2001). Digital particle image velocimetry/thermometry and application to the wake of a heated circular cylinder. *Exp .in Fluids* 30:327-338.

Pham, A.H., Hua, Y., Gray, N.B. (1999). Eddy current tomography for metal solidification imaging. In 1st World Congress on Industrial Process Tomography, Buxton. 451-458.

Plaskowski, A., Beck, MS., Thorn, R., Dyakowski, T. (1994). *Imaging Industrial Flows.* IPP, Bristol and Philadelphia.

Praisner, T.J., Sabatino, D.R., Smith, C.R. (2001). Simultaneously combined liquid crystal surface heat transfer and PIV flow-field measurements. *Exp. in Fluids* 300:1-10.

Prescott, P.J., Incropera, F.P. (1994). Convection transport phenomena and microsegragation during solidification of a binary metal alloy: Part1 & 2. *J. Heat Transfer (ASME)* 116:735-749.

Prescott, P.J., Incropera, F.P. (1996). Convection heat and mass transfer in alloy solidification. *Adv. in Heat Transfer* 28:231-338.

Quénot, G., Pakleza, J., Kowalewski, T.A. (1998). Particle Image Velocimetry with Optical Flow. *Exp. in Fluids* 25:177-189.

Raffel, M., Willert, C., Kompenhans, J. (1998). *Particle Image Velocimetry*. Springer Verlag, Berlin-Heidelberg-New York.

Rinkevichius, B.S. (1998). *Laser Diagnostics in Fluid Mechanics*. Begell House Inc., New York, Wallingford.

Roache, P.J. (1997). Quantification of uncertainty in computational fluid dynamics. *Ann. Rev. Fluid Mech.* 29:123-160.

Rucki Z., Szczepanik Z, Blogowska K. (2002). Measurements of the resistance filed in binary mixture with freezing (in Polish), *PW Prace Naukowe-Konferencje,* Conference XVIII Zjazd Termodyn., printed by Politechnika Warszawska, 22: 991-1000.

Sabatino, D.R., Praisner, T.J., Smith, C.R. (2000). A high-accuracy calibration technique for thermochromic liquid crystal temperature measurements. *Exp. in Fluids* 28:497-505.

Sielschott, H. (1997). Measurement of horyzontal flow in a large scale furnace using acoustic vector tomography. *Flow Meas. and Instr.* 8:191-197.

Sirrel, B., Holliday, M., Cambell, J. (1995). The benchmark test 1995. *In Proc. of the 7th Conference on Modelling of Casing, Welding and Advanced Solidification Process,* TMS Pub. 915-933.

Soeller, C. (1994). *Untersuchung der Konvektionsströmung in von oben gekühlten Behältern mittels tomographischer Speckle-Interferometrie und digitaler Bildverarbeitung*. Dissertation, University of Göttingen.

Straley, S.M.J. (1974). Physics of Liquid Crystals. *Review of Modern Physics* 46/4:617-704.

Takeda, Y. (1999). Ultrasonic Doppler method for velocity profile measurement in fluid dynamics and fluid engineering. *Exp. in Fluids* 26:177-178.

Van Buren, P.D., Viskanta, R. (1980). Interferometric measurement of heat transfer during melting from a vertical surface. *Int. J. Heat Mass Transfer* 23:568-571.

Viskanta, R. (1988). Heat transfer during melting and solidification of metals. *J. Heat Transfer (ASME)* 110:1205-1219.

Westerweel, J. (1993). *Digital Particle Image Velocimetry - Theory and Application*. Delft, Delft University Press.

Westerweel, J., Dabiri, D., Gharib, M. (1997). The effect of a discrete window offset on the accuracy of cross-correlation analysis of digital PIV recordings. *Exp. in Fluids* 23:20-28.

Xu, H., Fife, S., Andereck, C.D. (2002). Ultrasound thermal imaging (UTI) of convective opaque fluids. *Eurotherm Seminar 71 Proceedings* Reims (France), Oct. 28-30, 2002 (Edits. C. Padet & H. Burkhardt), University of Reims. 331-336.

Yarin, A., Kowalewski, T.A., Hiller, W.J., Koch, St. (1996). Distribution of particles suspended in 3D laminar convection flow. *Physics of Fluids* 8:1130-1140.

Modelling Methodologies for Convection-Diffusion Phase-Change Problems

F. Stella and M. Giangi

Dipartimento di Meccanica e Aeronautica, Università degli studi di Roma "La Sapienza"
Via Eudossiana 18 00184 Rome, Italy

1 Methodologies

1.1 Introduction

In recent years numerical simulation of phase-change problems have attracted much interest due to their significance for several technological process. Melting and solidification are typical examples of phase change met in the metallurgical industries or crystal growth technology. These processes involve complex phenomena of mass and heat transfer that determines the quality of the solid phase.

Temperature differences in the melt give rise to buoyancy forces that produce significant convective flow. It appears that there is a close relationship between the structure of the solid formed and the convective flow in the melt. Moreover the study of natural convection and the analysis of flow fields in the liquid are of great relevance for evaluating growth velocity in the solid and the shape of the liquid-solid interface.

For this reason, in the recent years, a number of studies have been performed using computational fluid dynamics (CFD). Due to the complexity of the problem, application of numerical method is not a trivial task. The mathematical models for the simulation of two phase problems are usually classified in terms of *moving grids* and *fixed grids* methods.

In the moving grids methods conservation equations are applied separately to each phase and heat flow conditions are invoked at the solid-liquid interface (Richtmyer and Morton, 1967, Crank, 1984, and Unverdi and Tryggvanson, 1992). The phase boundary is tracked in time using the classical Stefan formulation, i.e. balance of heat fluxes at the interface.

In the fixed grid methods (Fabbri and Voller, 1997, Bennon and Incropera, 1988, Voller, *et al.*, 1987, and Bennon and Incropera 1987) a unique set of equations and boundary conditions is used for the whole domain, including both solid and liquid phase avoiding the problem of tracking the liquid-solid interface. The heat and mass transfer conditions on the moving interface are incorporated into the governing equations using suitable source terms.

In this chapter a description of the most common mathematical models (moving grids and fixed grids methods) used in literature to study phase-change convection problems are presented. In Sec.2 details of a continuum mathematical model (enthalpy method) are presented. Dimensionless parameters to study solidification/melting problems are introduced in Sec.3, while numerical tecniques to discretized the continuum model (described in Sec.2) is presented in Sec.4. Finally applications of melting and solidification problems are discussed in Sec.6.

1.2 Stefan Problem

Let's consider a one-dimensional, (one-directional) solidification problem along a thin uniform rod with two phases (liquid and solid). The rod is assumed to be perfectly insulated along

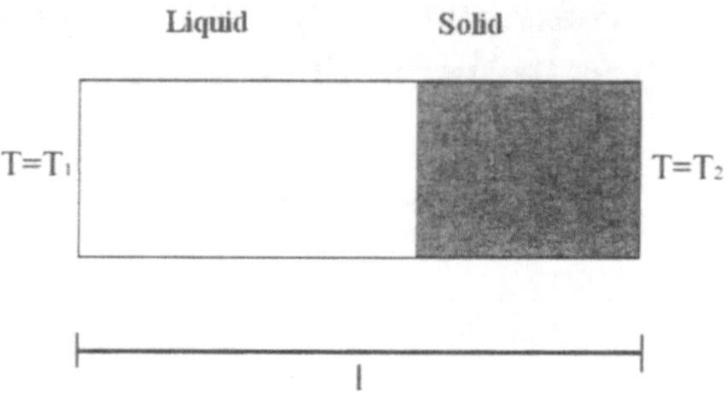

Figure 1.2.1 Definition sketch for the Stefan Problem.

its length so that the temperature changes occur through heat conduction along its length. Let L represent the length of the rod ($0 \le x \le L$) and T_m is the phase change (i.e. melting) temperature. The ends of the rods are at fixed temperatures so that during the solidification the interface $x = s(t)$ is fixed at $0 \le s(t) \le L$.

The study of this problem requires the equations for conservation of energy both in solid and in liquid phase. Under these conditions and assuming pure conduction on the interface, the equations can be written as:

$$\rho_s c_s \frac{\partial T_s}{\partial t} = \frac{\partial}{\partial x}\left(\lambda_s \frac{\partial T_s}{\partial x}\right) \tag{1.2.1a}$$

$$\rho_l c_l \frac{\partial T_l}{\partial t} = \frac{\partial}{\partial x}\left(\lambda_l \frac{\partial T_l}{\partial x}\right) \tag{1.2.1b}$$

in which T is the temperature, ρ is the density, λ is the thermal conductivity, c is the heat capacity of the solid $(.)_s$ and liquid $(.)_l$ phases.

Equation (1.2.1) is solved together with a proper set of boundary conditions of the type:

$$\begin{aligned} &T = T_1 \quad x = 0 \quad t > 0 \\ &T = T_2 \quad x = l \quad t > 0 \\ &T = T_0 \quad \forall x \ t = 0, s(0) = 0 \end{aligned} \tag{1.2.2}$$

with $T_1 > T_2$.

Furthermore on the interface $x=s(t)$ the following conditions must also apply:

$$T_l = T_s = T_m \qquad (1.2.3)$$

$$\lambda_s \frac{\partial T_s}{\partial x} - \lambda_l \frac{\partial T_l}{\partial x} = \rho_s L_f \frac{ds}{dt} \qquad \textit{Stefan Condition} \qquad (1.2.4)$$

where L_f is the latent heat of solidification.

Equation (1.2.3) represents the continuity condition for the temperature at the solid-liquid interface, while equation (1.2.4), usually called the *Stefan condition*, represents the conservation of energy at the interface. From equation (1.2.4) is also possible to determine the rate of change of the phase, and thus the velocity of the interface.

1.3 Moving Grids

The approach of phase change problems, by means of moving boundary methods, requires the definition on each physical phase (solid, liquid and/or gas) of the proper set of mathematical equations, with proper boundary conditions and with the thermal balance condition (Stefan condition) on the interface.

Since in moving grid models, the position of the interface has to be explicitly treated, different numerical techniques has been developed, the most commonly used are the *front tracking* and *front fixing* methods (Crank, 1984).

Front tracking methods. When a *front tracking* method is adopted, the position of the moving boundary has to be evaluated at each time step (see Figure 1.3.1). In such method the interface position is determined by means of additional computational elements (e) that are arbitrary posed inside the computational domain. The additional computational effort due to these extra elements is usually very small.

Figure 1.3.1. Advection of the interface.

The interface is advected by integrating in time the equation:

$$\frac{d\boldsymbol{x}^{(e)}}{dt} = \boldsymbol{u}^{(e)} \qquad (1.3.1)$$

where **x** are the coordinates of the element e and **u** the velocities of the interface points that are interpolated from the grid data (Unverdi and Tryggvason, 1992).

Generally speaking *front tracking techniques* is best suited for well-defined fronts. The basic idea of such methodology is discussed by Richtmyer and Morton (1967). The major drawback of front tracking is its complexity. Since it is generally necessary to reconstruct the interface grid during calculation procedure, often computational points have to be added in regions where the grid stretches, while it is usually required to eliminate points in regions where the interface is compressed. The situation is even more complicated when three-dimensional problems are investigated. Another problem for front tracking arise when one front interacts with a second one, in this case one has to be eliminated because of the merging of fronts.

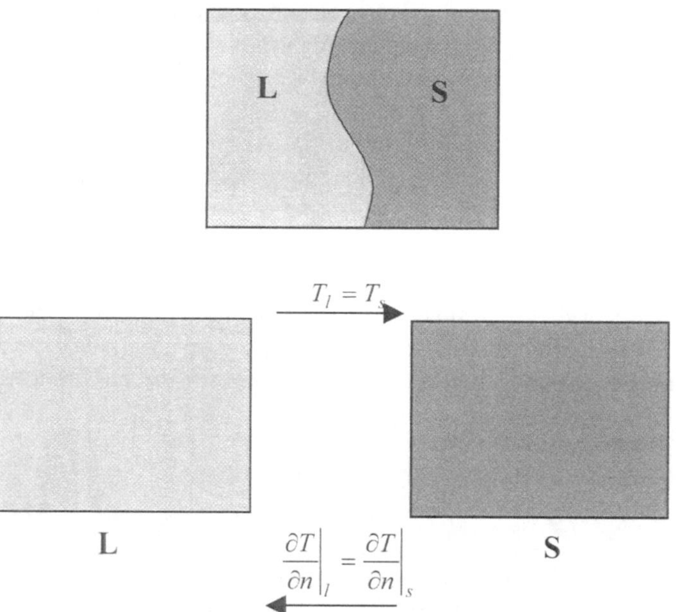

Figure 1.3.2. Advection of the interface.

Front fixing methods. When a *front fixing* method is adopted equations (1.2.1) - (1.2.4) are transformed by means of a new space coordinates, in such a way that the physical interface is transformed into a portion of the boundary of this new computational grid, resulting also fixed in time in the new system. With this approach the evolution of the interface is followed by adapting a coordinate transformation that changes with time, moving with the interface (see Figure 1.3.2). One advantage of this method is that the computational domain is fixed, so the grid can be refined near the interface. Main disadvantage of this method is that the equations are changed and in some case can became more complex than the original ones.

1.4 Fixed Grids

When this methodology is adopted, field equations are reformulated in a new, unique set of equations, in such a way that the Stefan condition is implicitly bound up in a new form of the equations, which applies over the whole of a fixed domain. The position of the moving boundary may be evaluated *a-posteriori* as a post processing of the numerical solution. Most commonly used methods are called: *phase field method* and *enthalpy methods*.

Phase field method. The *phase field method* for solidification phenomena is an open formulation (Shyy *et al.* 1996, Fabbri and Voller, 1997) based on a free-energy functional, $F(p,T)$. Where T is the temperature and $p(x,t)$ is a variable of phase assuming the value of $+1$ in the liquid and -1 in the solid phase respectively, the interface region, of length x, corresponds to intermediate value of p (see Figure 1.4.1). The local Helmholtz free energy density F, is postulated as a double-well function with two distinct minima, each corresponding to a phase, as shown in Figure 1.4.2. The total free energy of the system is:

$$\Psi = N\int_V F dV \tag{1.4.1}$$

where N is the number of molecules per unit volume.
The kinetics of the function p usually is related to the total free energy ψ as following (Shyy *et al.* 1996):

$$\frac{\partial p}{\partial t} = -A\frac{\delta\Psi}{\delta p} \tag{1.4.2}$$

where A is always positive and may depend on p and the temperature.
As F at equilibrium the functional ψ (1.4.1) has a minimum with respect to variations in p and thus satisfies the Euler-Lagrange equation

$$\frac{\delta\Psi}{\delta p} = 0 \tag{1.4.3}$$

By (1.4.1)-(1.4.3) the equation for the evolution of the phase field (see Shyy *et al.* 1996 for further details) can be written as:

$$\alpha\xi^2\frac{\partial p}{\partial t} = \xi^2\nabla^2 p - \frac{\partial F}{\partial p}$$

where α is a constant proportional to the interface thickness and ξ is a constant depend on the potential $F(p,T)$. So with the *phase field method (PF)* the explicit treatment of the interface conditions equations (1.2.3) - (1.2.4) is avoided by solving a coupled of non-linear system of evolution equations for the temperature T and the phase p (Shyy *et al.* 1996, Fabbri and Voller, 1997):

$$\rho c\frac{\partial T}{\partial t} - \nabla\cdot(\lambda\nabla T) = -\frac{1}{2}L_f\frac{\partial p}{\partial t} \tag{1.4.4a}$$

$$\alpha\xi^2\frac{\partial p}{\partial t} = \xi^2\nabla^2 p - \frac{\partial F}{\partial p} \tag{1.4.4b}$$

The Fourier heat conduction (1.4.4a) equation has a source term to account for the latent heat release at the moving interface, while the phase equation (1.4.4b) is governed by the imbalance between the excess interface free-energy and the potential $F(p,T)$.

The phase function equations are not unique, in the sense that one is free to choose the form of the potential hence it might be possible to generate different numerical results by choosing alternative forms for $F(p,T)$. Examples of the different possibilities include; Caginalp (1989), Caginalp and Socolovski (1991), Caginalp and Chen(1992), Kobayashi (1993), Wang *et al.* (1993) and Kobayashi (1994). One such example is shown in Figure 1.4.2 (Fabbri and Voller, 1997).

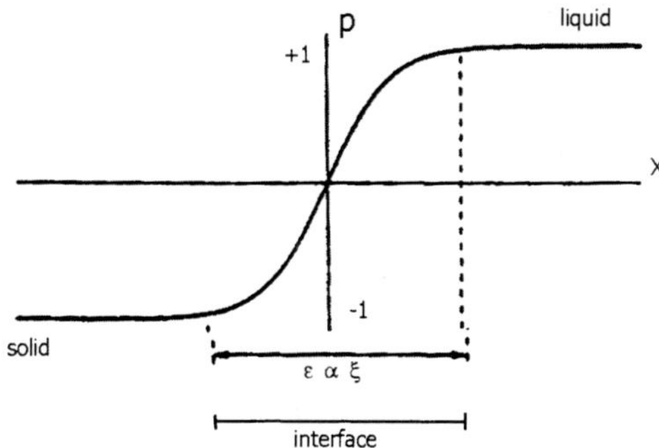

Figure 1.4.1. Variation of the phase variable, p.

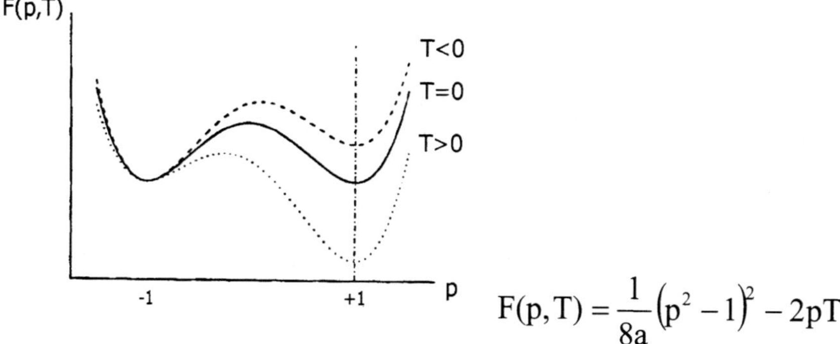

Figure 1.4.2. One possible Free-energy function.

Enthalpy method. In the enthalpy method, the Stefan condition, equation (1.2.4), is implicitly enclosed in the equations by introducing an enthalpy or total heat function $h(T)$. $h(T)$ is the total heat content, i.e. the sum of specific heat and the latent heat required for a phase change. The heat jump at the phase-change boundary is incorporated in the definition of $h(T)$ using:

$$h(T) = \int_{T_0}^{T} c(\vartheta)d\vartheta, \qquad if\ T < T_m$$

$$h(T) = \int_{T_0}^{T} c(\vartheta)d\vartheta + L_f, \qquad if\ T > T_m \tag{1.4.5}$$

$$\int_{T_0}^{T} c(\vartheta)d\vartheta \le h(T) \le \int_{T_0}^{T} c(\vartheta)d\vartheta + L_f, \qquad if\ T = T_m$$

where T_0 is some fixed temperature and T_m the temperature of phase change.

By the definition equation (1.4.5) the problem, given by equations (1.2.1), can be reformulated over the whole fixed domain occupied by the two phases as the single equation:

$$\rho\frac{\partial h}{\partial t} = \nabla\cdot(\lambda\nabla T) \tag{1.4.6}$$

The position of the interface is evaluated *a-posteriori* as the numerical solution in which,

$$\int_{T_0}^{T} c(\vartheta)d\vartheta \le h(T) \le \int_{T_0}^{T} c(\vartheta)d\vartheta + L_f\ . \tag{1.4.7}$$

Enthalpy porosity method. In the *enthalpy porosity method* (Crank, 1984) a control volume with a material undergoing to phase change is considered so a liquid mass fraction $f_l=f_l(T)$ is introduced. By assuming constant and equal specific heats (c) in the liquid and in the solid phases the enthalpy (h) can be written in the solid and in the liquid respectively as:

$$\begin{aligned} h_s &= cT \\ h_l &= cT + L_f \end{aligned} \tag{1.4.5}$$

The total enthalpy in to the whole control volume V (liquid and/or solid) can be defined as:

$$h = (1 - f_l)h_s + f_l h_l = cT + f_l L_f \tag{1.4.6}$$

So the heat equation (1.2.1) in to the whole domain (liquid and/or solid) becomes:

$$\rho c\frac{\partial T}{\partial t} - \nabla\cdot(\lambda\nabla T) = -L_f\frac{\partial f_l}{\partial t} \tag{1.4.7}$$

where the source term $\frac{\partial f_l}{\partial t}$ takes into account the latent heat due to the phase-change.

The zones where $T>T_m$ are entirely liquid and therefore $f_l=1$, whilst the zones where $T<T_m$ are entirely in the solid state and hence $f_l=0$. In the zones undergoing phase change $T=T_m$ and f_l varies between 0 and 1. By setting $f_l=(p+1)/2$, equation (1.4.7) becomes equivalent to the equation (1.4.4a), except that in the enthalpy method the liquid fraction f_l is a temperature function

that depend on the nature of the material so the enthalpy method can be extended to the study alloy solidification problems (Bennon and Incropera, 1988).

This method has also been developed for convection/diffusion phase change problems. In this case the enthalpy equation (1.4.4) has to be extended to include advection terms and must be coupled with the mass and momentum conservation equations (Voller *et al.*, 1987, Bennon and Incropera, 1987, and Scheudergger, 1963).

The equations on the whole domain (liquid and/or solid) are derived by writing separately the equations on the liquid and solid regions, then the equations are integrated on a control volume containing both solid and liquid phases, so the equations are written in each phase but with interface terms. Finally summing the equations for the liquid and solid phases and using averaged quantities the continuum equations are reached (Voller *et al.*, 1987, Bennon and Incropera, 1987, and Scheudergger, 1963). This method is discussed in detail in the following section.

2 Mathematical Models

2.1 Generalised Governing Equations

The homogenisation technique based on the averaging variables (velocity, enthalpy, species concentration) and physical properties over the whole domain can be used. This technique allows the construction of a new set of conservation equations for mass, momentum, energy and species (Bennon and Incropera, 1987, 1988).

The average density, thermal conductivity, velocity vector components, enthalpy and concentration are defined using,

$$\begin{aligned}\rho_m &= g_s\rho_s + g_l\rho_l \\ \lambda_m &= g_s\lambda_s + g_l\lambda_l \\ \mathbf{v}_m &= f_s\mathbf{v}_s + f_l\mathbf{v}_l \\ h_m &= f_s h_s + f_l h_l \\ C_m &= f_s C_s + f_l C_l\end{aligned} \quad (2.1.1)$$

where ρ is the density and λ is the thermal conductivity, $\boldsymbol{v}$ is the velocity, h is the enthalpy, and g and f are the volume and mass fractions of the phases. The subscripts $(\cdot)_l$, $(\cdot)_s$ and $(\cdot)_m$ indicate the liquid, solid, and mixture respectively.

By substituting equation (2.1.1) into the conservation equations they can then be rewritten as:

Continuity

$$\frac{\partial \rho_m}{\partial t} + \nabla \cdot (\rho_m \mathbf{v}_m) = 0 \quad (2.1.2)$$

Momentum

$$\frac{\partial(\rho_m \mathbf{v}_m)}{\partial t} + \nabla\cdot(\rho_m \mathbf{v}_m \mathbf{v}_m) = \nabla\cdot\left(\mu_m \frac{\rho_m}{\rho_l} \nabla \mathbf{v}_m\right) - \nabla p + \rho_m \mathbf{G} + \mathbf{D} \tag{2.1.3}$$

Energy

$$\frac{\partial(\rho_m h_m)}{\partial t} + \nabla\cdot(\rho_m \mathbf{v}_m h_m) = \nabla\cdot(\lambda_m \nabla T) - \nabla\cdot[\rho_m (h_l - h_m)(\mathbf{v}_m - \mathbf{v}_s)] \tag{2.1.4}$$

Concentration

$$\frac{\partial(\rho_m C_m)}{\partial t} + \nabla\cdot(\rho_m v_m C_m) = \nabla\cdot(\rho_m D \nabla C_m) - \nabla\cdot(\rho_m (C_l - C_m)(\mathbf{v}_m - \mathbf{v}_s)) \tag{2.1.5}$$

where μ is the dynamic viscosity, $\boldsymbol{G}$ is the buoyancy term, $\boldsymbol{D}$ is a source term and D is the diffusivity of the solute.

For a wide range of solid-liquid phase change systems, the multiphase region is characterized by a fine permeable solid matrix that is often being considered stationary (static solidification $\mathbf{v}_s=0$) or constrained to free body translation (continuous solidification $\mathbf{v}_s=const$). For such systems analogies can be drawn to flows through porous media.

Specially, *Darcy's law* suggests that phase interaction forces are proportional to the superficial liquid velocity relative to the velocity of the porous solid (Slattery, 1978):

$$\mathbf{D} = -\frac{\mu_l}{K}\frac{\rho_m}{\rho_l}(\mathbf{v}_m - \mathbf{v}_s) \tag{2.1.6}$$

in which $\boldsymbol{K}$ represents the permeability.

Equation (2.1.3) reduces, as it must, to the appropriate single phase limits as $\boldsymbol{K} \to 0$ (pure solid) and $\boldsymbol{K} \to \infty$ (pure liquid).

The permeability $\boldsymbol{K}$ is a porosity function g_l of media and by the *Carman-Kozeny equation* for flow in a porous media (Scheudergger, 1963), it may be written in the form:

$$\boldsymbol{K} = \boldsymbol{K}_0 \left[\frac{g_l^3}{(1-g_l)^2}\right] \tag{2.1.7}$$

in which $\boldsymbol{K}_0$ is a constant which depends on the solidification microstructures.

With the substitution of equations (2.1.6) and (2.1.7), equation (2.1.3) becomes,

$$\frac{\partial(\rho_m \mathbf{v}_m)}{\partial t} + \nabla\cdot(\rho_m \mathbf{v}_m \mathbf{v}_m) = \nabla\cdot\left(\mu_m \frac{\rho_m}{\rho_l} \nabla \mathbf{v}_m\right) - \nabla p + \rho_m \mathbf{G} + B(\mathbf{v}_m - \mathbf{v}_s) \tag{2.1.8}$$

where

$$B = -\frac{\mu_l}{K_0}\left[\frac{(1-g_l)^2}{g_l^3 + q}\right]\frac{\rho_m}{\rho_l} \tag{2.1.9}$$

in which q is a computationally small number used to avoid division by zero.

In practice the effect of B is as follows: in full liquid element ($g_l = 1$) B is zero and has absolutely no influence; in elements that are changing phase, the value of B will dominate over the convective and diffusive components of the momentum equation, thereby forcing them to imitate the *Carman-Cozeny* law; whilst in totally solid elements ($g_l=0$), the large value of B will swamp all other terms in the governing equations and force any velocity predictions effectively to zero.

There are alternative methods by which the behaviour of the velocities in the vicinity of the phase change can be modelled.

Garling (1980) takes the viscosity μ to be a function of temperature such that it rises to a large value in proximity of the interface. This has the effect of inhibiting the velocity so that any predicted velocities in the solid region are negligible. This method is know as the *enthalpy-viscosity method.*

In the *enthalpy model* developed by Morgan (1981) the velocities are imposed equal to zero when the amount of the latent heat results lower than a fixed value.

2.2 Enthalpy and Concentration Equations

Enthalpy equation. With the assumption that specific heat, c, is constant, the enthalpy can be written as the sum of the sensible and latent heat:

$$h = cT + f_l L_f \tag{2.2.1}$$

Substitution of equation (2.2.1) into the energy equation (2.1.4) yields

$$\rho_m c\frac{\partial T}{\partial t} + \rho_m c\nabla \cdot (\mathbf{v}_{\mathrm{m}}\mathrm{T}) = \nabla.(\lambda_m \nabla T) + S_T \tag{2.2.3}$$

where

$$S_T = -\frac{\partial}{\partial t}(\rho f_l L_f) \tag{2.2.4}$$

The source term (2.2.4) is used to account for latent heat release during phase change.

Equations (2.2.1)-(2.2.4) represent the fixed-grid approach for modelling heat transfer during a phase change process. An essential part of this approach is the derivation of an enthalpy-temperature-liquid fraction relationship.

Concentration equation. To account for the release of solute into the liquid during solidification of alloys a source term (Timchenko *et al.*, 1998,2000,2001) is added to the solute conservation equation,

$$\frac{\partial C_l}{\partial t} + \nabla.(\mathbf{v}_m C_l) = \nabla.(D\nabla C_l) + S_c \tag{2.2.5}$$

in which

$$S_c = \frac{\partial f_s}{\partial t}(1-k_p)C_l + f_s\frac{\partial C_l}{\partial t} \tag{2.2.6}$$

where k_p is the partition coefficient (C_s / C_l at the interface), $f_s = 1 - f_l$ is the local solid volume fraction. Note that S_c is zero throughout the liquid except in the partially solidified control volume containing the interface.

During solidification, the melting temperature can vary (in some cases) due to changes in solute concentration. With the assumption that phase change takes place under local thermodynamic equilibrium, the temperature at the interface, *i.e.*, the melting temperature T_m, can then be expressed as

$$T_m = T_{m0} + m_l C_I \tag{2.2.7}$$

where T_{m0} is the melting temperature of pure solvent, m_l is slope of the liquidus, assumed to be constant and obtained from the phase diagram and C_I is the interface solute concentration in the liquid.

2.3 Vorticity-Velocity Formulation

With the assumption that the density and viscosity are constant, the continuity (2.1.2) and the momentum equations (2.1.8) in a vorticity-velocity formulation (Stella and Giangi, 2000, and Guj and Stella, 1988) can be written as:

$$\rho_m \frac{D\omega_m}{Dt} = \mu_m \frac{\rho_m}{\rho_l}\nabla^2\omega_m + \nabla\times(\rho_m \mathbf{F}) + \nabla\times(B(\mathbf{v}_m - \mathbf{v}_s)) \tag{2.3.1}$$

$$\nabla^2 \mathbf{v}_m = -\nabla\times\omega_m \tag{2.3.2}$$

2.4 Vorticity-Stream Function Formulation

The governing time dependent equations describing mass and momentum transport in the vorticity-stream function formulation are (Timchenko *et al.*, 1998,2000,2001):

$$\rho_m \frac{D\omega_m}{Dt} = \mu_m \frac{\rho_m}{\rho_l}\nabla^2\omega_m + \nabla\times(\rho_m \mathbf{G}) + \nabla\times(B(\mathbf{v}_m - \mathbf{v}_s)) \tag{2.4.1}$$

$$\nabla^2\psi = -\omega_m \tag{2.4.2}$$

and the velocity is given by

$$\mathbf{v}_m = \nabla\times\psi \tag{2.4.3}$$

3 Dimensionless Equations

Study of solidification/melting problems can be conducted by introducing the following dimensionless parameters:

Prandtl number

$$\Pr = \frac{\nu}{k}$$

which is a measure of the diffusion of momentum with respect to the diffusion of heat.

Rayleigh number

$$Ra = \frac{g\beta\Delta T H^3}{\nu k}$$

which indicates the importance of the buoyancy-driven force respect to viscous force.

Stefan number

$$Ste = \frac{c\Delta T}{L_f}$$

which is a measure of latent heat during phase change.

Marangoni number

$$Ma = \frac{|\partial\sigma / \partial\theta| H\Delta T}{\mu k}$$

which indicates the importance of the termocapillary-driven force respect to viscous force.
Here β is the coefficient of thermal expansion, H is the cavity height where the phase-change problem is studied, ΔT is the temperature difference between the hot and the cold walls, κ is the thermal diffusivity, μ is the dynamic viscosity, ν is the kinematic viscosity, σ is the surface tension and L_f the latent heat of fusion.

Melting/solidification problems, in a close cavity. The adimentional governing equations in vorticity-velocity formulation (2.3.1-2.3.2)-(2.2.3), assuming the buoyancy force $\mathbf{G} = \rho\mathbf{g}$ (with ρ liquid density and $\mathbf{g}$ gravity acceleration), can be written in the following dimensionless form:

$$\frac{D\omega_m}{Dt} = \Pr\nabla^2\omega_m - Ra\Pr\nabla\times\left(\sum_{i=1}^{N}\gamma_i\theta^i\frac{g}{|g|}\right) + \nabla\times\left(B\left(\mathbf{v}_m - \mathbf{v}_s\right)\right) \tag{3.1}$$

$$\nabla^2\mathbf{v} = -\nabla\times\omega_m \tag{3.2}$$

$$\frac{D\theta}{Dt} = \nabla\cdot\left(\tilde{\lambda}\nabla\right)\theta - \frac{1}{Ste}\frac{\partial f_l}{\partial t} \tag{3.3}$$

The dimensionless form of the equation is based on a reference velocity:

$$\mathbf{v}^* = k/H$$

In the Equation (3.1)

$$\gamma_i = \frac{a_i}{a_1}\Delta T^{i-1}$$

where a_i are the coefficients of the liquid density expressed in Fourier series expansion as temperature function with order N; if N=1 (in Eq. (3.1)) a linear relation between density and temperature is considered.
In Eq.(3.1) the source term B is:

$$B = \frac{-C(1-f_l)^2}{\left(f_l^3 + q\right)}$$

where C is a large constant value and q is a computational small quantity used to avoid singularity in solid zone $\left(\approx 10^{-3}\right)$.

In the Equation (3.3) $\tilde{\lambda} = \dfrac{f_s\lambda_s + f_l\lambda_l}{\lambda_l} = f_s\left(\dfrac{\lambda_s}{\lambda_l}\right) + f_l$. In the liquid region $f_l = 1$ then $\tilde{k} = 1$, while in the solid region $f_s = 1$ and hence $\tilde{\lambda} = \lambda_s/\lambda_l$.

Melting/solidification problems in presence of a free-surface. To study phase-change problems with termocapillary convection, the Marangoni number has be introduced, so the governing equations in vorticity-velocity formulation (2.3.1-2.3.2)-(2.2.3) can be written in the following dimensionless form:

$$\frac{Ma}{\Pr}\frac{D\omega_m}{Dt} = \nabla^2\omega_m - \frac{Ra}{Ma}\nabla\times\left(\sum_{i=1}^{N}\gamma_i\theta^i\frac{g}{|g|}\right) + \nabla\times\left(B\left(\mathbf{v}_m - \mathbf{v}_s\right)\right) \tag{3.4}$$

$$\nabla^2\mathbf{v} = -\nabla\times\omega_m \tag{3.5}$$

$$Ste\frac{D\theta}{Dt} = \frac{Ste}{Ma}\nabla\cdot\left(\tilde{\lambda}\nabla\right)\theta - \frac{\partial f_l}{\partial t} \tag{3.6}$$

With the non-dimensional scheme adopted, the reference velocity is given by:

$$v^* = \frac{\left(|\partial\sigma / \partial\theta|\Delta T\right)}{\mu}$$

This scheme has the advantage that no parameters appear in the free boundary conditions, but has the disadvantage that the reference velocity changes when ΔT is changed. Since ΔT is usually the parameter used for changed Ma in the experiments, this requires that the velocity field be rescaled using different references values.

3.1 Boundary conditions

For **melting/solidification problems, in a close cavity** all the boundaries of the cavity can be assumed to be solid, impermeable and at rest (Giangi and Stella (1999), Giangi et al. (1999), Stella and Giangi (2000)). The top and bottom wall of the cavity are adiabatic while the left wall is maintained at a temperature $\theta_h > \theta_m$ and the right wall at a temperature $\theta_c < \theta_m$ (with θ_m phase change temperature). The velocity field on the boundary is assumed zero, a schematic representation of the boundary conditions is shown in Figure 3.1.1.

For **melting/solidification problems in presence of a free-surface** all the boundaries of the cavity, except the top free surface, are assumed to be solid, impermeable and at rest. The velocity boundary conditions on the rigid wall are **v**=0.
The top free surface can be assumed to be flat and the tension variation can be considered to be a linear function of the temperature only (Giangi et al. (2002)).
With the non-dimensional scheme adopted the Marangoni effect on the top free surface can be expressed as (see Figure 3.1.2):

$$\mathrm{v} = 0$$

$$\frac{\partial \mathrm{u}}{\partial y} = -\frac{\partial \theta}{\partial x}$$

where x and y are the horizontal and vertical axis respectively and u and v are the horizontal and vertical component of the velocity field.

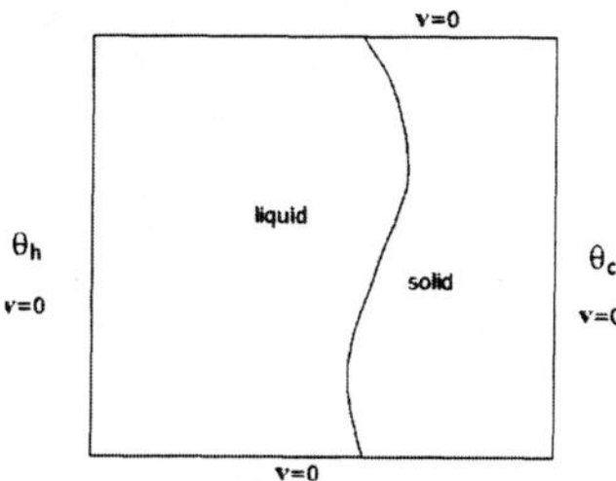

Figure 3.1.1. Melting/solidification in a close cavity: boundary conditions.

On the top and bottom wall a linear distribution of the temperature can be considered (see Figure 3.1.2), while the left and right vertical walls they are isothermal with the left wall maintained at a temperature $\theta_h > \theta_m$ and the right wall at a temperature $\theta_c < \theta_m$ (with θ_m phase change temperature).

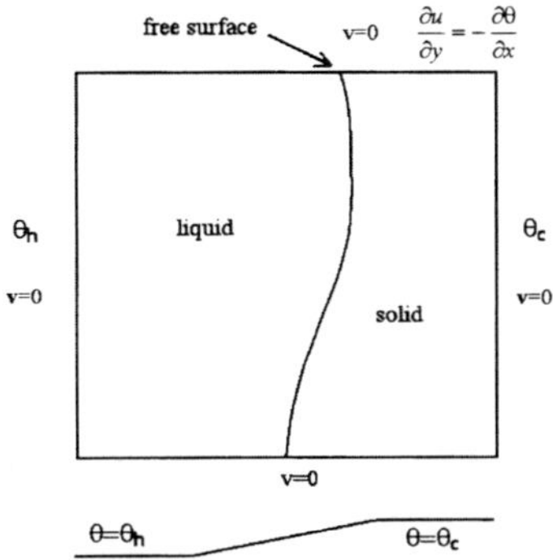

Figure 3.1.2. Melting/solidification in presence of a free-surface: boundary conditions.

4 Numerical Method

Variable location

In the numerical discretization usually the variable are located in the computational domain on a *staggered grid* (see Figure 4.1) so the mass conservation is preserved also in the discrete form.

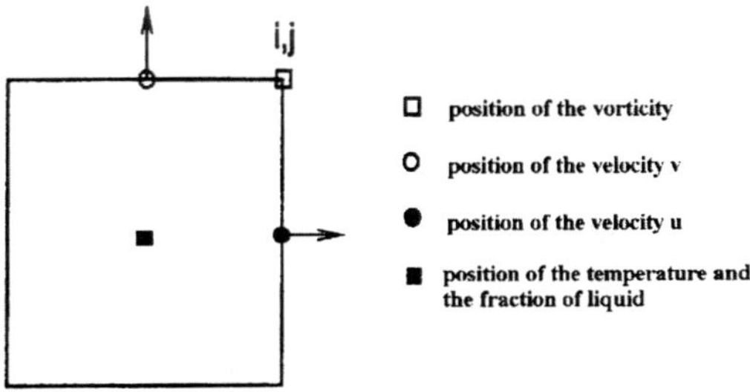

Figure 4.1. Staggered grid

The use of the staggered mesh requires some special treatment for boundary conditions:

East-West side

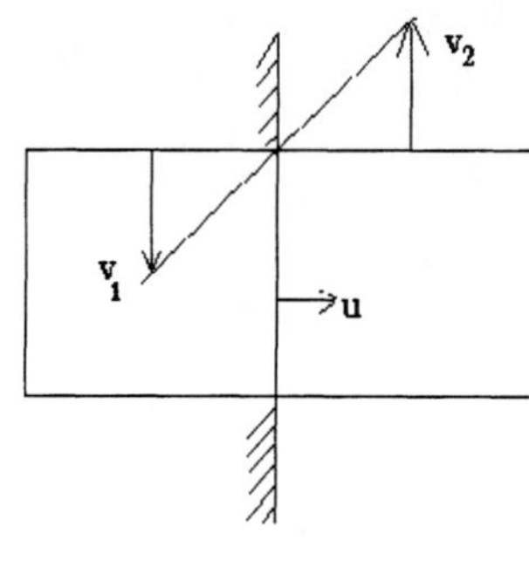

$$u = u_B = 0$$

$$\frac{v_1 + v_2}{2} = v_B = 0$$

North-South side

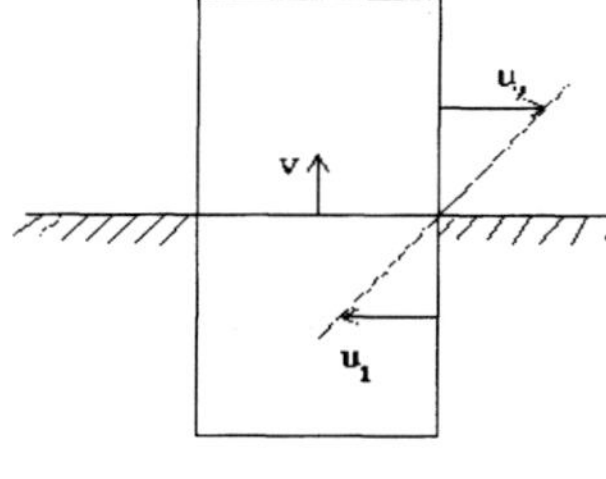

$$\frac{u_1 + u_2}{2} = u_B = 0$$

$$v = v_B = 0$$

where $(.)_B$ denotes the values on the boundary.

Numerical Discretization

The governing equations (3.4)-(3.6), (3.1)-(3.3) can be discretized using a finite volume method (Stella *et al.* (2000)); this method allows to preserve the flux trough the edges of the computational grid.

Let is consider as example the energy equation (3.3): integrating this equation in the control cell $\Omega_{i-1/2,j-1/2}$ (centred in i-1/2,j-1/2) with length Δx, height Δy and vertex A,B,C,D (see Figure 4.2):

$$\frac{\partial}{\partial t}\int_{\Omega_{i-1/2,j-1/2}} \theta d\Omega + \oint_{\Omega_{i-1/2,j-1/2}} (\mathrm{v}\theta)\cdot \mathbf{n} dS = \oint_{ABCD} \left(\tilde{\lambda}\nabla\theta\right)\cdot \mathbf{n} dS - \frac{1}{Ste}\frac{\partial}{\partial t}\int_{\Omega_{i-1/2,j-1/2}} f_l \, d\Omega$$

where **n** denotes the external normal to the control cell.

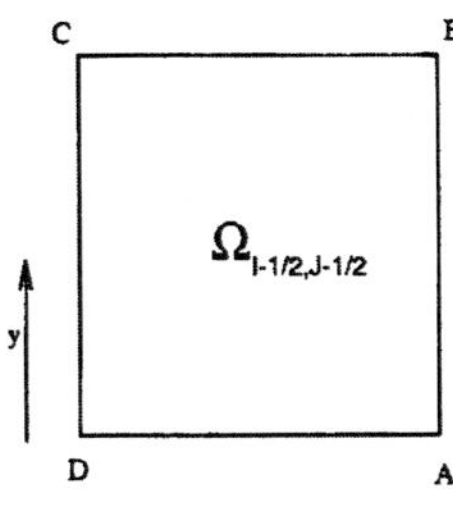

Figure 4.2. Finite volume discretization.

Results (Hirsch (1983)):

$$\left(\frac{\partial \theta_{i\text{-}1/2,j\text{-}1/2}}{\partial t}\right)\Delta x \Delta y+\left[(\mathrm{u}\theta)_{i,j-1/2}-(\mathrm{u}\theta)_{i-1/2,j-1/2}\right]\Delta y+\left[(\mathrm{v}\theta)_{i,j-1/2}-(\mathrm{v}\theta)_{i-1/2,j-1/2}\right]\Delta x=$$

$$\left[\left(\tilde{\lambda}\frac{\partial \theta}{\partial y}\right)_{i-1/2,j}-\left(\tilde{\lambda}\frac{\partial \theta}{\partial y}\right)_{i-1/2,j}\right]\Delta y+\left[\left(\tilde{\lambda}\frac{\partial \theta}{\partial y}\right)_{i,j-1/2}-\left(\tilde{\lambda}\frac{\partial \theta}{\partial y}\right)_{i,j-1/2}\right]\Delta x-\frac{1}{Ste}\left(\frac{\partial (f_l)_{i-1/2,j-1/2}}{\partial t}\right)\Delta x \Delta y$$

where:

$$(\mathrm{u}\theta)_{i,j-1/2}=\mathrm{u}_{i,j-1/2}\theta_{i,j-1/2}=\mathrm{u}_{i,j-1/2}\left(\frac{\theta_{i-1/2,j-1/2}+\theta_{i+1/2,j-1/2}}{2}\right)$$

$$(\mathrm{u}\theta)_{i-1/2,j-1/2}=\mathrm{u}_{i-1,j-1/2}\theta_{i-1/2,j-1/2}=\mathrm{u}_{i-1,j-1/2}\left(\frac{\theta_{i-3/2,j-1/2}+\theta_{i-1/2,j-1/2}}{2}\right)$$

$$(\mathrm{v}\theta)_{i-1/2,j}=\mathrm{v}_{i-1/2,j}\theta_{i-1/2,j}=\mathrm{v}_{i-1/2,j}\left(\frac{\theta_{i-1/2,j+1/2}+\theta_{i-1/2,j-1/2}}{2}\right)$$

$$(\mathrm{v}\theta)_{i-1/2,j-1}=\mathrm{v}_{i-1/2,j-1}\theta_{i-1/2,j-1}=\mathrm{v}_{i-1/2,j-1}\left(\frac{\theta_{i-1/2,j-1/2}+\theta_{i-1/2,j-3/2}}{2}\right)$$

$$\left(\tilde{\lambda}\frac{\partial \theta}{\partial x}\right)_{i,j-1/2}=\left(\tilde{\lambda}\right)_{i,j-1/2}\left(\frac{\theta_{i+1/2,j-1/2}-\theta_{i-1/2,j-1/2}}{\Delta x}\right)$$

$$\left(\tilde{\lambda}\frac{\partial \theta}{\partial x}\right)_{i-1,j-1/2}=\left(\tilde{\lambda}\right)_{i-1,j-1/2}\left(\frac{\theta_{i-1/2,j-1/2}-\theta_{i-3/2,j-1/2}}{\Delta x}\right)$$

$$\left(\tilde{\lambda}\frac{\partial \theta}{\partial x}\right)_{i-1/2,j-1}=\left(\tilde{\lambda}\right)_{i-1/2,j-1}\left(\frac{\theta_{i-1/2,j-1/2}-\theta_{i-1/2,j-3/2}}{\Delta y}\right)$$

If the thermal conductivity is equal in the liquid and in the solid $\tilde{\lambda} = 1$ while for problem with different liquid and solid conductivity to evaluate the terms $(\tilde{\lambda})_{i,j-1/2}$, $(\tilde{\lambda})_{i-1,j-1/2}$, $(\tilde{\lambda})_{i-1/2,j}$, $(\tilde{\lambda})_{i-1/2,j-1}$ the following procedure can be applied (Voller and Swaminathan (1993)).
By introducing a Kirchoff transformation

$$\phi = \int_{\vartheta_{rif}}^{\theta} \tilde{\lambda}(\alpha)d\alpha \tag{4.1}$$

where ϑ_{rif} is a reference temperature, results:

$$\tilde{\lambda}_x = \frac{\partial\phi/\partial x}{\partial\vartheta/\partial x}, \tilde{\lambda}_y = \frac{\partial\phi/\partial y}{\partial\vartheta/\partial y} \tag{4.2}$$

where $\tilde{\lambda}_x$ and $\tilde{\lambda}_y$ are the $\tilde{\lambda}$ horizontal and vertical components.
By Eq.(4.2):

$$(\tilde{\lambda})_{i,j-1/2} = \frac{\phi_{i+1/2,j-1/2} - \phi_{i-1/2,j-1/2}}{\theta_{i+1/2,j-1/2} - \theta_{i-1/2,j-1/2}} \qquad (\tilde{\lambda})_{i-1,j-1/2} = \frac{\phi_{i-1/2,j-1/2} - \phi_{i-3/2,j-1/2}}{\theta_{i-1/2,j-1/2} - \theta_{i-3/2,j-1/2}}$$

$$(\tilde{\lambda})_{i-1/2,j} = \frac{\phi_{i-1/2,j+1/2} - \phi_{i-1/2,j-1/2}}{\theta_{i-1/2,j+1/2} - \theta_{i-1/2,j-1/2}} \qquad (\tilde{\lambda})_{i,j-1/2} = \frac{\phi_{i-1/2,j-1/2} - \phi_{i-1/2,j-3/2}}{\theta_{i-1/2,j-1/2} - \theta_{i-1/2,j-3/2}} \tag{4.3}$$

Considering that in each control cell the thermal conductivity is constant by Eq.(4.1) and Eq.(4.3) results (Voller and Swaminathan (1993)) :

$$(\tilde{\lambda})_{i,j-1/2} = \frac{\theta_{i+1/2,j-1/2}\tilde{\lambda}_{i+1/2,j-1/2} - \theta_{i-1/2,j-1/2}\tilde{\lambda}_{i-1/2,j-1/2}}{\theta_{i+1/2,j-1/2} - \theta_{i-1/2,j-1/2}}$$

$$(\tilde{\lambda})_{i-1,j-1/2} = \frac{\theta_{i-1/2,j-1/2}\tilde{\lambda}_{i-1/2,j-1/2} - \theta_{i-3/2,j-1/2}\tilde{\lambda}_{i-3/2,j-1/2}}{\theta_{i-1/2,j-1/2} - \theta_{i-3/2,j-1/2}}$$

$$(\tilde{\lambda})_{i-1/2,j} = \frac{\theta_{i-1/2,j+1/2}\tilde{\lambda}_{i-1/2,j+1/2} - \theta_{i-1/2,j-1/2}\tilde{\lambda}_{i-1/2,j-1/2}}{\theta_{i-1/2,j+1/2} - \theta_{i-1/2,j-1/2}}$$

$$(\tilde{\lambda})_{i-1/2,j-1/2} = \frac{\theta_{i-1/2,j-1/2}\tilde{\lambda}_{i+1/2,j-1/2} - \theta_{i-1/2,j-3/2}\tilde{\lambda}_{i-1/2,j-3/2}}{\theta_{i-1/2,j-1/2} - \theta_{i-1/2,j-3/2}}$$

Solution procedure

A fully implicit method can been adopted for the vorticity and velocity equations (Eqs.(3.1)-(3.2)), while the temperature field (Eq.(3.3)) is solved separately in order to evaluate the variation in the local liquid mass fraction (Stella and Giangi (2000)). The two linearized algebraic systems are solved using a preconditioned BI-CGStab method (Van De Worst (1992)). The numerical algorithm is represented in Figure 4.3.

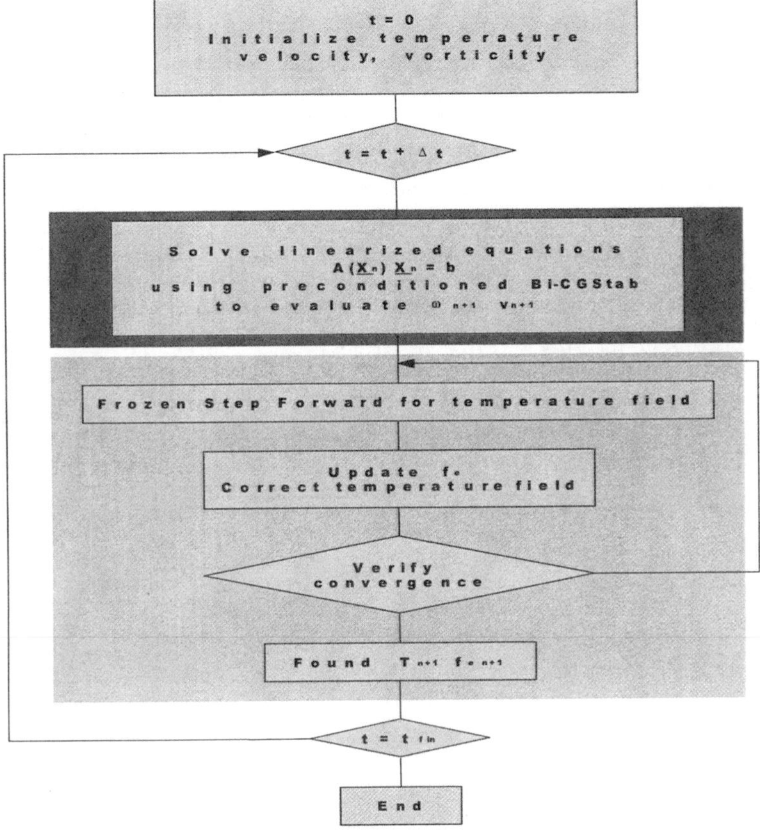

Figure 4.3. Numerical algorithm.

The solution procedure to solve the linearized vorticity-velocity equations consist of two part: the preconditioner and the iterative method (Stella and Bucchignani (1996)).

Preconditioner

It is well known (Golub and Van Loan (1990)) that for an iterative method of conjugate gradient type, the numerical residual at iteration k can be estimated as

$$\|r_k\| < \left(\frac{\sqrt{\mu(A)} - 1}{\sqrt{\mu(A)} + 1} \right)^{2k} \|r_0\|$$

where $\mu(A)$ is the condition number (i.e. the ratio between the maximun and the minimum eigenvalue) of the matrix $\boldsymbol{A}$. It is obvious that the best condition for the convergence of the method is when μ is small and close to 1. On the other hand, if μ is large, the iterative method would either converge very slowly or may even diverge, due to roundoff error.

The aim of the preconditioner is to convert the original linear system to an equivalent but better conditioned system. This consists of finding a real matrix $\boldsymbol{C}$ such that $\mu(\boldsymbol{C}^{-1}\boldsymbol{A}) < \mu(\boldsymbol{A})$. In this way the linear system,

$$\boldsymbol{C}^{-1}\boldsymbol{A}x=\boldsymbol{C}^{-1}\boldsymbol{b}$$

has (by design) better convergence and stability characteristics than the original system. It is obvious that the matrix $\boldsymbol{C}$ must be chosen carefully: it should be close to the inverse of $\boldsymbol{A}$, but easy to invert to avoid increased computational cost. An excellent review of preconditioners is given in (Golub and Van Loan (1990)) .

One of the most widely used preconditioners is the Incomplete Cholesky factorization (***ILU***). In this case, the matrix $\boldsymbol{C}$ is defined as the product $\boldsymbol{LU}$ of a power ($\boldsymbol{L}$) and an upper ($\boldsymbol{U}$) triangular matrix. $\boldsymbol{L}$ and $\boldsymbol{U}$ are generated by a variant of the Crout factorization algorithm (Ralston and Rabinowitz (1978)): only the elements that are originally non-zero in the matrix $\boldsymbol{A}$ are evaluated and stored in the $\boldsymbol{LU}$ factorization; the other elements are assumed to be zero. This avoid the generation of nonzero elements are assumed to be zero, so that $\boldsymbol{L}$ and $\boldsymbol{U}$ factors have the same sparsity patterns as the original matrix $\boldsymbol{A}$, with an enormous saving in both CPU time and central memory.

The algorithm can be summarizzed as follows: the column r of $\boldsymbol{L}$ (r=1,...,n) is generated by equation

$$l_{ir} = a_{ir} - \sum_{k=1}^{r-1} l_{ik} u_{kr} \quad i=r,\dots,n$$

followed by row r of $\boldsymbol{U}$, given by

$$u_{rj} = a_{rj} - \left(\frac{\sum_{k=1}^{r-1} l_{rk} u_{kj}}{l_{rr}} \right) \quad j=r+1,\dots,n$$

with a further condition that if a_{ij} is equal to zero, the corresponding l_{ij} element (or u_{ij} element) is also set equal to zero. All the elements of the main diagonal of $\boldsymbol{U}$ are set equal to unity.

Iterative method

One of the most efficient iterative methods for symmetric positive-definite systems is the conjugate gradient method. Different forms of the conjugate gradient method have been proposed to deal with unsymmetric problems, like those generated in CFD. The most widely used are GMRES (Saad and Schultz (1986)), CGS (Sonneveld (1989)), and Bi-CGSTAB (Van der Vorst (1992)).

The Bi-CGStab method may be summarized as follows (Van der Vorst (1992)).

1. An arbitrary initial solution x_0 is chosen and the residual r_0 is evaluated:

$$r_0=b-\boldsymbol{A}*x_0$$

2.

$$\text{if } r_0 \le \varepsilon \text{ stop} \quad \text{else:}$$

$$p = r_0 \qquad \hat{\zeta}_0 = 1 \qquad \varphi = 1 \qquad \alpha_0 = 1$$

$$w_0 = 0 \qquad q_0 = 0$$

$$k = 1$$

3.

$$\hat{\varphi} = p * r_{k-1} \qquad \zeta_k = \frac{\hat{\varphi}}{\varphi} * \frac{\hat{\zeta}_{k-1}}{\alpha_{k-1}} \qquad \varphi = \hat{\varphi}$$

$$q_k = r_{k-1} + \zeta_k * (q_{k-1} - \alpha_{k-1} * w_{k-1})$$

$$w_k = A * q_k$$

$$\hat{\zeta}_k = \frac{\hat{\varphi}}{p * w_k}$$

$$s = r_{k-1} - \hat{\zeta}_k * w_k$$

$$\tau = A * s$$

$$\alpha_k = \frac{\tau * s}{\tau * \tau}$$

$$x_k = x_{k-1} + \hat{\zeta}_k * q_k + \alpha_k * s$$

$$r_k = s - \alpha_k * \tau$$

if $r_k \le \varepsilon$ stop

else $k = k + 1$ and go to item 3

Here p, q, w, s, τ are work vectors with n elements (n is the number of equations), while $\zeta, \hat{\zeta}, \varphi, \hat{\varphi}, \alpha$ are work scalars.

Iterative procedure for the thermal field (temperature θ and liquid fraction f_l).
At each time step the liquid fraction f_l and the temperature field θ in Equation. (3.3) are solved by using the following iterative procedure.
At the new time step the initial iterative fields are set to the previous time step (n) values:

$$\theta^* = \theta^n, \theta^0 = \theta^*, f_l^0 = f_l^*$$

then, the following iterative system, Eqs.(4.1- 4.3), is solved:

$$f_l^k = f_l^{k-1} + Ste\left(\theta^{k-1} - \theta_m\right) \tag{4.1}$$

subject to the following constraint:

$$f_l^k = \max\left[0, \min\left(f_l^k, 1\right)\right] \tag{4.2}$$

$$Ste\left(\frac{\theta^k - \theta^*}{\Delta t}\right) + Ste\nabla \cdot \left(\mathrm{v}\,\theta^k\right) = Ste\nabla^2\theta^k + \frac{f_l^* - f_l^k}{\Delta t} \tag{4.3}$$

where k is the index of iteration level, Δt the time step, * indicates the previous time step values and θ_m is the phase change temperature.
Steps (4.1) to (4.3) are repeated until

$$\left\| f_l^k - f_l^{k-1} \right\| < \varepsilon \quad \text{and} \quad \left\| \theta^k - \theta^{k-1} \right\| < \varepsilon \tag{4.4}$$

During the computations $\varepsilon \cong 10^{-8}$ is usually adopted (Stella and Giangi (2000)).

5 Comparison of Moving and Fixed Grid Models

In this section a comparison of the two frequently used computational techniques for solving phase-change problems (*moving grid models* and *fixed grid methods*) is presented.

5.1 Example of application: melting.

A large number of papers are related in the literature to a problem where phase change is driven by laminar thermal convection in the melt.

In particular melting problems have enormous importance in applied problems, for example, in metallurgical processes or in microchip industry when reduction of macrosegregation is approached using zone refining.
For this reason, in recent years, a number of studies have been performed using computational fluid dynamics (CFD) to study melting problems.

Melting of pure gallium in a rectangular container with aspect ratio 1.4 has been studied by Brent *et al.* (1988) using a finite volume approach and by Dantzig (1989) using a finite element method. It is interesting to note that in the second paper Dantzig found a solution that during the quite long initial transient is dominated by a flow structure consists of small rolls that disappear after very long time (see Figure 5.1.1).

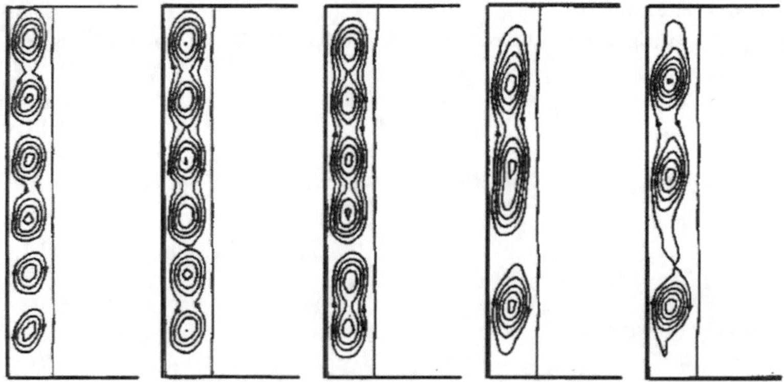

Figure 5.1.1. Evolution of multiple cell patterns in melted region from 42 to 50s (J. Dantzig 1989).

In contrast, Brent *et al.* (1988) found a solution that is never dominated by multiple rolls structure, that is from the very early stage the structure presents a single recirculating cell. They found a single cell structure and, as consequence, a much flatter solid/liquid interface (see Figure.5.1.2).

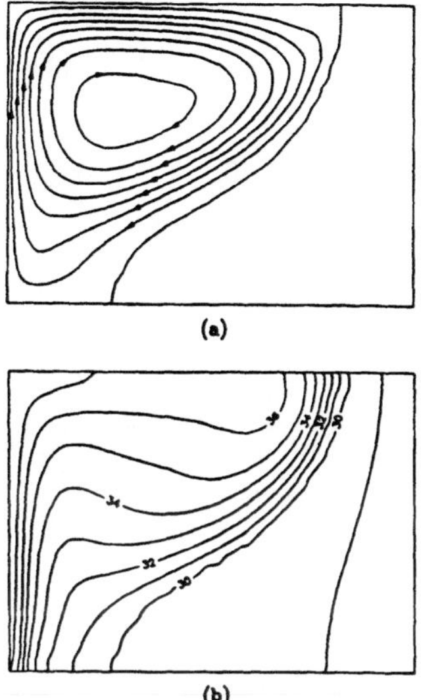

Figure 5.1.2. Streamlines (a) and temperature fields (b) at 19 min (A.D. Brent *et al.* 1988).

The reason of this strong difference has probably to be found in the limitation of computational resources of the time (1988) and in the coarse mesh 45×35 adopted as a consequence, that was not fine enough to resolve the multiple cell flow structure. Furthermore, an indication that a reduction of mesh points may cause a reduction of the number of recirculation cells.

As a further demonstration of the interest aroused in the scientific community by melting problems and the strong need of a reference solution, Gobin and Le Quere promoted the development of a benchmark solution on melting driven by natural convection. A first set of results have been presented by Bertrand *et al.*, (1999). In that comparison, many authors have found a multiple cell structure at the early stage, while many others have not.

Because of the difficulties of the problem a deep mesh sensitivity analysis has been performed by Stella and Giangi (2000) in order to clarify mesh independence of the flow structure and the solution found. Mesh sensitivity has been repeated during the evolution of the melting process in order to evaluate the proper mesh for the specific flow structure under study in that specific context. As consequence, an optimal solution strategy has been adopted, using different meshes during the various phases of the transient, in order to reduce the enormous requests of computation resources maintaining the required numerical accuracy.

Results shown that only the use of a very fine mesh allows the observation of the multicellular flow structure described (see Fig.5.1.3).

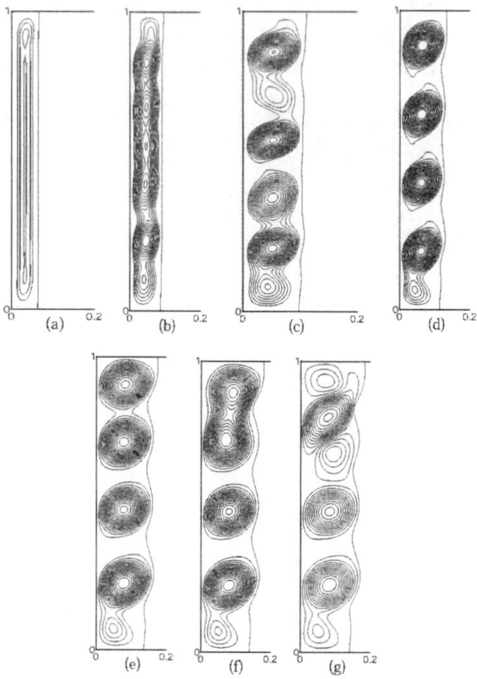

Figure 5.1.3a. Evolution of the solid-liquid interface and the corresponding streamlines (F. Stella and M. Giangi 2000) .

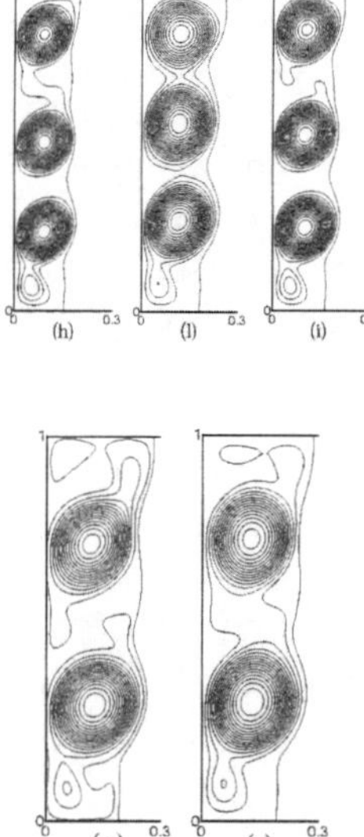

Figure 5.1.3b. Evolution of the solid-liquid interface and the corresponding streamlines (F. Stella and M. Giangi 2000) .

A comparison of a front-tracking or transformed grid (T-grid) procedure to a fixed grid or enthalpy method for melting of gallium in a rectangular cavity with isothermal side walls and adiabatic top and bottom walls has been also presented by Viswanath and Jaluria (1993).

In this paper the streamlines and isotherms obtained from the enthalpy method (see Fig.5.1.4) are compared with the results from the T-grid method (see Fig.5.1.5) at four time intervals: 6 min., 10 min., 15 min.,19 min.

In the T-grid approach the centre of the eddies lies closer to the midplane of the cavity (at x=0.55) whereas in the enthalpy method it is located higher up (at x=0.7). In the T-grid method there is an inflection in the interface profile close to the bottom of the cavity a t=15 min. The finer flow structure is better predicted by the T-grid method. A still finer resolution is probably necessary at the phase front for the enthalpy method. The advantage of the T-grid method for this problem is clear: the grid can be preferentially concentrated near the phase front, where important and interesting phenomena occur, and very few points may be used in the solid where only diffusion is the dominant process, obtaining a more sharp interface.

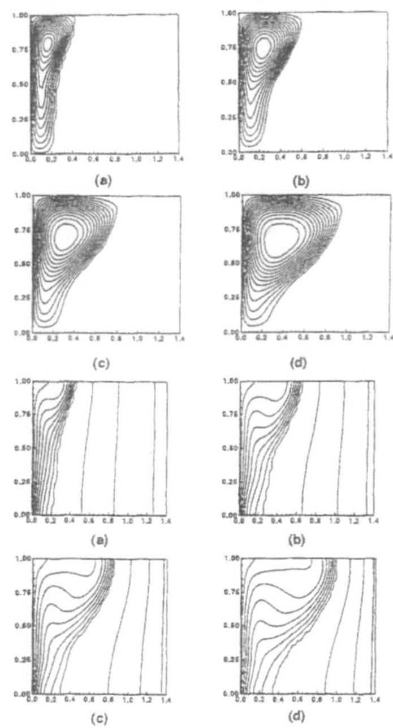

Figure 5.1.4. Streamlines and isotherms using the enthalpy method (R. Viswanath and Y. Jalura 1993).

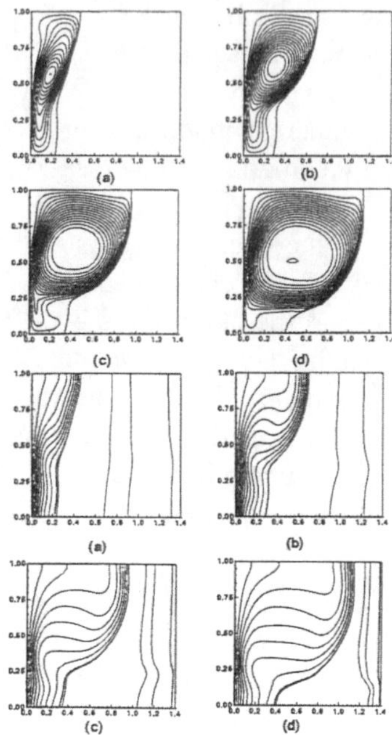

Figure 5.1.5. Streamlines and isotherms using the transformed grid method (R. Viswanath and Y. Jaluria 1993).

6 Applications

6.1 Solidification of Pure Materials

Solidification of Water. A transformed grid finite differences model (Kowalewski and Rebow, 1998), a fixed grid finite volume model (Giangi *et al.*, 1999) and a fixed grid finite element model (Banaszek *et al.*, 1998) have been applied to the problem of water freezing in a cube shape cavity.

The aims of this works have been to verify several assumptions used for modelling natural convection. The results obtained for both methods, presented in terms of velocity and temperature fields and interface position, have been compared with experimental results.

Differences between the methods are manly limited to the recirculation at the bottom cold corner (Fig.6.1.1a).

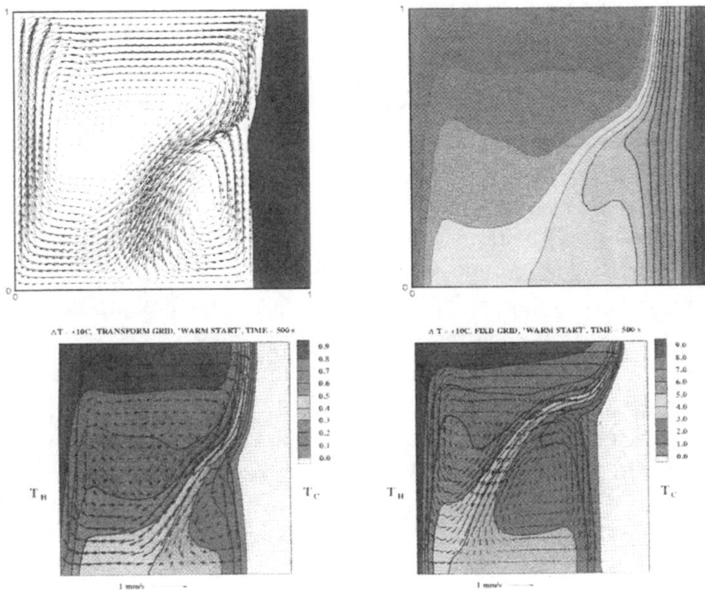

Figure 6.1.1a. Comparison of flow pattern and temperature field for freezing of water at time t=500s. The upper row: *Fixed grid volume model* (Giangi *et al.* 1999) ; the lower row: *transformed grid finite differences model* (Kowalewski and Rebow 1998) - left side, fixed grid finite element model (Banaszek *et al.* 1998) – right side.

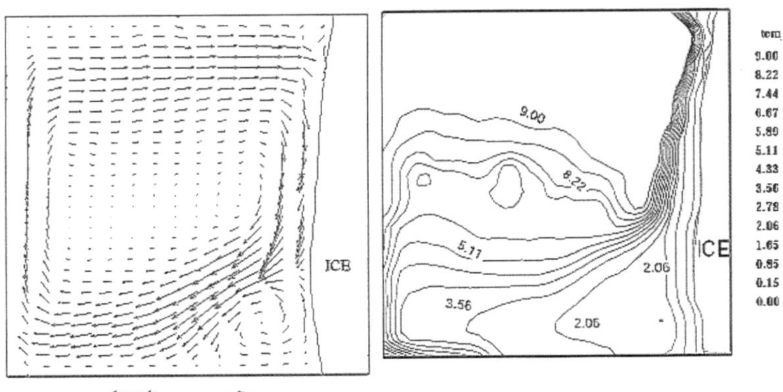

Figure 6.1.1b. Experimental results (T.A. Kowalewski 1998).

The solution of the transformed grid model seems closer to the experimental findings (Rebow and Kowalewski, 1998) (Fig.6.1.1b), when the flow pattern is compared, but it evidently overestimates convective heat flux at the bottom part of the interface (Fig.6.1.1a). The results indicate that the numerical methods used need several improvements.

Melting of a Pure Metal In the present section a study of melting of pure gallium in a rectangular container (1.4 × 1) is presented (Stella and Giangi (2002)). The selected problem has been already been studied by Brent *et al.* (1988) using a finite volume approach and by Dantzig (1989) using a finite difference method. It is interesting to point out that in the second paper by Dantzig found a solution that had quite a long initial transient dominated by a flow structure consisting of small rolls that disappears at longer times, on the contrary Brent *et al.* (1988) found a solution that is never dominated by multiple rolls structure, i.e. also at the very early stage the structure presents a single cell (see Figure 6.1.2). This result found by Brent *et al.* (1988) seems to be confirmed by the experimental findings of Gau and Viskanta (1986). The flow structures found during the present research, also using a very fine mesh, and the physics of the problem seems to indicate that the flow at the very early stage presents this multiple cell structure.

The numerical simulation of this problem was conducted in a two-dimensional cavity (height H=6.35 cm; width L=8.89 cm) with one heated wall. The left-hand wall is maintained at a temperature $T_h=38°C > T_m$ and the right-hand wall at $T_c=28.3°C < T_m$ (with $T_m=29.78°C$ the phase change temperature). At time t=0 only solid is present in the cavity (see Figure 6.1.3) so the initial condition for liquid fraction and temperature are $f_{li}=0$ and $T_i=T_c$.

The physical property were taken from one of Gau and Viskanta's experiments (1986) and corresponded to $Pr = 0.0216$, $Ste=4.6 \times 10^2$ and $Ra=7\times 10^5$.

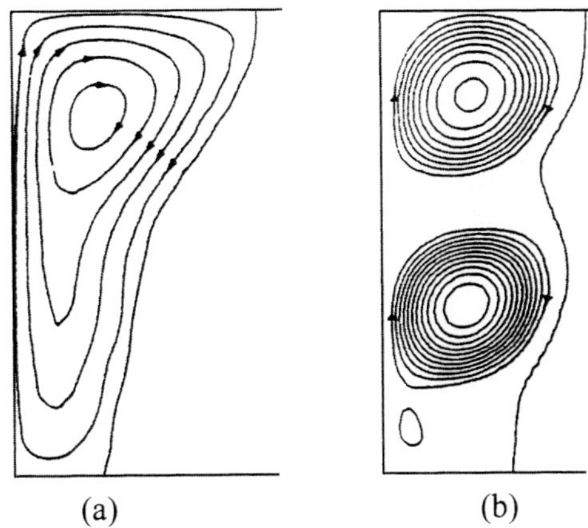

Figure 6.1.2. Comparison of single and multi cell structures at t≅ 6 min.
(a) Voller (1988) (b) Dantzig (1989).

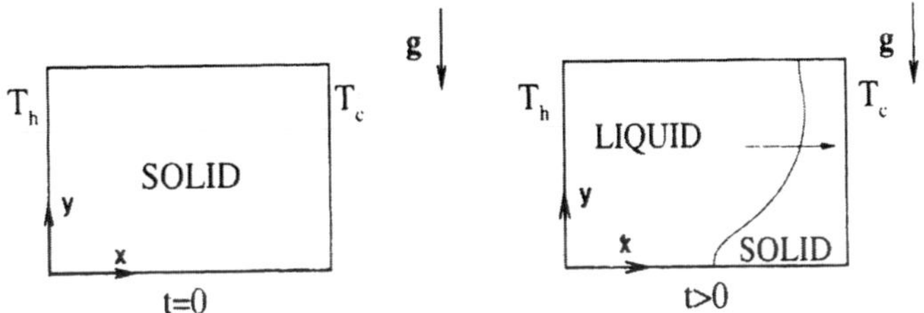

Figure 6.1.3. Problem definition. Aspect ratio of 1.4.

Three different computational meshes have been considered during the sensitivity analysis as reported in Figure 6.1.4. Obviously, since the fluid phase is present only on a limited portion of the physical domain, the evaluation of the vorticity and velocity equations is performed only on a small region that completely contains all the fluid zone. On the remaining part of the domain, only the energy equation has been solved (see Figure 6.1.5). This region has been enlarged during the melting procedure, in order to completely contain all the liquid zone. In this way the enormous requirement of CPU time has been reduced, allowing the solution of the very fine meshes proposed in a limited elapsed time.

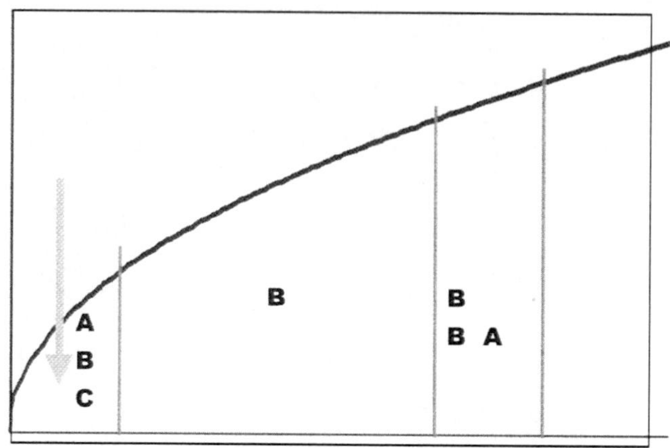

Mesh	Δx	Δy	Equivalent mesh
A	1/160	1/80	225x81
B	1/320	1/160	449x161
C	1/640	1/320	897x321

Figure 6.1.4. Mesh sensitivity.

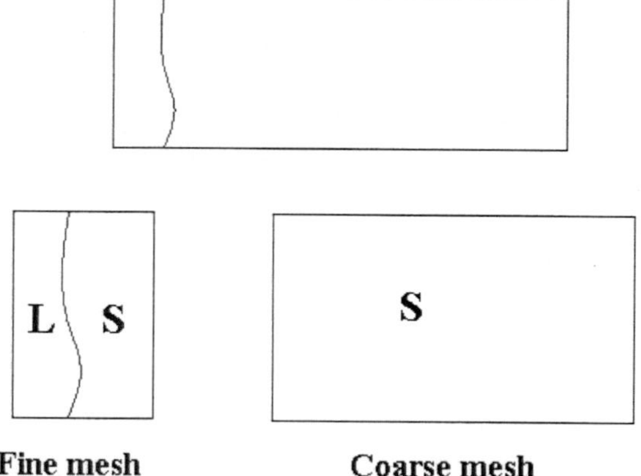

Figure 6.1.5. Mesh adoption.

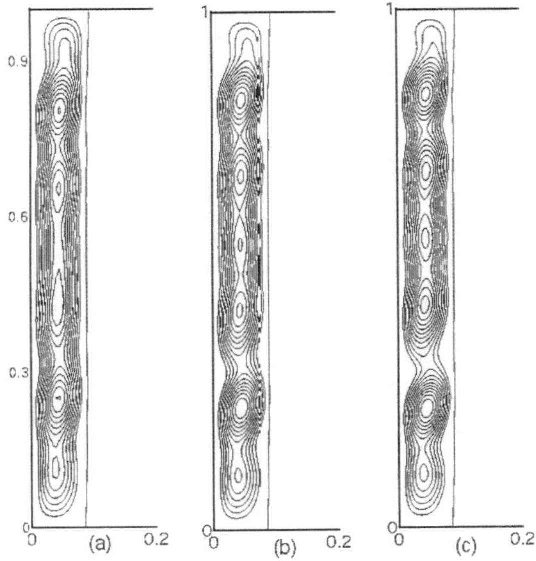

Figure 6.1.6. Streamlines at t=32s. Mesh: (a) **A**, (b) **B**, (c) **C.**

The first analysis has been conducted considering the transient solution from the initial state (t=0 s) to the solution at time t=32 s, this physical time has been chosen because presents the larger number of recirculations, as described in the work of Dantzig (1989). From the comparison of the streamlines (Figure 6.1.6) obtained with the three meshes has been observed, we observe that mesh **A** is not accurate enough for solving the several recirculations created from the flow field. This conclusion is even more evident when the horizontal velocity profiles are compared at x=0.029, (i.e. near the middle of the recirculations) (Figure 6.1.7).

A further analysis of Figure 6.1.7 shows that there is a maximum difference of 10% between the values obtained with mesh **B** and **C**. Also the positions of all the maximum and minimum of the velocity are correctly represented. So meshes **B** and **C** seem to be adequate in the interval t=0 s - t=32 s for the purposes of the present section.

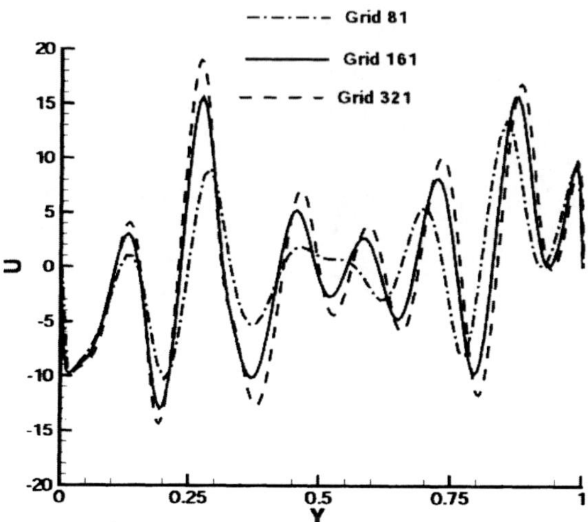

Figure 6.1.7. Horizontal velocity profiles near the middle of the recirculation cell (x=0.029) at τ=32 s (t=0.109).

Since thickness of the melted zone grows, the aspect ratio of the fluid zone decreases. This induces a reduction of the number of recirculations, so starting from t=32 s it is possible to evaluate the use of a mesh even coarser than **B**.

The second test has been conducted considering a transient solution from t=120 s. to t=150 s., in this case only meshes **B** and **A** have been considered. As initial solution for mesh **A**, the projected solution from mesh **B** at t=120 s. has been taken. So the two simulations start from the same initial state.

Results obtained in terms of stream function are presented in Figure 6.1.8. It is easy to see that mesh **A** is enough accurate for the prediction of the flow pattern at t=150 s. Results in terms of horizontal velocity near the middle of the recirculation cell (x=0.11) are presented in Figure 6.1.9 showing a good agreement between the two meshes used. The maximum difference on peak velocity is of the order of 5%. The same type of comparison have been made on the vertical velocity showing similar or even better agreement, results have not been reported here for seek of brevity.

Summarizing the results obtained in the mesh sensitivity analysis we can conclude:

- mesh **C** is adopted from t=0 s. to t=32 s.,
- mesh **B** is adopted from t=32 s. to t=150 s.,
- mesh **A** is adopted from t=150 s..

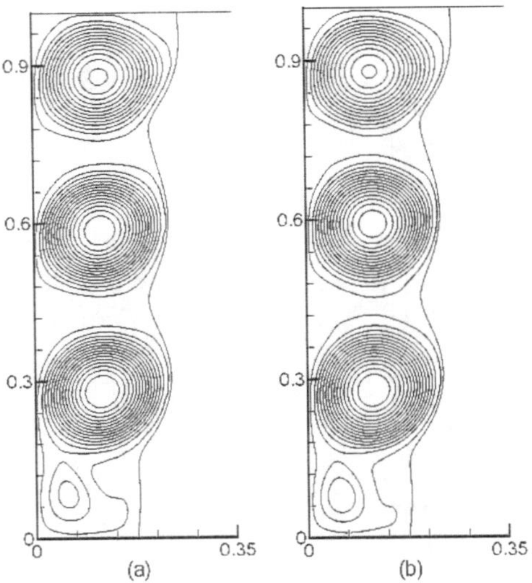

Figure 6.1.8. Streamlines at t=150s. Mesh: (a) **A**, (b) **B**.

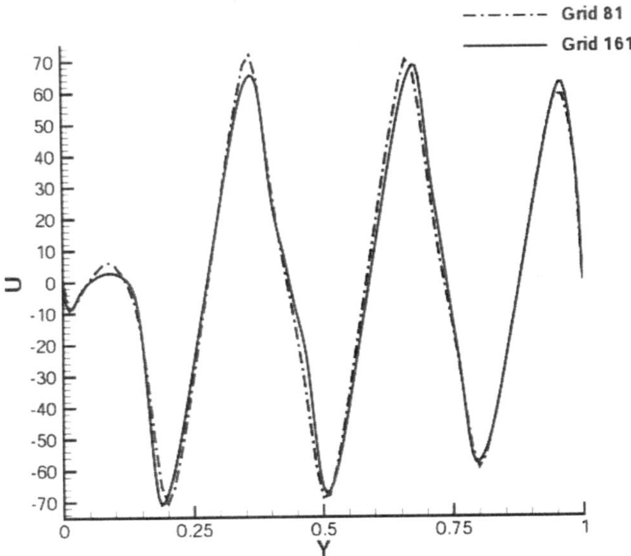

Figure 6.1.9. Horizontal velocity profile near the middle of the recirculation cell (x=0.11) at τ=150 s (t=0.511).

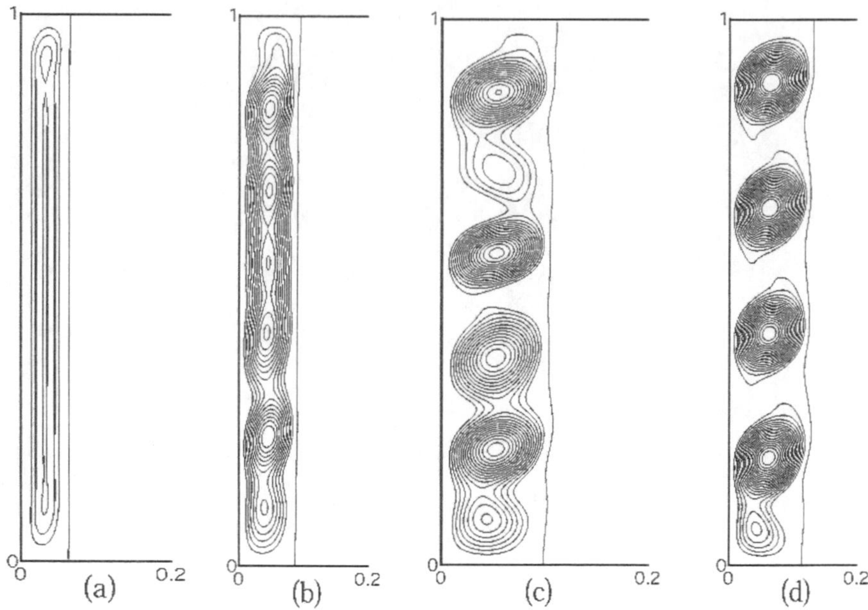

Figure 6.1.10. Evolution of the liquid-solid interface and corresponding streamlines a) t = 16 s; b) t = 32 s; c) t = 42 s; d) t = 58 s.

Flow pattern evolution during the melting process can be subdivided into two main phases, with different and contrasting behaviours. The first one consists of the early stage of melting, that in our case goes approximately from t=0s to t=42s, is characterized by a flow field confined into a very narrow vertical layer. During this period we observe formation of several recirculations. The second one, in which melting process has been already developed, presents a larger fluid region with aspect ratio more close to unity. During this phase we observe the re-merging of many of the recirculations developed in the first phase into larger ones.

In the Figure 6.1.10, the evolution of the liquid/solid interface and the corresponding streamlines are reported (from t=16s to t=42s). At time t=32s the liquid region is characterized by small rolls with an approximate averaged wave number $\alpha = 3.14$.

A qualitative comparison of results obtained with those of Lee and Korpela (1983) in a vertical slot is tried, by adopting an appropriate *equivalent Grashof number* (Gr^*) based on the thickness of the melted layer δ and the temperature difference ΔT between the hot wall and the phase-change temperature, (Gr^*) can be evaluated at each time step (Table 6.1.1). In particular the flow configurations at time t=16s and t=32s (Figure 6.1.11) are in qualitative agreement, in terms of wave number shape and overall flow structure, with those obtained in the mentioned work for Grashof equal to 5000 and 15000 respectively. Anyway, we must observe that results by Lee and Korpela (1983) has been obtained for Pr=0, with a rigid flat boundary and at steady state whereas in the melting problem the flow configurations are transient. The presence of this multicellular flow structure induces on the solid/liquid interface a wavy shape. The results obtained are in good agreement, in terms of shape and position of the solid/liquid interface, with those of Dantzig (1989).

Table 6.1.1. Time, nondimensional thickness of the melted layer (δ), equivalent Grashof number (Gr_e).

τ	δ	Gr_e
16 s	0.064	7219
32 s	0.090	20076
42 s	0.10	27540

On the contrary very poor agreement is found with numerical results obtained by Brent *et al.* (1988). In that work Brent *et al.* (1988) did not find, also at very early stage, the multi cellular flow structure obtained here. On the contrary they found a single cell structure and, as consequence, a much flat solid/liquid interface. The reason of this strong difference has probably to be found in the limitation of computational resources of the time (1988) and in the coarse mesh (45 ×35) adopted as consequence, not fine enough to resolve the multiple cell flow structure. Differences also arise when results are compared with the experiment of Gau and Viskanta (1986) although they do not report the flow structure of the flow field the quite straight shape of the solid/liquid interface they report for early melting stage is an indication of the absence of flow recirculations. The subject is obviously controuvers, since in a recent benchmark promoted by Gobin Le Quéré with the contribution of several authors, for the simulation melting problems in a similar geometry (Bertrand *et al.* (1999)) solutions obtained from the different authors with various methods, in some case present recirculations in some case don't. In particular one of the author (P.Le Quéré) describes a behaviour very similar to the one presented here with a multiple cell structure at very beginning that is reorganized by re-merging of the cells when melting goes further.
Going on with the melting process the size of the fluid zone increases and, as consequence, its apsect ratio decreases. In this way the number of cells decreases determining a less wavy front (Figs.6.1.12-6.1.13). Correspondingly the isotherms near to the liquid/solid interface are smoother (Fig.6.1.13). Also in this case, good agreement is also found when the shape and the liquid/solid position are compared with those of Dantzig (1989). In particular in the Table 6.1.2 a comparison between the adimensional thickness A_i=L_i/H (where H is the height of the cavity and L_i (i=1,2,3) the width of the liquid region at y=0, y=0.5, y=1) is reported. These values are calculated at different time from t=387.5s to t=500s. As in the previous case a more jagged solid/liquid interface is found with respect to Brent *et al* (1988) and Gau Viskanta (1986). This is obviously a consequence of the differences found at the earlier stages. Anyway, we observe that, as the liquid/solid interface advanced in the time, the differences between the various solution decreases (Fig.6.1.14). This is due to the fact that in this phase merging of the flow structures take place, reducing the influence of the differences on the initial flow structure.

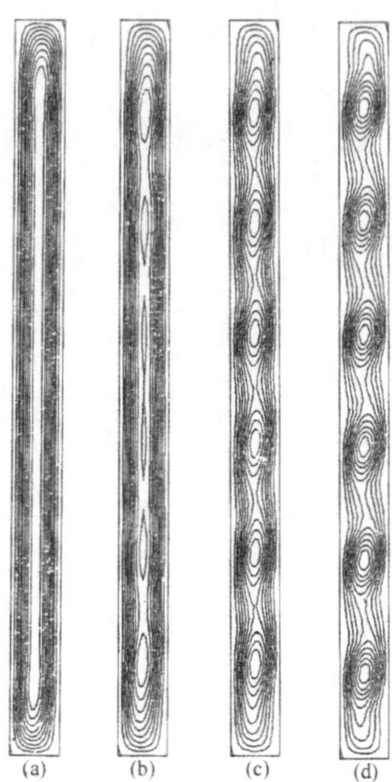

Figure 6.1.11. Streamlines in tall cavities. Lee and Korpela (1983)
(a) Gr = 5000, (b) Gr = 8000, (c) Gr = 10000, (d) Gr = 15000.

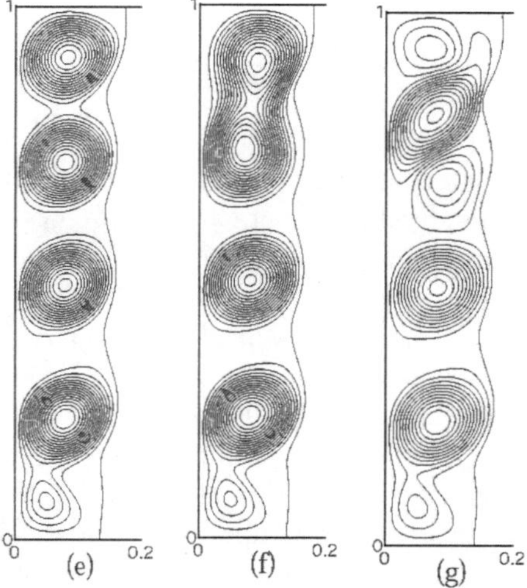

Figure 6.1.12. Mesh sensitivity e) t = 85 s; f) t = 88 s; g) t = 90 s.

Table 6.1.2. Adimensional thickness of liquid phase from τ=387.5 to τ=500 s. Comparison with Dantzig.('89).

τ	y	L/H	Dantzig ('89)	Δ
τ=387.5 s	0	0.280	0.305	8%
	0.5	0.476	0.444	7%
	1	0.488	0.486	0.4%
τ=425 s	0	0.293	0.333	12%
	0.5	0.509	0.447	12%
	1	0.512	0.513	0.1%
τ=462.5 s	0	0.305	0.361	15%
	0.5	0.543	0.500	8%
	1	0.536	0.542	0.11 %
τ=500 s	0	0.317	0.388	18%
	0.5	0.567	0.555	2%
	1	0.555	0.555	0

6.2 Solidification of Alloys

The study of solidification processes is very important for the growth of high quality crystals. The performance of electronic devices strongly depends on the presence of microscopic crystal defects and compositional variations caused by convection effects. The MEPHISTO[1] experiment was a cooperative US-French-Australian research effort directed towards gaining a detailed understanding of crystal growth with reference to the solidification behaviour of Bi-1 atomic % Sn alloy. It combined ground-based experiments and a series of experiments conducted in a micro-gravity environment so that convection was decreased to a level at which crystal growth was largely diffusion controlled. The latest MEPHISTO experiment was performed on board the US Space Shuttle *Columbia* during the USMP-4[2] mission in November-December 1997. The MEPHISTO-4 apparatus, shown schematically in Figure 6.2.1, consists of three parallel tubes or ampoules (only one is shown in the figure), each containing some sample material, around which are placed two "furnaces", each comprising a pair of heating and cooling jackets. Between each heating and cooling jacket is a nominally adiabatic or insulated zone. One furnace is fixed, and acts to generate a reference state; the other can be moved over the tubes. If it moves in the direction from the cooling to the heating jacket (*i.e.*, to the right in Figure 6.2.1), the material will be progressively solidified from left to right; when it moves in the opposite direction, melting will take place.

[1] Matériel pour l'Etude des Phenomènes Intérrasants de la Solidification sur Terre et en Orbite

[2] United States Microgravity Payload

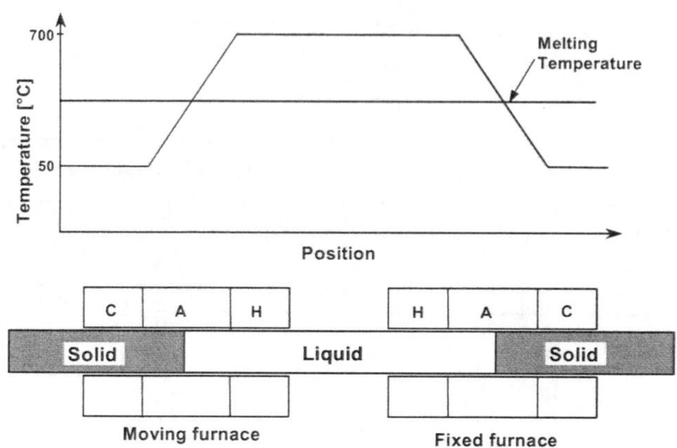

Figure 6.2.1. Schematic diagram of the MEPHISTO apparatus. H and C denote the hot and cold sections of the furnaces; A denotes the adiabatic zone.

The in-flight experiment, which lasted for almost 14 days, consisted of a series of "events", most of which contained three phases: solidification, holding and melting. The solid/liquid (s/l) interface advanced progressively along the ampoule over a period of about a week, during which a number of events at different solidification and melting speeds ranging from 0.74 to 40 μm/s, and lasting for periods of up to several hours occurred. During the following week, the procedure was reversed, culminating in a rapid quenching of the alloy to preserve the interface composition and shape for post-flight analysis. Part of the MEPHISTO program was the numerical modelling of the solidification process , which was performed by researchers at the CFD Research Laboratory at The University of New South Wales, Sydney, Australia. The numerical simulations allowed the investigation of the effects of natural convection on the interface shape, the prediction of planar (*i.e.*, non-dendritic) front instabilities and the calculation of the segregation or redistribution of solute during solidification.

For modelling transient phase change processes, a fixed-grid single domain approach is widely accepted to be simpler and lower in computational cost than front tracking methods. However, the standard fixed-grid enthalpy formulation has a major weakness: it produces spurious numerical oscillations in the predicted temperature and interface positions. To overcome this oscillatory behaviour, Voller and Fabbri (1995) suggested selecting only predictions that occur when the interface crosses over a node point. Tacke (1985) employed a special discretization of the heat fluxes which allowed the removal of the numerical oscillations for a *one-dimensional* phase change problem. Laouadi *et al.* (1998) developed a numerical method based on averaging the microscopic conservation equations of a two-phase composite medium over a control volume with the assumption that the phases may coexist at a temperature different from the melting temperature.

The problem becomes more complex for isothermal plane front (*i.e.*, non-dendritic) solidification of binary alloys. The severe discontinuity of solute concentration at the solid/liquid interface and the sharp gradients of concentration near the interface induced by the low values of both the diffusion coefficient for tin in bismuth and the partition coefficient of tin require special treatment of the diffusion flux at the interface to obtain accurate solutions.

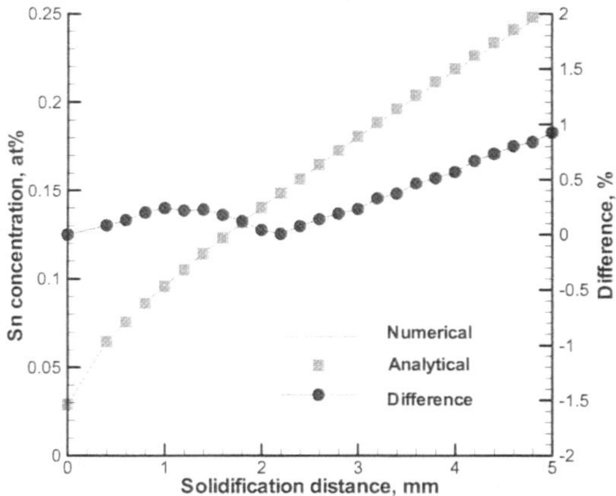

Figure 6.2.2. Analytical and numerical solute concentration in the solid for 5 mm of solidification.

Additional difficulties arise when the effect of solute concentration on melting temperature is taken into consideration and the s/l interface position becomes dependant on concentration as well as temperature. Accurate estimation of interface concentration during plane front solidification becomes essential to obtain a smooth history of the interface position.

The model is applied to the simulation of some of the MEPHISTO events. The problem involves heat conduction in the solid alloy and in the walls of the ampoule containing the alloy; thermal and solutal convection; and diffusion in the liquid. Solute diffusion in the solid is neglected. The effects of concentration-dependent melting temperature on the phase change processes are incorporated.

Comparisons with analytical solutions. To validate the physical and mathematical models, Timchenko *et al.* 1998, performed a comparison of numerical results (C_n) with the analytical solution (C_a) of Smith *et al.*, 1955 for one-dimensional, diffusion-controlled plane front solidification was performed. No convection was included. Results of this comparison are shown in Figure 6.2.2. It can be seen that the computed solute concentrations in the solid at the mid-height of the ampoule are very close to the analytical, diffusion controlled values. The relative difference $(C_a - C_n)/ C_a$ is less than 1%. As noted below, segregation occurs and the concentrations away from the mid-height would differ from these one-dimensional values.

Modelling of MEPHISTO experiments. For the solutions presented, A grid size of 0.2 mm in the x and 0.3 mm in z directions was used with time step of 0.5 s. This mesh size were found to be adequate by extensive testing against results obtained using finer meshes.

The model has been applied to the simulation of the experiments performed during the 1997 flight of MEPHISTO-4. In the example described below (the events identified in the flight schedule as 11E and 11F, Abbaschian *et al.*, 2001) solidification at a pulling speed (the speed of the moving furnace) of 3.34 μm/s occurred for 0.333 h. The furnace was then stopped for 3.7 hours (an "extended hold") during which time almost complete rehomogenization of the liquid occurred. Solidification at a speed of 1.85 μm/s followed for 0.6 h. The melting temperature was calculated according to (21) with m = -2.32 K/at%. The magnitude of the gravity vector was taken to be 1 μg, *i.e.*, 9.81×10^{-6} m s^{-2}, acting in a direction normal to the axis of the ampoule. The variation of the thermal conductivity between the solid and liquid phases was taken into account with k_l = 12.4 W/m K and k_s = 6.5 W/m K. Properties values for pure liquid bismuth (Timchenko *et al.*, 1998) taken at the reference temperature of 271.3 °C (the equilibrium melting temperature of Bi) were used. The partition coefficient k_p for Sn in Bi was taken to be 0.029. Reported values of the diffusion coefficient D for dilute Sn in Bi near 271.3 °C vary from $1.76 \times 10^{-9} m^2/s$ (Buell and Shuck, 1970) to $2.7 \times 10^{-9} m^2/s$ (Niwa *et al.*, 1957).

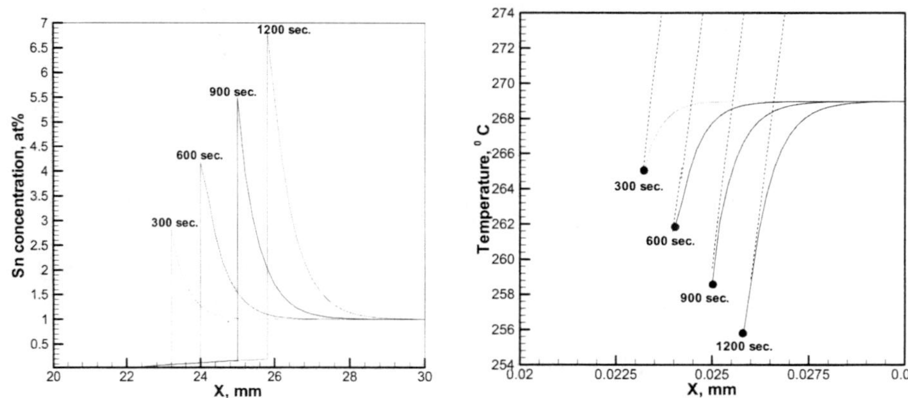

Figure 6.2.3. a) Distribution of solute concentration at the mid-height of the ampoule during event 11E (b) Distribution of melting temperature (solid lines) and the actual temperature distribution (dashed lines) at the mid-height of the ampoule during event 11E.

In these calculations $D = 2.0 \times 10^{-9} m^2/s$ was chosen after a comparison of numerical solutions with post-flight microprobe results for solute concentration in the solid.

The moving temperature profile imposed on the outer walls of the ampoule consisted of a cold zone (T_c = 50 °C), an adiabatic zone and a hot zone (T_h = 700 °C). The length of the adiabatic zone was 20 mm, leading to an internal temperature gradient in the liquid of approximately 20 K/mm.

Figure 6.2.3(a) shows the distribution of solute concentration at the mid-height of the ampoule during the 11E solidification event. Figure 6.2.3(b) shows the distribution of melting temperture (solid lines) and the actual temperature distribution (dashed lines) at the mid-height of the ampoule during the same event. Because of the increase of solute concentration at the interface during solidification, the melting temperature decreases according to the slope of liquidus in the phase diagram. During the hold, solute concentration at the interface decreases due to diffusion of solute into the liquid as shown in Figure 6.2.4.

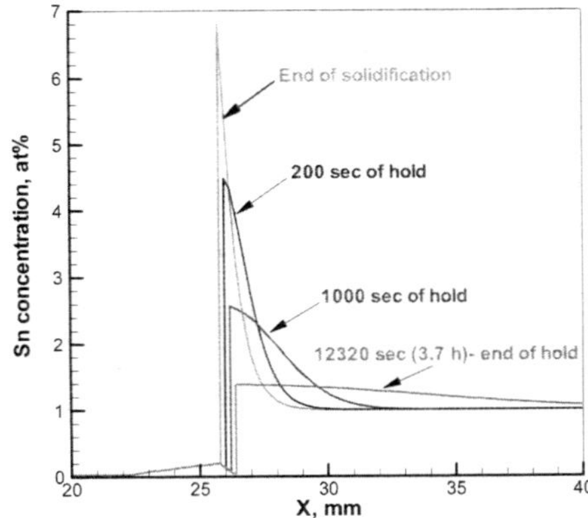

Figure 6.2.4. Distribution of solute concentration at the mid-height of the ampoule during event 11E.

Figure 6.2.5 shows temperature contours and the velocity field in the liquid phase (a) at the start of solidification, (b) in the end of event 11E and (c) after event 11F. It can be seen that both temperature and velocity fields are moving along the sample as time progresses accord ing to the moving boundary temperature profile.

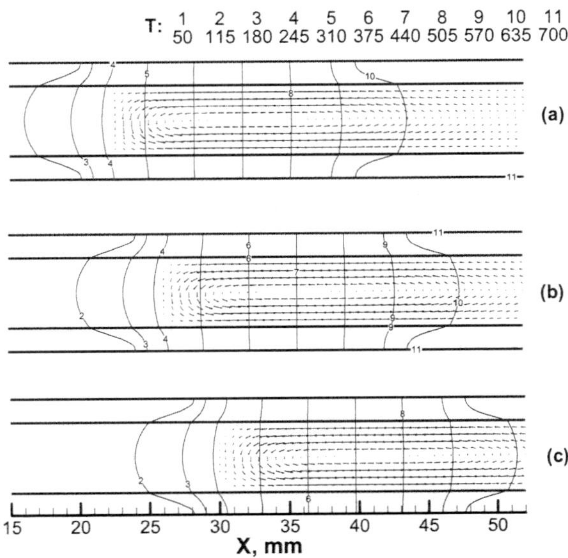

Figure 6.2.5. Temperature contours and velocity field (a) at the start of solidification, (b) at the end of event 11E and (c) after event 11F

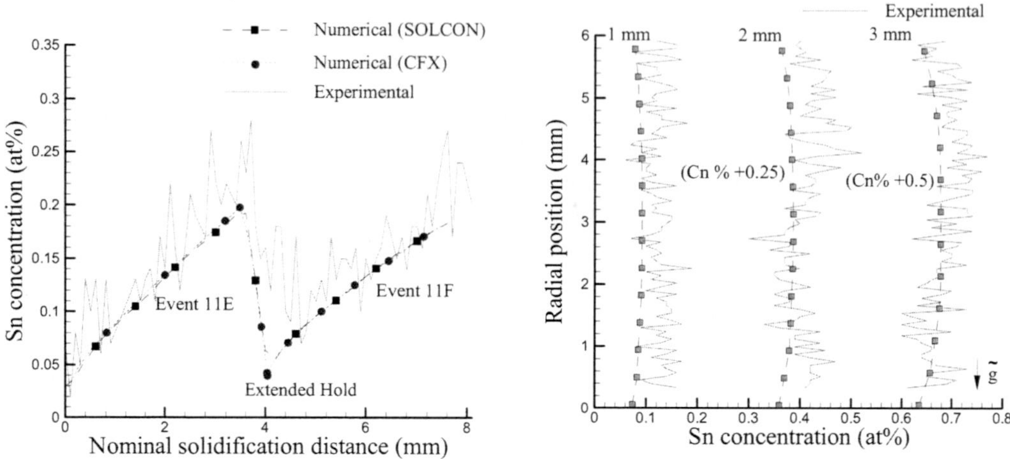

Figure 6.2.6. Solute concentration in the solid (a) along the ampoule centre line and (b) across the ampoule.

The distribution of solute concentration in the solid along the centre line of the sample is shown in Figure 6.2.6a. Numerical solutions are presented together with the microprobe results obtained after the flight from the experimental samples. Figure 6.2.6b shows the distribution of solute concentration across the solid. Both axial and transverse numerical distributions are in very good agreement with the experimental results.

Figure 6.2.7 shows the interface shape after the last event of solidification (11F) in the MEPHISTO experiment . The computed interface shape is in excellent agreement with that observed in the actual experiment.

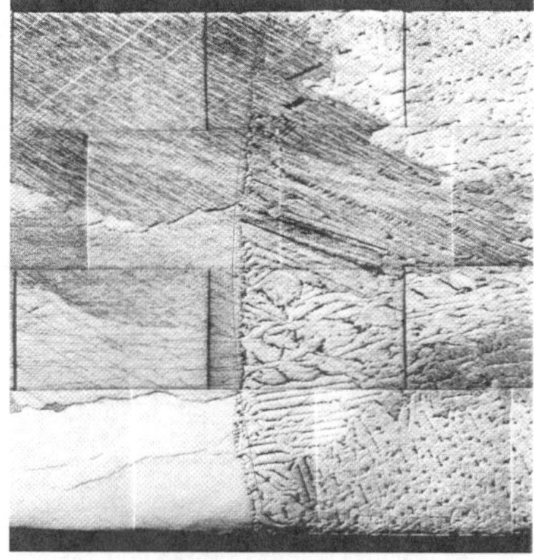

Figure 6.2.7. Comparison of the computed and measured interface shape.

6.3 Marangoni and Solidification

The problem studied in this Section (Giangi *et al.* 2002) is a simplification of the MEPHISTO-4 apparatus discussed in the previous section. Although the MEPHISTO-4 experiment was with a Bi 1 at% Sn alloy, to evaluate thermocapillary convection, this study will be simplified by considering only a pure substance, since Marangoni convection is not affected from the presence of such a small quantity of impurity. The fluid considered in this study is pure Bismuth ($Pr = 0.02161$), for which the physical properties are given in Table 6.3.1. During the solidification of this aterial the Stefan number is $Ste = 1.8$.

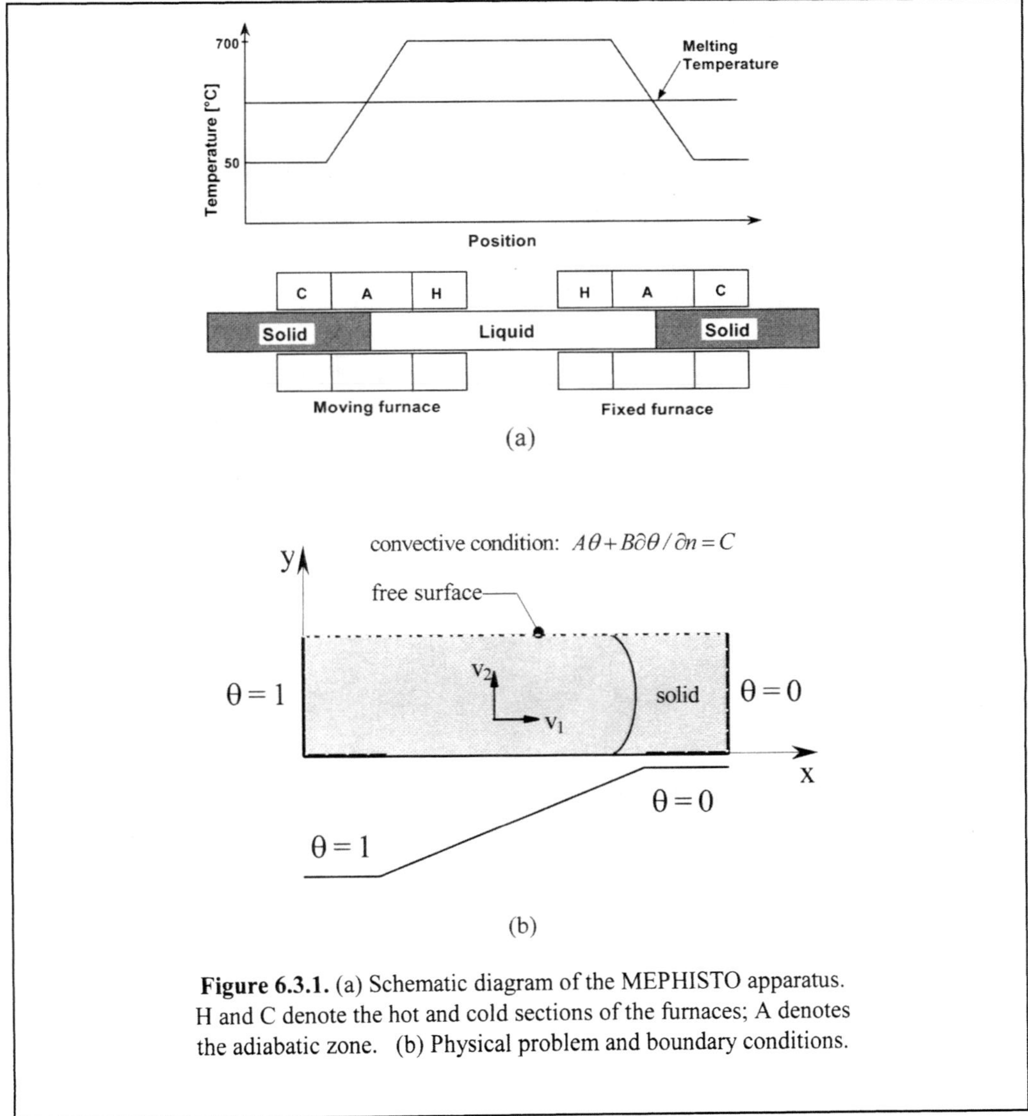

Figure 6.3.1. (a) Schematic diagram of the MEPHISTO apparatus. H and C denote the hot and cold sections of the furnaces; A denotes the adiabatic zone. (b) Physical problem and boundary conditions.

Table 6.3.1. Physical properties of Bismuth.

Property	Symbol	Value
Density	ρ	10,070 kg/m^3
Specific heat	C_p	144.87 J/kg K
Thermal conductivity (liquid)	k_λ	12.4 W/m K
Thermal conductivity (solid)	k_s	6.5 W/m K
Thermal diffusivity	α	8.500×10^{-6} m^2/s
Kinematic viscosity	ν	1.837×10^{-7} m^2/s

Numerical simulations were conducted for values of Marangoni number up to 16,120. Since typical values for the gravity level in space can vary from 0.1 up to 2,500 μg, a value of 45 μg that corresponds to a Ra = 5 has been assumed. This realistic value is big enough to include the residual effects due to gravity on the flow dominated by Marangoni, but not too big to mask the Marangoni effects, which we wish to investigate. The MEPHISTO experimental conditions correspond to Ma=16,120 but, to clearly explain the effects of increasing of Ma, numerical simulation for values from Ma=1,612 up to Ma=16,120 have been presented. During the numerical simulations a uniform mesh on the whole computational domain has been used. For $Ma \leq 4\,836$ a 161x41 mesh was used and for $Ma \geq 8\,060$ a 641x161 mesh was used. For the imposed thermal gradient and as a consequence of the Marangoni effects, high velocity gradients occur near the free surface and strong recirculating flows are then generated by these surface gradients; whilst in the remaining part of the flow, the velocity are very small because of the weak gravitational field. Two particular cases was been considered:

- Case A Constant and equal liquid and solid conductivities. Negligible resistance due to the external heat transfer coefficient.
- Case B Different liquid and solid conductivities. Convective boundary condition with $h = 20$ W/K m^2.

Case A

Steady state streamlines and the solid/liquid interface positions are presented in Fig. 6.3.2. The liquid/solid interface occurs at the $\theta = 0.32$ isotherm, corresponding to the phase change temperature.

The flow patterns observed for all Ma have similar base structures i.e. a clockwise recirculation main flow near the solid/liquid interface, which dominant the whole flow field. This recirculating cell is due to the Marangoni effect which accelerates the flow on the free surface from left to right. As Ma increases this flow becomes progressively stronger and consequently increases the interface curvature. There is a marked change in the curvature as Ma is increased from 1,612 to 3,224 (compare Fig. 6.3.2a and 6.3.2b).

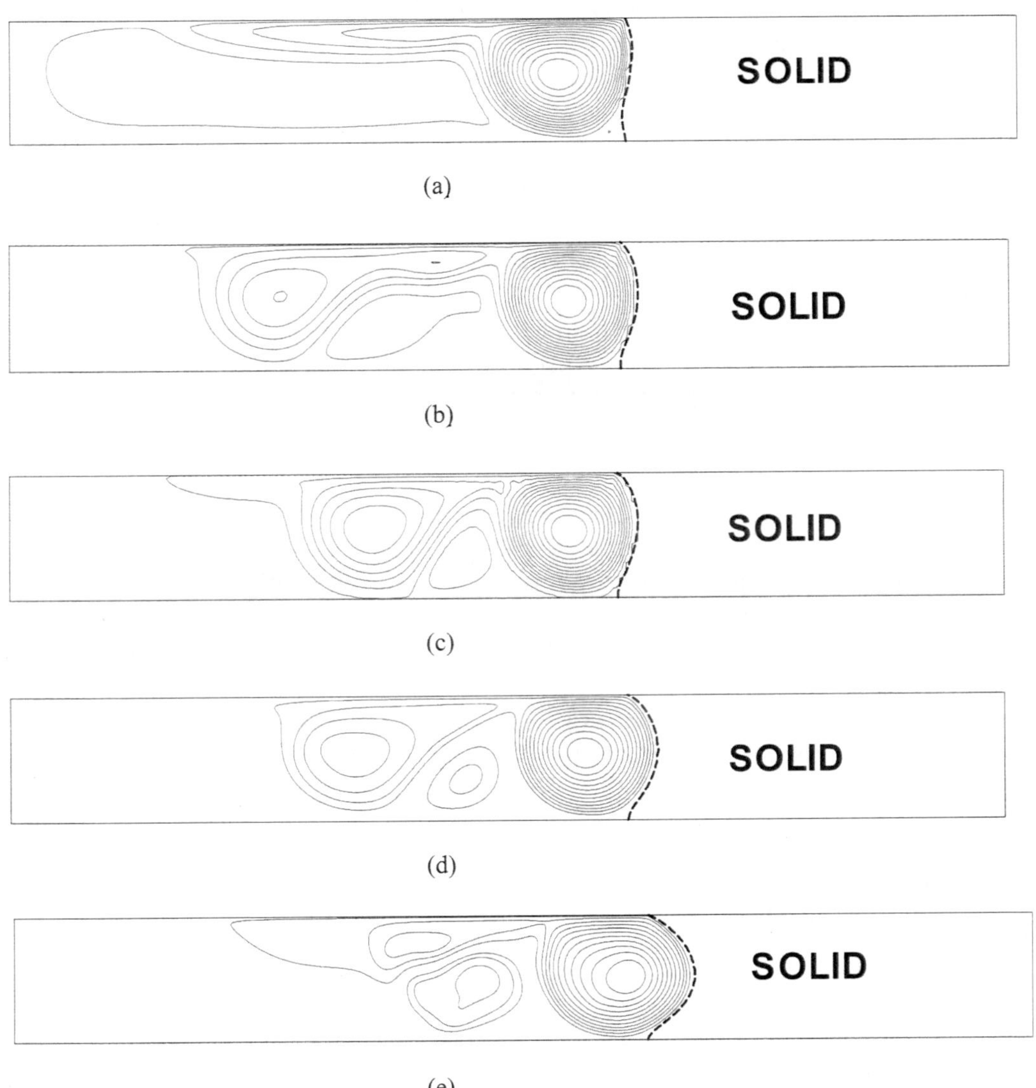

Figure 6.3.2. Stream-function (ψ) and solid/liquid interface (dashed line) for Ra=5.
(a) Ma = 1612 (ψ_{min} = -2.35×10^{-3}, ψ_{max} = 1.97×10^{-4}, $\Delta\psi$ = 1.8195×10^{-4});
(b) Ma = 3224 (ψ_{min} = -2.66×10^{-3}, ψ_{max} = 3.07×10^{-4}, $\Delta\psi$ = 2.1174×10^{-4});
(c) Ma = 4836 (ψ_{min} = -2.21×10^{-3}, ψ_{max} = 3.14×10^{-4}, $\Delta\psi$ = 1.8015×10^{-4});
(d) Ma = 8060 (ψ_{min} = -1.64×10^{-3}, ψ_{max} = 2.33×10^{-4}, $\Delta\psi$ = 1.3708×10^{-4});
(e) Ma = 16120 (ψ_{min} = -0.355×10^{-3}, ψ_{max} = 2.239×10^{-4}, $\Delta\psi$ = 0.307×10^{-4}).

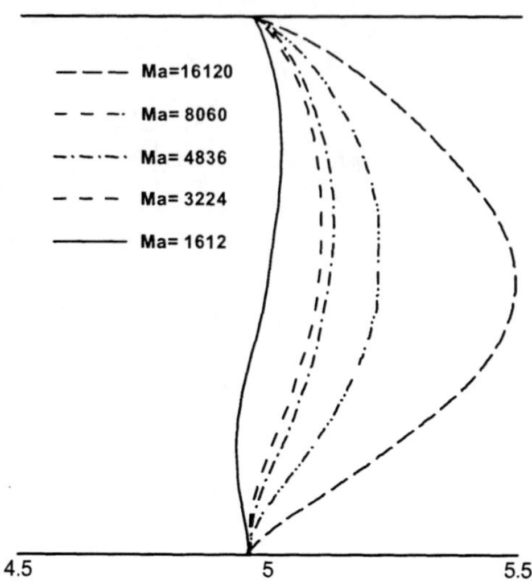

Figure 6.3.3. Solid/liquid interface at different values of *Ma*.

It can be seen in Fig. 6.3.2 that a secondary recirculation zone is induced in the flow field. This recirculation is located on the left of the main recirculating zone and is rotating in an anti-clockwise direction. As *Ma* increases, this secondary recirculating zone is strongly influenced by the strengthening main Marangoni cell.

For $Ma \geq 3{,}224$ this recirculation is surrounded by a Marangoni driven recirculation (Fig.6.3.2b), while for $Ma \geq 4{,}836$ the recirculation becomes progressively stronger (Fig.6.3.2c). Thermocapillary convection strongly influences the solid/liquid interface shape. In all the cases investigated long time histories have been followed until steady state is reached. The steady state solid/liquid interface shape which develops for different Marangoni numbers is presented in Fig. 6.3.3. It can be seen that both buoyancy and thermocapillary convection result in increased curvature of the interface. However it is clear that the Marangoni effect dominates the distortion of the interface, with the maximum curvature at $Ma = 16{,}120$. The change in the curvature is particularly significant as *Ma* is increased from *1,612 to 3,224* and also when it changes from *8,060* to *16,120.*

Fig. 6.3.4 shows the effect of Marangoni number on the temperature field. The isotherms exhibit increased distortion near the solid/liquid interface due to the strong primary cell which forms due to the surface tension effects. For higher *Ma* the effect of the secondary cells is also quite evident (see Fig. 6.3.4d and 6.3.4e).

The horizontal component of the velocity on the free surface for all Ma considered is presented in Fig. 6.3.5. For better comparison, the velocities have been reported in dimensional form. The reference velocity $\mathbf{v}^* = Ma(\alpha / H)$ has been evaluated using data from MEPHISTO (*H*= 6 mm and α= 8.5 x 10^{-6} m^2/s). It can bee seen that substantial free surface velocities develop with a maximum horizontal velocity of 176 mm/s at $Ma = 16120$. These high velocities induce significant recirculation regions as already seen in Fig.6.3.3, resulting in the interface curvature.

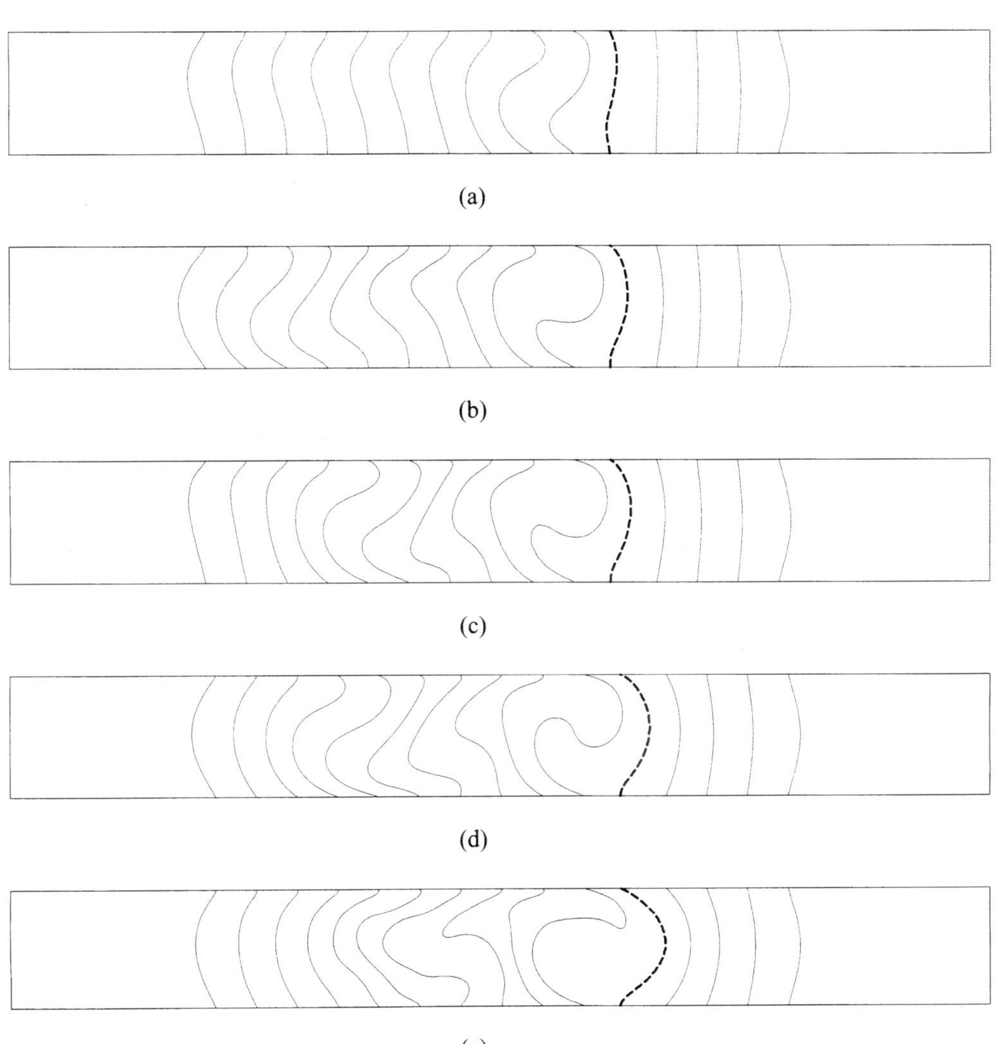

Figure 6.3.4. Isotherms ($\theta_{min} = 0.0625$, $\theta_{max} = 0.9375$, $\Delta\theta = 6.73076 \times 10^{-2}$) for Ra=5 for different values of Ma. (a) $Ma = 1612$; (b) $Ma = 3224$; (c) $Ma = 4836$; (d) $Ma = 8060$, (e) $Ma = 16120$. The liquid/solid interface (dashed line) subdivides the liquid region (left zone) from the solid region (right zone).

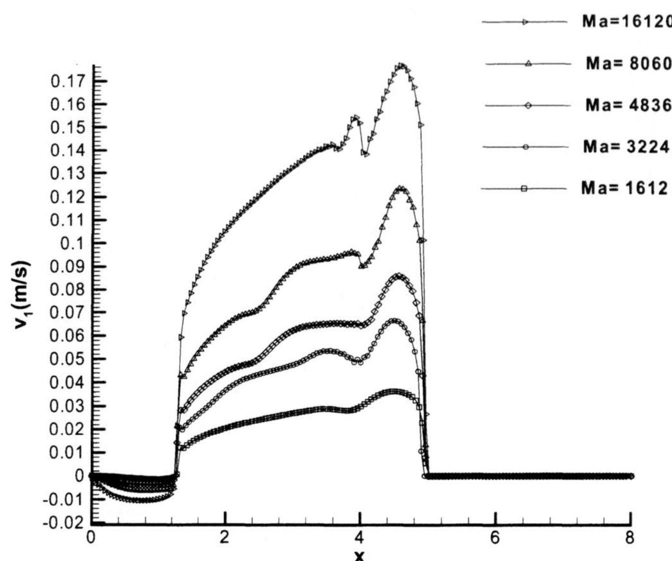

Figure 6.3.5. Velocity profiles on the free surface at different values of *Ma*. Velocities are presented using dimensional values.

Case B

The influence of the gap near the free surface and the effect of different thermal liquid and solid conductivities on the interface shape has been investigated for *Ma = 8,060* and *Ma = 16,120*. Streamlines and the solid/liquid interface positions are presented in Fig. 6.3.6a-b. As was seen in Case A, a strong clockwise recirculating main flow, due to the Marangoni effect, is present near the solid/liquid interface.

Due to the argon conductivity the temperature distribution near the top free surface is substantially modified compared to that of case A and so the Marangoni effect, imposed by the condition $\partial v_1/\partial y = -\partial\theta/\partial x$ is increased, resulting in the main clockwise recirculation being much stronger than Case A (see Fig. 6.3.6). It can be seen in Fig. 6.3.6a-b that two small recirculations are induced in the flow field. The isotherms are highly distorted due to the strong convection (see Fig. 6.3.6c-d).

The effect of the argon layer and the different liquid and solid conductivities also results in significant curvature of the interface. For *Ma = 8,060* the interface is more curved and an asymmetry is introduced. This effect becomes dominant for *Ma = 16,120* (see Fig. 6.3.7).

As the Marangoni number is increased from 8,060 to 16,120 the amount of material solidified is reduced. The significant effect of the argon layer on the free surface velocity is clearly visible in Fig. 6.3.8.

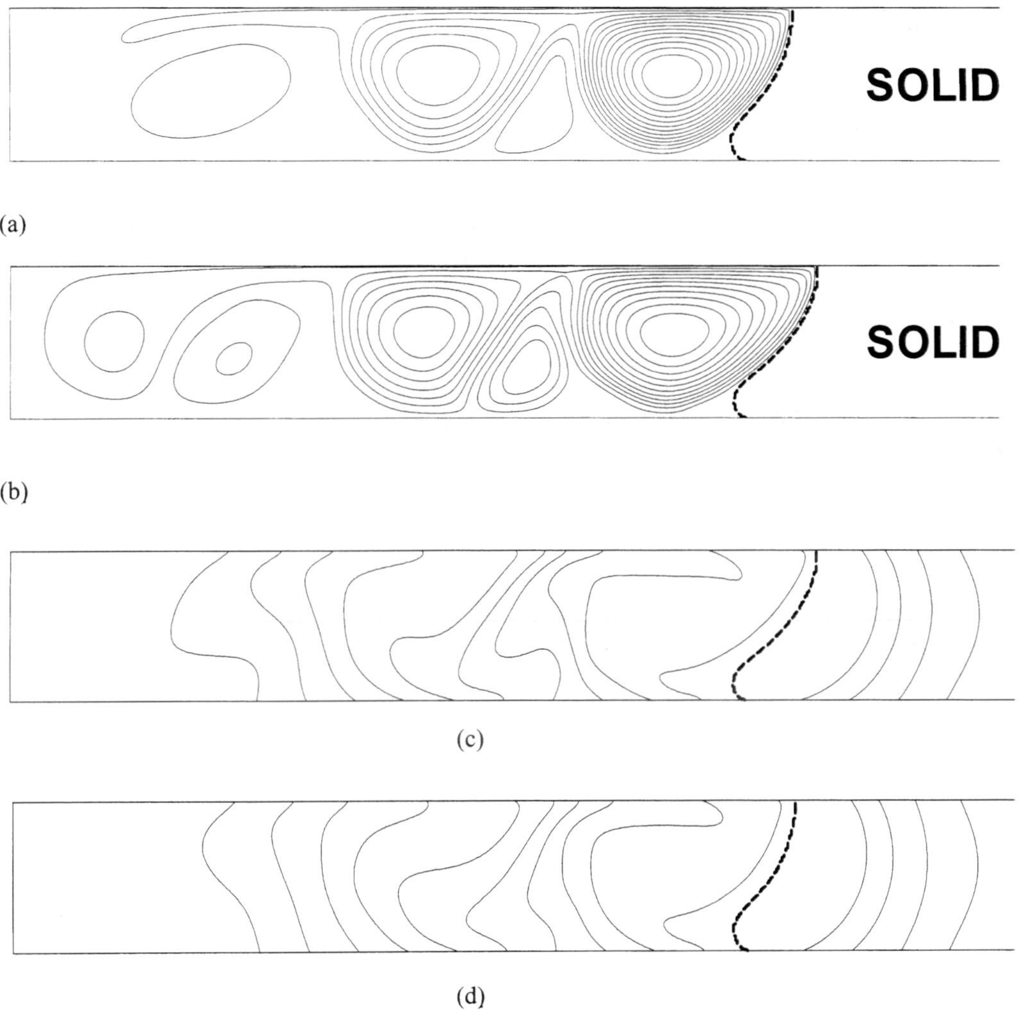

Figure 6.3.6 Stream-function (a & b) and Isotherms (c & d) for Ra=5.
(a) Ma = 8,060 (ψ_{min}=-1.30×10^{-3}, ψ_{max}=0.43×10^{4}, $\Delta\psi$=1.0367×10^{-4});
(b) Ma =16,120 (ψ_{min}=-0.835×10^{-3}, ψ_{max}=2.09×10^{-4}, $\Delta\psi$=0.8041×10^{-4});
(c) Ma = 8,060 (θ_{min}=0.063, θ_{max}=0.938, $\Delta\theta$=6.730×10^{-2});
(d) Ma = 16,120 (θ_{min}=0.063, θ_{max}=0.938, $\Delta\theta$=6.730×10^{-2}).

7 Conclusions

Much attention has been paid during the last few years to understanding of phase change phenomena. The reason for this is the enormous applied importance that this problem has in various industrial processes such as casting, welding, melting and purification of metal.

The central problem with modelling phase change problems is the treatment of the moving solid-liquid interface. Since the interaction of fluid flow with moving boundaries leads to a highly coupled, non linear system, such problems have remained analitically intractable. Two alternative approaches are used in literature: moving grid models and fixed grids models.

From the comparison of results appears as for fixed temperature phase change problems (pure metals with a single melting temperature) the moving grids models give better predictions of the interface position and the fluid flow, compared to the enthalpy method. The enthalpy method would require a prohibitively fine mesh to achieve the same accuracy.

But with transformed grids methods at each time step new grids have to be generated, this limits the flexibility of the method and substantially slows down the computations.

The fixed grid methods, as the enthalpy method, are ideal for non-isothermal phase change problems and mushy region with dendrites, which are present in many practical phase change processes (for example alloys), because the flow of liquid can be best simulated using suitable porous-flow equations.

Generally speaking, the application of the numerical method to the practical problems involves a compromise between the complexity of the model (time necessary to solve the problem) and accuracy of the results obtained.

Acknowledgements

We gratefully acknowledge Prof. E. Leonardi for the help during the development of this work.

Nomenclature

A	aspect ratio	[H/l]
C	large computational quantity	
c	heat capacity	[J/Kg K]
D	diffusivity of solute	
f	mass fraction of the phases	
F	free energy density	
G	buoyancy term	
g	volume fraction of the phases	
g	gravity acceleration	[m/s^2]
h	specific enthalpy	[J/Kg]
K	permebility	
Gr	Grashof number Ra/Pr	
H	height	[m]
k_p	partition coeffiecient	
l	length	[m]
L_f	latent heat	[J/kg]

Ma	Marangoni number $(\partial\sigma / \partial T)\Delta Tl / (k\mu)$
Nu	Nusselt number $\alpha l/\lambda$
Pe	Péclet number Re*Pr
Pr	Prandtl number [ν/k]
q	small computational quantity
Q	heat quantity [J]
Ra	Rayleigh number $g\beta\Delta TH/k\nu$
Re	Reynolds number [ul/ν]
S	Source term
Ste	Stefan number $c\ \Delta T/L_f$
t	time [s]
T,ΔT	temperature, temperature difference [K]
u,v,w	velocity components in Cartesian coordinate x,y,z [m/s]
V	Volume [m^3]

Greek letters

α	heat transfer coefficient	[$W/(m^2K)$]
β	coefficient of cubic expansion	[1/K]
k	thermal diffusivity	[m^2/s]
λ	thermal conductivity	[W/(m K)]
μ	dynamic viscosity	[Pa s]
ν	kinematic viscosity	[m^2/s]
ρ	density	[kg/m^3]
σ	surface tension	[N/m]
Ψ	total free energy density	

Subscripts

c	cold wall
h	hot wall
l	liquid phase
s	solid phase

References

Abbaschian, R., de Groh III, H.C, Leonardi, E., de Vahl Davis, G., Coriell, S., and Cambon, G., 2001. 'Final report for the shuttle flight experiment on USMP-4: In Situ Monitoring of Crystal Growth Using MEPHISTO', NASA Technical Publication, NASA/TP-2001-210825.

Banaszek, J., Rebow, M. and Kowalewski, T. A. (1998). Fixed grid finite element analysis of solidification, In de Vahl Davis, G. and Leonardi, E., eds., *Advances in computational heat transfer CHT-97*, 471-478, Cesme, Belgell House.

Bennon, W. D. and Incropera, F. P. (1987). A continuum model for momentum, heat and species transport in binary solid-liquid phase change systems I, Model formulation, *Int. J. of Heat and Mass Transfer* **30**, 2161-2170,

Bennon, W. D. and Incropera, F. P. (1988). Numerical analysis of binary solid-liquid phase change using a continuum model. *Num. Heat Transfer* **13**, 207-216.

Bertrand, O., Binet, B., Combeau, H., Coutier, S., Dellanoy, Y., Gobin, D. Lacroix, M., Lequere, P., Medale, M., Mencinger, J. , Sadat, H. and Veira, G. (1999). Melting driven by natural convection. A comparison exercise: first results, *Int. J. Thermal Sci.* **38**, 5-26.

Brent, A. D., Voller, V. R. and Reid, K. J. (1988).Enthalpy-porosity technique for modelling convection-diffusion phase-change: application to the melting of a pure metal, *Num. Heat Transfer* **13**, 297-318.

Buell, C. and Shuck, F., 1970. Diffusion in the liquid Bi-Sn system, *Metallurgical Transactions,* **1**, 1875-1880.

Caginalp, G. and Socolovski, E. A., (1991). *J. Comput. Phys*., **95**, 85.

Caginalp, G., (1989). *Phys. Rev. A*. **39**, 11, 5887.

Caginalp, G. and Chen X. (1992). *On the evolution of Phase Boundaries*, Spring Verlag New York.

Chen, P.Y.P., Timchenko, V., Leonardi, E., de Vahl Davis, G. and de Groh III, H.C., 1998. A numerical study of directional solidification and melting in microgravity, HTD-Vol. 361-3/PID-Vol. 3. *Proc. Of ASME Heat Transfer Divn*., HTD-Vol.361-3/PID-Vol.3, *Proc. of the ASME Heat Transfer Division* - Volume 3, R.A. Nelson Jr., L.W. Swanson, M.V.A. Bianchi, C. Camci (eds), ASME, New York, 75-83.

CFX-4.2:Solver, 1997. Harwell Laboratory, Didcot, Oxfordshire OX11 ORA, U.K.

Crank, J. (1984). *Free and moving boundary problems*. Oxford Science Publications.

Dantzig, J. (1989). Modelling liquid-solid phase change with melt convection, *Int. J. Num. Meth. In Eng.* **28**, 1769-1785.

Fabbri, M. and Voller, V. R. (1997). The phase-field method in the sharp interface limit: a comparison between model potential. *J. Comp. Physics* **130**, 256-265.

Gau, C. Viskanta, R. (1986), Melting and solidification of a pure metal on a vertical wall, *Transaction of the ASME* **108**, 174-181.

Garling, D. K. (1980). Finite element analysis of convective heat transfer problems with change of phase, *Computer Meth. Fluids*, 257-284.

Giangi, M., Stella, F. and Kowalewski, T. A. (1999). Phase change with free convection: fixed grid simulation, Comp. Visual. Sci. **2**, 123-130.

Giangi, M., Stella, F. (1999). Analysis of natural convection during solidification of a pure metal, *Int. J. Comp. Fluid Dynamics*, **11**, 341-349.

Giangi, M., Stella, F., Leonardi E. and de Vahl Davis G. (2002). A numerical study of solidification in the presence of a free surface under microgravity condition, *Num. Heat Transfer: Part. A*, **41**, 579-595.

Golub G. and Van Loan C. (1990). *Matrix computations*, Johns Hopkins University Press, Baltimore, MD.

Guj, G. and Stella, F. (1988). Numerical Solution of High-*Re* Recirculating Flows in Vorticity-Velocity Form, *Int. J. Num. Meth. in Fluids*, **8**, 405-416.

Hirsch C. (1983). *Numerical computation of internal and external flows*, Spring Verlag New-York,1, Fondamental of numerical discretization.

Kobayashi R. (1993), *Physica D*., **63**, 410.

Kobayashi, R. (1994), *Exp. Math*. 3, 59.

Kowalewski, T. A. and Rebow, M. (1998). An experimental benchmark for freezing water in the cubic cavity, In de Vahl Davis, G. and Leonardi, E., eds., *Advances in Computational Heat Transfer CHT-97*, 149-156, Cesme, Belgell House.

Laouadi, A., Lacroix, M. and Galanis, N., 1998. A numerical method for the treatment of discontinuous thermal conductivity in phase change problems, *International Journal of Numerical Methods for Heat and Fluid Flow,* **8,** 265-287.

Lee, Y. and Korpela S.A. (1983).Multicellular natural convection in a vertical slot, *J. Fluid Mech.*, 126, 91-121.

Morgan, K. (1981). A numerical analysis of freezing and melting with convection, *Comp. Meth. Applied in Eng.* **28**, 275-284.

Niwa, K., Shimoji, M., Kado, S., Watanabe, Y. and Yokokawa, T., 1957. Studies of diffusion in molten metals, *AIME Tran.*, **209**, 96-101.

Ralston A. and Rabinowitz P. (1978). *A first course in numerical analysis*, McGraw-Hill International Editions, Singapore.

Raw, W.Y. and Lee, S.L., 1991. Application of weighting function scheme on convection-conduction phase-change problems, *Int. J. Heat and Mass Transfer,* **34**, 1503-1513.

Rouzaud A., Favier J.J. and Thevenard D., 1988. A space instrument for fundamental materials science problems: the MEPHISTO program, Adv. Space Res., Vol.8, No.12, 49-59.

Richtmyer, R. D. and Morton, K. W. (1967). *Difference methods for initial value problems.* Interscience, New York.

Saad Y. and Schultz M.H. (1986). Gmres: A generalized minimum residual algorithm for solving non-symmetric linear systems, *SIAM, J. Sci. Stat. Comput.*, **7**, 856-869.

Scheudergger, A. E. (1963). *The physics of flow trough porous media*, Oxford University Press, London.

Shyy W., Udaykumar H.S., Madhukar M.Rao, Smith R.W. (1996). *Computational fluid dynamics with moving boundaries*, Taylor & Francis, London.

Slattery, J. C. (1978) *Momentum Energy and Mass Transfer in Continua*, Krieger, New York.

Smith, V.G., Tiller, W.A. and Rutter, J.W., 1955. A mathematical analysis of solute redistribution during solidification. Canadian J. of Physics, 33, 723-743.

Sonneveld P. (1989). CGS, A fast lanczos type type solver for nonsymmetric linear systems, *SIAM J. Sci. Stat. Comput.*, 1**0**, 36-52.

Stella, F. and Giangi, M. (2000). Melting of a Pure Metal on a Vertical Wall: Numerical Simulation, *Numerical Heat Transfer: Part. A*, **38**, 2, 193-208.

Stella F. and Bucchignani E. (1996). True transient vorticity-velocity method using preconditioned Bi-CGSTAB, *Numerical Heat Transfer: Part. B*, **30**, 315-339.

Tacke, K., Discretization of the explicit enthalpy method for planar phase change, 1985. *Int. J. Num. Methods in Eng.,* **21,** 543-554.

Timchenko, V., Chen, P.Y.P., de Vahl Davis, G. and Leonardi, E., 1998. Directional Solidification in Microgravity, *Heat Transfer 98* , J.S. Lee (ed.), Taylor & Francis, 241-246.

Timchenko, V., Chen, P.Y.P., de Vahl Davis, G., Leonardi, E and Abbaschian, R., 2000. A computational study of transient plane front solidification of alloys in a Bridgman apparatus under microgravity conditions, *Int. J. Heat and Mass Transfer*, **43**, 963-980.

Timchenko, V., Chen, P.Y.P., de Vahl Davis, G., Leonardi, E and Abbaschian, R., 2001. A numerical study of the MEPHISTO experiment, *CHT'01 Advances in Computational Heat Transfer II*, G. de Vahl Davis and E. Leonardi (eds), Begell House Inc., New York, 1089-1096.

Unverdi, S. O. and Tryggvason, G. (1992). A front-tracking method for viscous incompressible, multi-fluid flows. *J. Comp. Physics* **100**, 25-37.

Van De Worst H. (1992). A fast and smoothly converging variant of Bi-Cgstab for the solution of non-symmetric linear sistems, *SIAM, J. Sci. Stat. Comput.*, **132**, 631-644.

Viswanath, R. and Jaluria, Y. (1993). A comparison of different solution methodologies for melting and solidification problems in enclosures, *Num. Heat Transfer*, Part B, **24**, 77-105.

Voller, V. R., Cross, M. and Markatos, N. C. (1987). An enthalpy method for convection-diffusion phase change. *Int. J. Num. Meth. Eng*. **24**, 271-284.

Voller, V. R., Swaminathan C.R. (1993). Treatment of discontinuous thermal conductivity in control volume solutions of phase change problems. *Numerical Heat Transfer*, Part B, **24**, 161-180.

Voller, V.R., Fabbri, M. (1995). Numerical solution of plane-front solidification with kinetic undercooling, *Numerical Heat Transfer*, Part B, **32**, 467-486.

Voller, V.R., Brent, A.D. and Prakash, C., 1989. The modelling of heat, mass and solute transport in solidification systems. *Int. J. Heat Mass Transfer*, **32**, 1719-1731.

Wang S.L. *et al.*(1993), *Physica D*. **69**, 189.

Zeitfracht Medien GmbH
Ferdinand-Jühlke-Straße 7
99095 Erfurt, Deutschland
produktsicherheit@kolibri360.de